Future City

Volume 18

As of 2008, for the first time in human history, half of the world's population now live in cities. And with concerns about issues such as climate change, energy supply and environmental health receiving increasing political attention, interest in the sustainable development of our future cities has grown dramatically.

Yet despite a wealth of literature on green architecture, evidence-based design and sustainable planning, only a fraction of the current literature successfully integrates the necessary theory and practice from across the full range of relevant disciplines.

Springer's *Future City* series combines expertise from designers, and from natural and social scientists, to discuss the wide range of issues facing the architects, planners, developers and inhabitants of the world's future cities. Its aim is to encourage the integration of ecological theory into the aesthetic, social and practical realities of contemporary urban development.

More information about this series at https://link.springer.com/bookseries/8178

George Cristian Lazaroiu • Mariacristina Roscia
Vasile Sebastian Dancu
Editors

Holistic Approach for Decision Making Towards Designing Smart Cities

Springer

Editors
George Cristian Lazaroiu
University POLITEHNICA of Bucharest
Bucharest, Romania

Mariacristina Roscia
University of Bergamo
Bergamo, Italy

Vasile Sebastian Dancu
University of Bucharest
Bucharest, Romania

ISSN 1876-0899 ISSN 1876-0880 (electronic)
Future City
ISBN 978-3-030-85565-9 ISBN 978-3-030-85566-6 (eBook)
https://doi.org/10.1007/978-3-030-85566-6

The cover image represents the holistic approach of future cities that integrate the Environmental, Social, and Governance (ESG) themes, never forgetting that the Nature is the "upside-down" of Smart Cities, the vital connection that, without it, the human being, the plants and the animals would not exist. The ESGs can represent the key elements also for the companies that will adequately invest for creating long-term value, not only for the economic assets, but also for the environmental and social domains of Smart Cities. Photo Credit: Ciprian Butnaru

This Springer imprint is published by the registered company Springer Nature Switzerland AG
The registered company address is: Gewerbestrasse 11, 6330 Cham, Switzerland

To my beloved mother Margareta,
To my beloved father Claudio,
To my sociology students.

Preface

A city is not gauged by its length and width,
but by the broadness of its vision
and the height of its dreams.

Herb Caen

Nowadays, more than 50% of the world population live in urban areas, consuming more than 75% of natural resources, generating approximately 75% of carbon emissions and 50% of global wastes, and consuming between 60 and 80% of energy consumption. The European Union has established a target to become climate-neutral by 2050, and the clean development and resilient planning of cities represent a solution towards the sustainable future of our society.

In the last decades, the renewable energy revolution and ICT-based sustainable solutions determined the boost of clean electricity and pollution reduction, improvement of citizens' lives, transformation of urban environment and regulatory structures, and transformation of cities towards smarter cities where citizens and sustainability are at the core. New business models are burgeoning in smart cities, fostering economic prosperity, social wellness, and environmental sustainability so as to comply with the EU targets for current cities. Such ambitions towards smart cities are not free of challenges, as designing a new smart city is much simpler than going to incorporate new technologies into an already built urban space.

This book focuses on a holistic approach for decision-making towards designing smart cities that require energy and sustainable development, smart infrastructure, and social involvement. The book provides an integrated, sustainable, and multidisciplinary approach for transition of urban areas into smart cities.

This book consists of 14 chapters grouped in 3 parts.

The first part of the book is dedicated to energy and sustainable development and consists of five chapters. Chapter 1 introduces the readers to smart cities, the subject matter of this book, and approaches the role of citizens as main drivers towards the transformation of cities into smart ones. This can be achieved by adapting the cities to the ongoing changes of inhabitants' needs through a holistic and resilient vision which harmoniously unites the actors in the urban context. In Chap. 2, the integration of photovoltaic installations in a new smart city landscape is analyzed considering the near-zero energy buildings and positive energy districts. The building-integrated photovoltaics at building level and at community/district level are analyzed, together with the challenges of photovoltaic installations integration. Chapter 3 describes active participation of consumers in smart cities in demand-side response programs and potential sustainability strategies. Consumer participation is required for sustainability to use smart cities technologies to holistically cope with their necessities. The chapter ends with the relationship between energy transition and involvement of smart consumers, and due to the concentration of inhabitants, the cities tend to be representative of the energy sustainability for the whole country. Chapter 4 provides an analysis of security and privacy challenges in smart grids. The architectures of smart grids for smart cities are described. The threats and attacks on Internet of Things–enabled smart grids are presented. Digital substation in smart grids is investigated from the point of view of possible attacks. A security pyramid is presented for analyzing the threats on a smart grid substation in Norway.

In Chap. 5, the smart lighting systems for smart cities are presented. The smart lighting should lead to energy savings, without reducing the quality of lighting services, and preservation of traffic and pedestrian security as well as customer well-being. A transformation is ongoing towards a human-centric lighting system, with optimum balance between efficiency and quality of light.

The second part of the book, consisting of five chapters, is dedicated to the smart infrastructure of smart cities. Chapter 6 describes the role of geoinformation and communication technologies in managing a smart city. The use of available data for developing innovative geoinformation services can lead to real-time information about the needs of the inhabitants in the city. Chapter 7 provides the framework for investigating future forms of citizen involvement within urban planning activities. The detailed characterization of existing urban planning software is realized and highlights the potential of these participatory tools to generate social and economic impacts. Chapter 8 summarizes the key indicators for developing smart cities, under pandemic threats, using artificial intelligence. A model to be used in policy-making processes as starting point for discussion among various stakeholders and citizens is proposed. Chapter 9 provides new business models for the charging of electric vehicles with smart infrastructures, a key component of smart cities. The blockchain technology creates scenarios of high interest in the electric mobility industry applied to the different actors participating in this market. Chapter 10 resumes the holistic view of transportation system in Africa cities, together with renewable energy sources in the future smart cities, towards improvement of developing countries in this territory. Smart cities, with the help of sustainable mobility, will foster the improvement of quality of life of populations in Africa.

The third part of the book, consisting of four chapters, describes the social involvement in smart cities. Chapter 11 describes the factors that influence the development of smart cities and the results of the empirical study on a large number of European cities. The levels of the smart cities are related to the geographical location of the cities and the gender of the governors. Chapter 12 describes the impact of social media in smart cities, the technologies interfering into social life, and assessment of public governance role regarding social media. Both quantitative and qualitative analyses of digital social media in smart cities are described in detail. Chapter 13 provides a framework for evaluating the real participation of citizens in smart cities, and the case study of Washington D.C. provides interesting information. Content and discourse analysis are used for assessing citizen-centric rhetoric. Chapter 14 describes the results of sociological surveys on the measurement of indicators related to the use of smart city technologies and governance in Romania for identifying how smart government ideologies are correlated with the satisfaction of inhabitants and prioritization towards improved quality of life.

This book analyzes the current developments and active technology advance in main aspects of smart cities, both in engineering and social aspects, with contributions from qualified experts in the field of study, with many applications for professionals and decision-makers in urban communities, local governments, industry, and students/researchers in academia.

This book can become a focal point not only for the scientific community, thus identifying beyond state-of-the- art modification for rendering the projects concrete, but also for the ones that would have to take decisions to make current cities smart cities, considering also the cultural and social particularities.

The content of this book can be arranged in different ways to cope with the requirements and interests of graduate teaching in development and transformation of cities towards smart cities.

The Smart City should not be seen as a technological project, based only on enabling technologies, but as a synergy between conscious citizenship and inclusive governance.

To create sustainable cities, a clear holistic vision is needed with an eye to the future, conceived under the paradigms of equity, sharing, participation, sustainability, and livability.

The concept behind this book is to give the readers, such as decision-makers, urban planners, and in particular citizens, a systemic, holistic vision, capable of holding together the various challenges that the urban context is facing: economic, environmental, and also social.

Bucharest, Romania — George Cristian Lazaroiu

Bergamo, Italy — Mariacristina Roscia

Bucharest, Romania — Vasile Sebastian Dancu

Acknowledgments

The editors would like to acknowledge the contribution of reviewers who read the subsequent chapters and made valuable suggestions: Elisenda Jové-Llopis (University of Barcelona, Spain), Antoni Sudrià i Andreu (Universitat Politècnica de Catalunya, Spain), Ogerta Elizaj (University of Tirana, Albania), Santiago Sanchez-Acevedo (SINTEF Energy Research, Norway), Vincenco Puri (Universita' degli Studi di Milano, Italy), Marco Antônio Dalla Costa (Universidade Federal de Santa Maria, Brazil), George Cristian Lazaroiu (University POLITEHNICA of Bucharest, Romania), Stuart Mucklejohn (Ceravision Ltd, UK), Joanna Bac-Bronowicz (Wroclaw University of Science and Technology, Poland), Mieczysław Muraszkiewicz (Politechnika Warszawska, Poland), Matteo Caglioni (Université Côte d'Azur, France), Eda Beyazit (Istanbul TU, Turkey), Cengiz Kahraman (Istanbul Technical University, Turkey), Vasile Dancu (University of Bucharest, Romania), Patricia Milanés Montero (Universidad de Extremadura, Spain), Mariacristina Roscia (Universita di Bergamo, Italy), María Jesús Barroso Méndez (Universidad de Extremadura, Spain), Samal Tazhiyeva (Al-Farabi Kazakh National University, Kazakhstan), Kirill Ratnikov (Finance University of the Government of the Russian Federation, Russia), John McNutt (University of Delaware, USA), Paolo Cardullo (Maynooth University, Ireland), and Nicolae Golovanov (University POLITEHNICA of Bucharest, Romania).

Substantial help and assistance during this entire period were received from Markus Spaeth and Solomon George of Springer Nature, to whom our thanks are kindly addressed.

The editors specially thank all authors for their effort and great work to make this book a success. Collaborating with these stimulating colleagues has been a privilege.

George Cristian Lazaroiu
Mariacristina Roscia
Vasile Sebastian Dancu

Contents

Editors and Contributors

About the Editors

George Cristian Lazaroiu is professor at University POLITEHNICA of Bucharest, Romania. He obtained his bachelor's and master's degrees from University POLITEHNICA of Bucharest, Romania, in 2002 and 2003, respectively. In 2006 he obtained his PhD in electrical engineering from Politecnico di Milano, Italy, and in 2009, his PhD in energy engineering from POLITEHNICA University of Bucharest, Romania. He obtained his habilitation in energy engineering in 2015. Dr. Lazaroiu developed large international cooperation with highly prestigious universities from all over the world taking part in PhD commissions and supervising foreign postdocs and PhD students. He acts as evaluator in research grants for many research funding bodies like Romanian UEFISCDI, the Italian MIUR, and the Kazakhstan NCSTE, and as expert for evaluating research projects in energy domain within HORIZON 2020 of EU European Commission. Dr. Lazaroiu authored or coauthored 5 books and more than 150 papers in prestigious international journals and national/international conferences. He is Senior Member of the IEEE, member and secretary of the Energy Commission, Engineering Sciences Section of the Romanian Academy. He was the co-chair of IEEE 16th International Conference on Harmonics and Quality of Power (ICHQP 2014) and general co-chair of 2019 IEEE PES Innovative Smart Grid Technologies Europe (ISGT-Europe 2019). He is active in IEEE Working Groups in the areas of power quality. He obtained the "Constantin Budeanu" Best Paper Award at IEEE ICHQP 2006 the Best Poster Award at ICRERA 2014, the Business Excellence Award at Int. Conf. Business Excellence 2017, the Best Paper Award of *Inventions* journal in 2018, and the IEEE PES Chapter Outstanding Engineering Award in 2018. He is editor-in-chief of the *International Journal of Energy and Environmental Engineering*, energy executive editor of University POLITEHNICA of Bucharest *Scientific Bulletin Series C-Electrical Engineering and Computer Science*, editorial board member of *Renewable Energy*, and editorial board member of Smart Cities. His research interests include energy systems development, smart cities, monitoring and control of power quality, electricity markets, and optimization of power systems. University POLITEHNICA of Bucharest, Bucharest, Romania,

cristian.lazaroiu@upb.ro

Mariacristina Roscia is the mother of two awesome daughters: Chantal and Micol. She obtained her bachelor's and master's degrees in electrical engineering from the University of Naples "Federico II" as well as her PhD, from the same university, with a dissertation entitled "Possible energy scenarios and sustainable development." Since 2019, she is associate professor of the courses "Electrical systems," "Smart Home into Smart City," and "Smart Electrical Infrastructure" at the University of Bergamo, Italy. She was adjunct professor at Politecnico di Milano, Italy, for more than 10 years teaching "Automation of transport systems" and "Electric transport systems I." Dr. Roscia has authored more than 90 papers in Scopus international journals and international conferences. She acts as reviewer for more than 30 different Scopus journals and conferences as well as international research programs. She is expert evaluator for HORIZON 2020 Work Program 2018-2020, Energy challenge 2020, and EU European Commission. She contributed as technical support for the city of Naples for: optimization of energy services, SMART CITY projects, SEAP (Sustainable Energy Action Plan), and participation in the CDP (Carbon Disclosure Project). Sustainable development has been the first key element of the research activity, towards the development, definition, and implementation of Smart Cities. Her research fields are smart grids in the presence of distributed generation, for the control and management of energy flows between the grid, and smart home and building. The systems related to electric mobility and intelligent lighting through the aspects of sustainability were considered jointly and integrated; these issues then saw in parallel the study of the related power quality issues. She has given several speeches on smart city aspects and several articles have received the Best Poster Award. University of Bergamo, Bergamo, Italy

cristina.roscia@unibg.it

Vasile Sebastian Dancu is a professor at University of Bucharest, Romania, with a rich teaching and research career in the field of social sciences and journalism. He obtained his doctorate in sociology from Babeş Bolyai University in Cluj-Napoca. He taught at prestigious universities in Romania such as Babeş Bolyai University Cluj-Napoca, the University of Bucharest, and the National Academy of Information Mihai Viteazul. He is a member of prestigious professional associations in Romania and abroad: ESOMAR (European Society for Opinion and Marketing Research), ISA (International Sociological Association), and SSR (Society of Sociologists in Romania). Throughout his teaching career, he founded several faculties, research centers, and master's programs in public and private universities of Romania in Bucharest and Cluj-Napoca. He founded two of the most important public opinion and social research companies in Romania: the Institute of Social Studies, Marketing and Advertising Metro Media Transilvania (1994) and the Romanian Institute for

Evaluation and Strategy – IRES (2009). He is author of numerous courses, works, and volumes on sociology and sociological research, and has also authored eight volumes on journalism, wherein he radiographs contemporary society. For his work, he has been rewarded with numerous awards and distinctions, the most recent being in November 2015: the *Cross of Merit in the Rank of Knight of the Order of Merit* of the Federal Republic of Germany by the President of the Federal Republic of Germany (November 2015) and *Professor Honoris Causa* of Babeş Bolyai University in Cluj-Napoca (November 2015). Since 2000, Dr. Dancu is actively involved in public life in Romania, occupying various positions as member of the Romanian Government, holding the Public Information portfolio, heading the Agency for Government Strategies, and is Minister of Regional Development and Public Administration. His research interests focus on sociology of public opinion, sociology of media and mass communication, security and risk management, culture and institutions, and research on interethnic relations. University of Bucharest, Bucharest, Romania,

vasile.dancu@sas.unibuc.ro

Contributors

Mohamed Abomhara Department of Information Security and Communication Technology, Norwegian University of Science and Technology, Trondheim, Norway

Doney Abraham Department of Information Security and Communication Technology, Norwegian University of Science and Technology, Trondheim, Norway

Laura Aelenei Laboratório Nacional de Energia e Geologia, Amadora, Portugal

Thomas S. Benson University of Delaware, Newark, DE, USA

Laurent Canale Univesrité Toulouse 3, Toulouse, France

Luísa Cagica Carvalho Instituto Politécnico de Setúbal (CICE-IPS) & CEFAGE, Universidade de Évora, Évora, Portugal

Mihnea Costoiu University POLITEHNICA of Bucharest, Bucharest, Romania

Elie Daher Luxembourg Institute of Science and Technology, Esch-sur-Alzette, Luxembourg

Fisnik Dalipi Department of Informatics, Linnaeus University, Växjö, Sweden

Vasile Sebastian Dancu University of Bucharest, Bucharest, Romania

Nina P. David University of Delaware, Newark, DE, USA

Régis Decorme 2R2M Solution, Paris, France

Rachel Desmaris 2R2M Solution, Paris, France

Virgil Dumbrava University POLITEHNICA of Bucharest, Bucharest, Romania

Pascal Dupuis Univesrité Toulouse 3, Toulouse, France

María Pache Durán Facultad de Empresa, Finanzas y Turismo, Universidad de Extremadura, Badajoz, Spain

Alemayehu Gebremedhin Department of Manufacturing and Civil Engineering, Norwegian University of Science and Technology, Trondheim, Norway

María Teresa Nevado Gil Facultad de Empresa, Finanzas y Turismo, Universidad de Extremadura, Badajoz, Spain

Dariusz Gotlib Faculty of Geodesy and Cartography, Warsaw University of Technology, Warszawa, Poland

Jacques Jansen University of Pretoria, Pretoria, South Africa

Sergei Kamolov Federal State Autonomous Educational Institution of Higher Education "Moscow State Institute of International Relations (University) of the Ministry of Foreign Affairs of the Russian Federation", Moscow, Russia

Korhan Kayisli Gazi University, Ankara, Turkey

Sylvain Kubicki Luxembourg Institute of Science and Technology, Esch-sur-Alzette, Luxembourg

George Cristian Lazaroiu University POLITEHNICA of Bucharest, Bucharest, Romania

Mohaddeseh Maktabifard 2R2M Solution, Paris, France

Dušan Milošević Faculty of Electronic Engineering, University of Niš, Niš, Serbia

Mimica Milošević Faculty of Business Economics and Entrepreneurship, Belgrade, Serbia

Robert Olszewski Faculty of Geodesy and Cartography, Warsaw University of Technology, Warszawa, Poland

Inna Sousa Paiva Instituto Universitário de Lisboa (ISCTE-IUL), Business Research Unit (BRU-IUL), Lisbon, Portugal

Burak Pak KU Leuven, Leuven, Belgium

Nazim Pigenet Kawantech SAS, Toulouse, France

Claudia Laurenta Popescu University POLITEHNICA of Bucharest, Bucharest, Romania

Mihai Octavian Popescu University POLITEHNICA of Bucharest, Bucharest, Romania

Mariacristina Roscia University of Bergamo, Bergamo, Italy

Alessandra Scognamiglio ENEA Centro Ricerche Portici, Naples, Italy

Alexandra Catalina Sima University POLITEHNICA of Bucharest, Bucharest, Romania

Alexandra Smagina Federal State Autonomous Educational Institution of Higher Education "Moscow State Institute of International Relations (University) of the Ministry of Foreign Affairs of the Russian Federation", Moscow, Russia

Ana Stanojević Faculty of Civil Engineering and Architecture, University of Niš, Niš, Serbia

Andreas Sumper Centre d'Innovació Tecnològica en Convertidors Estàtics i Accionaments (CITCEA-UPC), Universitat Politècnica de Catalunya (UPC), Barcelona, Spain

Manuel Villa-Arrieta Escola Tècnica Superior d'Enginyeria Industrial de Barcelona (ETSEIB), Universitat Politècnica de Catalunya (UPC), Barcelona, Spain

Sule Yildirim Yayilgan Department of Information Security and Communication Technology, Norwegian University of Science and Technology, Trondheim, Norway

Georges Zissis Univesrité Toulouse 3, Toulouse, France

Part I
Energy and Sustainable Development

Chapter 1
Smart Citizens for Realizing Smarter Cities

Mariacristina Roscia, Mihnea Costoiu, and George Cristian Lazaroiu

Abstract The cities of the future will be urban ecosystems in which cultures and identities can interact thanks to intelligent infrastructures and interconnections, equipped with widespread intelligence, through devices capable of constantly and continuously collecting data.

However, cities are not only technology, transport, economy, and data exchange, they are above all people, who live in the city and which will be intelligent only in relation to what it can simplify and improve the quality of life of those who live there. Therefore, before building smart buildings, sustainable mobility systems, and infrastructures, it is necessary to "build" a smart citizen, who is able to use technology to create well-being, culture, social inclusion, and participatory safety, in a holistic approach.

The role of citizens becomes so fundamental for the city of the future that is intelligent and sustainable, citizens who become many "nodes" of a network, where the exchange of data, linked to its behavior, movements, and habits, can finally become an added value and not just a simple exchange of information useful to a few, but to the whole community represented by the cities themselves.

Cities are, in fact, in an advanced growth phase centered on the service economy characterized by the diffusion of digital technologies and up-to-date ingenious organizational models. These favor the implication in policy processes, through the creation of data sharing structures, to outline the policies for intervention.

The pandemic has highlighted that the urbanization phenomenon is extremely important related to global and local connections between population, ambient, and development, to the point that any reflection on elaborating possible models of sustainable development causes questioning of the future of cities globally.

M. Roscia (✉)
University of Bergamo, Bergamo, Italy
e-mail: cristina.roscia@unibg.it

M. Costoiu · G. C. Lazaroiu
University POLITEHNICA of Bucharest, Bucharest, Romania
e-mail: mihnea.costoiu@upb.ro; cristian.lazaroiu@upb.ro

G. C. Lazaroiu et al. (eds.), *Holistic Approach for Decision Making Towards Designing Smart Cities*, Future City 18,
https://doi.org/10.1007/978-3-030-85566-6_1

The governance in the new cities should be founded on a strategic development vision that accounts both the overall flows space and the local space of physical spaces, in order to find a dynamic equilibrium between the contradictory expressions of values and the interests of the many subjects who live and operate there.

This is the reason why data accessibility and reliability are of particular importance in the smart city, which can be provided in a conscious way by citizens with a view to collaboration and sharing of objectives, in such a way as to discover and enhance the city itself, thus involving civic culture and social capital.

In any case, the city cannot and must not be considered only the application of technology, and above all it must not dominate the interest and well-being of citizens, who must be the subjects capable of shaping the city without being overwhelmed; it will be the Smart Citizens who are changing cities and not the other way around, adapting them to the needs that will change over time, this will only be possible if it is possible to have a holistic and resilient vision, which brings harmony between all the components of an urban context.

Keywords Holistic and resilient vision · Smart citizens · Urban policy · Smart cities · Citizen participation · Urban planning

Introduction

The desire for aggregation and sharing is the basis of man, and given the forecast that in 2050 the population living in urban zones is envisaged to arrive at 6.7 billion, equal to about 66% of the world population, it is fundamental to rethink urban contexts.

Urbanization represents an opportunity: cities attract talent and investments, and the concentration of people favors a faster expansion of knowledge and a higher rate of innovation and encourages the development of infrastructure.

The same cultural customs are influenced by life in the city, but at the same time, urbanization also has a negative side: congestion and pollution are growing, due to smog and the production of waste, and above all there is desocialization and an increase in costs. In the absence of planning and rules, urban growth becomes the cause of multiple vulnerabilities, and cities have grown without an efficient order.

Hence, there is a need for a new city model, defined as a "smart city," but before being considered smart, given its definition, it is also necessary to rethink the concept of the city itself, in order to define what can be truly smart. It is obviously not the simple technology applied to an urban context. For too long, different disciplines have been kept at a distance, as if the engineering aspect could not be touched by the social, inclusive, and economic aspect, so every part of the whole remained disconnected.

This is why there is a need to review the city in a holistic way; the purpose of this book is to provide a general harmony to all the parts that make up the life of man in cities. The first requirement is to increase the quality of human life in a sustainable city context.

The city has been considered by different authors, from different perspectives, whether they are engineers, architects, economists, psychologists, politicians, citizens, etc., as a fundamental element of community life.

The concept of "ideal city" can already be found in Plato's ideas, where absolute geometries and radial and checkerboard urban structures come together to create real cities. We must go back to the Italian Renaissance to speak of the definition of "ideal city," where the aim is to make the organization of urban spaces more rational: man was placed at the center of the system, and the environment in which he lived was also revolutionized. For the first time in history, the concept of "ideal city" became a real city, projecting its enormous influence on contemporary urban planning.

The search for efficient, organized, and well-functioning cities has been carried on through the centuries by many great modern and contemporary architects around the world. We remember some great exponents of world culture who report the concept of the city, who defines "*the city as a projection of society on the territory, the city is a social totality or a society considered as a totality, including its culture, its institutions, its ethics, its values, in short its superstructures, including its economic basis and the social relations which constitute its proper structure*" (Biagi 2019, 13). "*The city is 'a space-time' and by means of this dimension we can model an ideal type*."

Through the concept of the "right to the city," we want to create the "city for citizens" by revitalizing participatory urban civil societies (Douglass and Friedmann 1998, 156).

While several works investigate the issues and dilemmas associated with such a grassroots urban participation policy, others also recommend its construction, spreading, or transformation in intensifying restructuring processes (Brenner et al. 2009, 176).

For architect Le Corbusier, his ideal city is understood as a new urban environment called the Ville Verte, (green city), a city in nature, but full of inhabitants and full of activities and social relations. Architecture and urban planning must necessarily provide answers to industrialized society; they must recreate a relationship between economic needs and human needs that contributes to the finding of a territorial balance for the city.

Today, the ideal city has been transformed into a smart city, which translates the concept of the ideal city into something that, thanks to technology, can be achieved, remembering that the ultimate goal must be the well-being of the citizen who lives it and capable of changing himself, adapting to the new lifestyles over time.

View of Smart Citizenship as an Active Key of Smart Cities

Many works compare Lefebvre's concept of the "right to the city" (Brenner et al. 2011, 54). This catch-phrase resonates as the desire to create the "city for citizens" by energizing the participatory urban civil societies (Douglass and Friedmann 1998, 156).

Smart city deploys and efficiently uses the local resources (including but not limited to the social, cultural, and financial capital, the natural resources, and information and technology) to:

- Improve the life quality of its population, commuter workers, students, and other visitors.
- Significantly improve its resource-efficient use and decrease impact on the ambient and augmenting resilience.
- Build a green and innovative-driven economy.
- Promote a good local democracy.

Although there is no univocal definition of the concept of smart city, aspects such as digitization, inclusion, quality of life, and sustainability are fully included in this term. Therefore, a city is smart when it is capable to support a good quality of life and a social increase of citizens and businesses, optimizing the resources and spaces. The starting point is the construction of a strategic, planned, and organic vision, linked to the ability of an institutional body to capture the potential of a territory and evaluate it from a long-term perspective. A smart city is therefore not only able to reduce emissions, increase the efficiency of buildings, reduce waste of resources, and produce clean energy but can also be digital because it processes data and shares information; it is able to create new knowledge. It also has a history and a culture, and its inhabitants have a proactive role since they must possess an ecological sensitivity that leads them to adopt virtuous behaviors, consistent with the goal of sustainability. The smart city manages resources in an intelligent way, envisages to become economically sustainable, and is careful with the life quality and citizen needs, also thanks to the level and type of services offered. In short, it is an attractive and sustainable territorial space that keeps pace with innovations and the digital revolution.

Smart city systems have emerged as an important response to the challenges of resource-efficient use, city economic development, and sustainability. This new way of experiencing the city will change the intersections between physical and social resources. Cities are more and more required to provide more services, to concur in a worldwide economy and to provide well-being to the population in a sustainable way. Thus, cities need to become smarter. However, it seems clear that the vision of a smart city in a purely technological key is not feasible and is indeed deeply connected to the dimension of smart citizenship.

It will be necessary to make possible the democratic and participatory process of citizens, through algorithms applicable to platforms, a key element of technology based on smart cities, created specifically to integrate the code of behavior of smart citizens, in order to obtain not only efficient services but also effective in terms of improving the quality of life in an urban environment.

The increasingly evident use of new communication and information technologies has profoundly changed the processes of citizen participation, with consequent risks of homologation, isolation, flattening of contacts, and exchanges, as evidenced even in times of pandemics.

It should be remembered that in addition to the global network, there is a network of people who, with their relationships, contribute to social interaction, a key element of the Smart City (Taylor and Lodato 2019, 35).

So technology, IoT, and ICT will be an opportunity to review not only the way in which decisions or votes are made, but also the way in which we inform and interact, in order to make decisions collaboratively, based on rules that guarantee not only the majorities but also the minorities and the continuity of the democratic structure.

Hence, a new concept of intelligent citizenship is born, developing in a conscious context in which the rules that intelligent citizenship will give itself will no longer be lowered from above but incorporated into the code of the platforms that will emerge from the citizens' initiative and will have important consequences on the quality of smart cities under construction.

However, much attention must be paid to the inclusion of citizens in the process of making decisions and planning processes for smart cities, since a citizen, who is capable of using technology, responsible, and capable of the right information to allow their participation in an authentic way, is needed.

With this in mind, it must be remembered that end users, in urban areas and beyond, are no longer mere consumers but prosumers who must be increasingly enabled to interoperate with smart grids, smart home systems, smart metering, and blockchain. These new methods of energy transition are slowed down by the inaction of the prosumers, who at the moment remain passive elements in a system that is constantly evolving and which instead requires active communities.

Smart Education for Citizens

The success factor of a technological innovation therefore also lies in the degree of acceptance by the consumer. This degree of acceptance can be fostered through an active participation of the end user in the initial stages of realization of the technological device, as demonstrated by studies based on user-oriented and participatory design, which confirm the importance of involving the consumer right from the stage planning and drawing of the devices, in order to make them easier to use (Davidoff et al. 2006, 19).

Based on these considerations, it is evident that the understanding of the individual and group behaviors within the living environments and the home is fundamental, as the failure to consider these factors could lead the consumer to perceive the introduction of innovation in daily life as a "threat," which will affect their daily life to the point of touching the dimension of individual freedoms (Hargreaves et al. 2010, 6111).

The term smart citizens has now become known and widespread, and we cannot ignore this key element to have smart cities that are really usable, but we must ask ourselves the question on how to train them and how to educate a citizen who is aware and socially useful. .

It is not a simple task because, in fact, there is nothing precise or standardized, and each city has different needs, in relation to its geographical position, ethnic traditions, economic development, etc. Cities are constantly evolving, and new approaches are needed. Training the citizens of the future will not only mean changing physical tools and teaching them how to use emerging technology; the education system must be able to prepare citizens for future challenges, offering the development of skills and knowledge (Sutterlin et al. 2011, 8137).

It is necessary to implement the teaching experience with themes increasingly linked to the use of IT tools, such as cloud computing, open source, and 3D animations, also in relation to the issues of privacy, big data, open data, participation in the democratic process, and awareness of information, which must be the basis of new forms of learning.

We must be careful not to underestimate the question of time, since the world is linked to new pervasive technologies such as IoT, ICT, smart city, and electric mobility, which change rapidly, and there is the risk of chasing obsolete knowledge such as the Achilles paradox and Zeno's tortoise.

The risk is not to provide sufficient tools to make the citizens of the future truly smart, able to be the discriminating element of the cities that must necessarily be created, since they will become the focus of most of the population, who will concentrate precisely in them.

It is necessary to start from education at all levels, to prepare the urban community for the cultural and social change that technology lends itself in bringing and integrating learning with the process of creating smart citizenship, making both the public and the public sector interact privately to pursue a common goal. If you want to create a useful Smart City, there must be smart citizens able to operate on it, an intelligent community created from education at all levels to ensure maximum social inclusion and the determination of a new urban community. From this perspective, a further mission of the school can be glimpsed, which will have to train digital, connected, and collaborative citizens (and the pandemic has shown how this is the only element that allows not to create social disparities, isolation, cultural depression, alienation), determining a new way of being a community, where the strength will be in the high number of elements carrying skills, ideas, abilities, and experiences that can lead to innovative processes for an urban future suitable for new generations.

The latter is an element that is often overlooked; we often think of the present generation without considering the young people, who will represent the users of what is being achieved today. Young people are almost automatically excluded from any decision-making process, while it appears essential to integrate them into the process of planning and conception of future smart cities. This must be remedied quickly, through teaching and collaboration between private and public sectors and the involvement of the entire community starting from childhood.

Smart Citizens Indicator

In recent years, there have been several approaches in order to establish what a smart city is, through standards, codes, indicators, definitions, matrices, and key elements.

What can be the strategies and/or the standard model, based on guidelines, to create the smart urban community? Would it be possible to think that the process of creating smart citizens could be homologated?

Indicators are used for decision making. The results of indicators, be they single indicators or multi-indicator-based assessments, should reach relevant decision makers.

It is clear that progress over time is important to users of city indicators. Therefore, the city indexes should be elaborated to be quickly included in the city program for the collection of regular surveys and statistics. The result of the indicators' collection process should be included in the city planning processes.

Other groups like research institutes and companies use indicators for projects and for city. For the citizens, the indicators' values can lead to the better understanding of complex projects and their impact. Thus, indicators to quantify and to protect the human rights and the performance of social basis of the cities are developed (Marsal-Llacuna 2017, 563).

In these same years, many initiatives have been developed at national and local level, such as the Canadian Index of Wellbeing (CIW), the measurement of the Gross National Happiness Index in Buthan, and the Measures of Australia's Progress, while in the United Kingdom the Measuring National Well-being (MNW) program, which has developed "a shared and reliable set of indicators that citizens can turn to to understand and monitor national well-being"; these models can be the starting point for the definition and construction of smart cities through the degree of satisfaction and livability of citizens.

(a) Canada has introduced an index, the Canadian Index of Wellbeing (CIW), which serves to provide a parameter that regularly certifies the quality of life of Canadians – at national, provincial, and local levels – to contribute to social change, which has for objective the well-being of the population at the center of political decisions. An initial analysis and calculation of the CIW was carried out in the years preceding the COVID event, and the next report will be in 2021; certainly the pandemic will have determined considerable variations. It will be interesting to verify in which category or domain, the COVID event, but this comparison will certainly be of great support to be able to understand what has changed in citizens' expectations and therefore move in that direction, helping policy makers to make conscious choices with respect to the definition of a new standard of "quality of life" for Canadians. In particular, the CIW has been structured in order to track changes in eight categories or domains of the quality of life, including:

1. *Community vitality*

This domain relates the quality of life to the urban communities to which it belongs.

It serves to give an index on what is happening in the neighborhoods, for example, on how safe you can feel, if you are integrated or engaged in community activities or rather if you are socially isolated.

The vitality of the community has two main dimensions: social relationships with the sub-dimensions on social participation, social support, community security, and social norms and values with the subdomain on attitude on the community.

2. *Democratic commitment*

This category quantifies the participation in the democratic processes of the various political organizations. Obviously, this factor is crucial as the population of an urban context, which has a large amount of democratic involvement and being able to participate freely in political activities, can formulate their opinions and share political knowledge.

In this way, through the participation of citizens, democratic values are supported at all levels of government; it allows to build a climate of relationships and trust, which leads to a true democracy, based not on pre-election commitment but continuous throughout the mandate.

The sub-dimensions are the following:

- Participation.
- Communication.
- Leadership.

3. *Education*

As already mentioned in the paragraph on training smart citizens, education is a key fundamental element for every citizen worldwide and is a mirroring of our capacity to interact and have resilience in the society. This indicator is the basis of other indicators such as democratic participation, living standards, and health; it is not seen only as a domain linked to the acquisition of notions and classical education, but rather in consideration of informal learning; this is because smart communities encourage the thirst for knowledge at all ages and stages, and therefore some less conventional indicators of lifelong education are used.

The sub-dimensions are the following:

- Social and emotional expertise.
- Basic educational comprehension and expertise.
- Global academic accomplishment and involvement.

4. *Environment*

This serves to identify trends in the existence and utilization of natural resources in Canada's ambient. This domain is not only analyzing the environment sustainability but investigates the natural local resources, their variation in time, and the impact on the environment. The sense of this choice arises from the consideration that the well-being of human beings also depends on the state of the environment. Finally, accessible energy and materials, clean air and water, surface of wild areas, variety of species, and the resources play a fundamental role, so it appears to be an indicator that acts on various aspects.

5. *Healthy population*

This domain deals with measuring the state of health, lifestyle, and behavior of citizens and the factors that affect the health system; this is to evaluate whether the various aspects of human well-being are improving or worsening; and after the pandemic event, it will likely have a significantly changed value and importance, compared to pre-COVID.

The dimensions of the state of health were considered:

- Personal well-being.
- Physical and mental health conditions.
- Life duration and mortality.
- Functional well-being.

The dimensions as lifestyle and behavior and public health and healthcare quantify the factors influencing our well-being and are often influenced by public policies and health actions.

6. *Leisure and Culture*

The sphere of leisure and culture serve to define the interaction with the extracurricular activities, arts, and culture. Active participation has a positive effect on comfort by adding to better body health and raising opportunities to socialize, relax, and learn new things. Another important domain is understanding the benefits of these activities: why people involve in the activities and what needs are fulfilled. The third is the free time utilization or its significance for people related to quality of life. The last component involves the supplying of cultural possibilities.

7. *Living Levels*

This domain quantifies the value and spreading of income and well-being, analyzing poverty shares, income variations, job market safety and profession quality, and subsistence safety. It also considers basic necessities such as food, security, and reasonable housing.

The main income indicators are grouped into four components:

- Medium and median income and well-being.
- Income and spreading of well-being.
- Variability of income.
- Economic safety.

8. *Time Utilization*

The time utilization domain indices measures the spending of time, how this use is felt, what influences the use of time, and how this time is influencing our well-being. It is composed of four dimensions, which together provide the sensory experience of time:

- Time.
- Timing.
- Rhythm.
- Temporality.

This dimension is composed of indexes that accounts the period used for certain activities and the specific moment during the day. This implies how people can decide on the utilization of their time and the intensity in using it.

An important factor in the process of constructing the indicators was the "bottom up" procedure, for the creation of the eight domains and the criteria for choosing the indicators: in fact, the researchers started by listening to candidate citizens to identify what is important for the quality of their life.

(b) Another interesting experience was developed by the Center for Bhutan Studies (CBS), which proposed the development of indicators and collected in the Gross National Happiness (GNH), to quantify the progress and happiness of local society (Gross National Happiness. 2017). There were nine areas identified that contribute to people's well-being:

  (a) Earnings.
  (b) Health.
  (c) Education.
  (d) Governance.
  (e) Environment.
  (f) Time utilization.
  (g) Quality of life.
  (h) Cultural attractiveness.
  (i) Community strength.

The GNH index measures well-being holistically and looks at each person's GNH profile, using key well-being indicators grouped into nine domains. Each of the nine domains has the same weight. Within the domains, the weights of the indicators are different and identified, as shown in the examples in Table 1.1.

A profile is realized for each person, and the indicators are evaluated leading to a GNH score measuring the satisfaction level of the specific person.

(a) Statistics were obtained in 2002 in Australia collecting data on the society, economy, and ambient to release information on the local progress and adjust the decision making for social, economic, and environmental areas' improvement (Australian Bureau of Statistics 2012).

Table 1.1 Gross National Happiness (GNH), Center for Bhutan Studies

Domain	Indicators	Indicator weight
Psychological well-being	Life satisfaction	1/3
	Positive emotion	1/6
	Negative emotion	1/6
	Spirituality	1/3
Health	Self-reported health status	1/10
	Number of healthy days	3/10
	Disability	3/10
	Mental health	3/10
Time utilization	Working hours	1/2
	Sleeping hours	1/2
Education	Literacy	3/10
	Schooling	3/10
	Knowledge	2/10
	Values	2/10
Cultural diversity and resilience	Artisan skills	3/10
	Cultural participation	3/10
	Native language speaking	2/10
	Code of conduct	2/10
Good governance	Political participation	4/10
	Services	4/10
	Governance performance	1/10
	Fundamental rights	1/10
Community vitality	Donation (in time and money)	3/10
	Safety	3/10
	Community relationship	2/10
	Family	2/10
Ecological diversity and resilience	Wildlife destruction	2/5
	Urban issues	2/5
	Environment responsibility	1/10
	Pollution	1/10
Living standard	Income per capita	1/3
	Assets	1/3
	Housing quality	1/3

(b) In Germany, the federal government considers well-being as a guiding principle for the definition of the policies it pursues, just like the economic, social, and ecological objectives. The government believes that in order to be effective, the policy must contain the various aspects of well-being and the interrelationships between them, but also of businesses and industry, of society, and, last but not least, of each individual citizen, as the well-being that is not synonymous with the individual pursuit of happiness.

The welfare indicator system in Germany uses twelve dimensions and forty-six indicators to assess key parts of welfare. There are also other topics that include liberty, justice, migrant integration, welfare ratio between urban and rural, and sustainability. In principle, all dimensions and indexes are considered balanced important. All dimensions of welfare are important from the point of view of good governance (Helliwell et al. 2015, 40).

(c) The Hong Kong Quality of Life Index was first developed by the Chinese University of Hong Kong in 2003 for quantifying the well-being of the local population using twenty one indicators across social, economic, and ambient.
(d) In Europe, Switzerland developed sixteen indicators to determine if the country is realizing improvements toward sustainable development.

These are just some examples of the possibility of being able to measure citizen's smartness through the concept of well-being; the latter has been parameterized and evaluated in different ways, with different elements in common, such as education, participation, and inclusion rather than just technological skills. .

In reality, it remains fundamental that it is necessary to measure what interests the people themselves, in order to obtain indicators, which, in relation to the type of citizen, can give the correct and appropriate indications to create a smart city as the citizens themselves wish.

Conclusion

If the citizen will be regarded just as a customer or consumer, then the work of politicians will result in an efficient, low-cost service provider. The government can be influenced by other interests and citizens, losing trust, and the democratic relationship with institutions will get used to passivity and ignorance.

It must become clear that not focusing on people can lead to less utilized systems. Furthermore, the most stimulating and relevant part would be missing: the connection of people with high technology influences their well-being directly or not, so if you focus exclusively on technology, without a holistic, complementary view on the social, it is entirely possible that you will arrive at solutions that are not really as effective – nor as fair – as they should be.

As long as political considerations and choices are translated into a technological solution, with the dystopian vision that can solve problems autonomously, ignoring human capital, there will be no smart city, no well-being, and no evolution.

Unfortunately, there is a lot of commercial impact and the apparent inability for decision makers to correctly visualize this situation. Perhaps we need to point out that intelligence lies in the degree of citizen involvement, which cannot be provided through technology alone.

The creation of technology platforms will contribute to the equity of urban social systems. This, however, is a large period that would increase participatory planning process that involves experimental, educational, and people mobilization, and research and policy activities within a project will be shared.

A smart city without an intelligent community is unthinkable, but "smart" definitely does not imply solely the possibility to quickly solve engineering and mathematical problems as human needs are of a completely different kind. For example, if you want to solve the problem of pollution or waste, it is not enough to have efficient systems or innovative separate collection systems if only a small group of people use them.

The concept therefore returns that technology is necessary but not sufficient and must be subservient to the well-being of the community and with the strong element of inclusive participation.

The way citizens' voices can be heard (e.g., through well-being indicators) and make a difference in planning processes, called "collaborative turning point", can be the true basis of smart citizens for smart cities.

The responses of citizens of different countries, in relation to smart cities, can be very different, also in relation to the degree of progress of the urban context in which one lives. It is clear that when the transport systems already work well, it will not represent a fundamental element to find your own well-being.

In general, however, it can be concluded by saying that the most important guidelines for smart citizens include realization of innovation, improved public transport, friendly environment, increased education development, clean energy, infrastructure and digital services, better city control, realization of local businesses, and new employment and preservation of natural resources.

Cities need to be reviewed in a broad, non-sectorial perspective, in a holistic approach and, as Lefebvre wrote, "dealing with everyday life means changing it" and that "the city must be understood as a work of art," just like its citizens.

References

Australian Bureau of Statistics (2012) Measures of Australia's Progress: aspirations for our nation: a conversation with Australians about progress. Report https://www.nintione.com.au/?p=5336. Accessed 10 January 2021

Biagi F (2019) "Henri Lefebvre e la città"come opera d'arte. Note di teoria critica urbana. The Lab's Quarterly 12(3):7–22

Douglass M, Friedmann J (1998) Cities for citizens. Wiley:151–164

Brenner N, Marcuse P, Mayer M (2009) Cities for people, not for profit. City 13(2–3):176–184

Brenner N, Marcuse P, Mayer M (eds) (2011) Cities for people, not for profit: critical urban theory and the right to the city. Routledge

Davidoff S, Lee MK, Yiu C, Zimmerman J, Dey AK (2006) Principles of smart home control. In: Dourish P, Friday A (eds) UbiComp 2006: Ubiquitous Computing. UbiComp 2006. Lecture notes in computer science, vol 4206. Springer, Berlin, Heidelberg

Gross National Happiness (2017) A compass towards a just and harmonious society 2015 GNH survey report http://www.grossnationalhappiness.com/wp-content/uploads/2017/01/Final-GNH-Report-jp-21.3.17-ilovepdf-compressed.pdf. Accessed 10 January 2021

Hargreaves T, Nye M, Burgess J (2010) Making energy visible: a qualitative field study of how householders interact with feedback from smart energy monitors. Energy Policy 38(10):6111–6119

Marsal-Llacuna ML (2017) Building universal socio-cultural indicators for standardizing the safeguarding of citizens' rights in smart cities. Soc Indic Res 130:563–579

Helliwell JF, Richard L, Sachs J (eds) (2015) World happiness report 2015. Sustainable Development Solutions Network, New York

Sutterlin B, Brunner TA, Siegrist M (2011) Who puts the most energy into energy conservation? A segmentation of energy consumers based on energy-related behavioural characteristics. Energy Policy 39(12):8137–8152

Taylor S, Lodato T (2019) Actually existing smart citizens. City 23(1):35–52

Chapter 2
Integrated Photovoltaics: An Opportunity for Urban Decarbonization

Laura Aelenei and Alessandra Scognamiglio

Abstract The need for shifting to urban renewable energy integration and use is critical when the goal is decarbonizing the global energy system. Due to cities' integrating framework, the energy transition in urban environment towards renewable energy can be achieved in all end-use sectors, not only in power but also in heating, cooling and transport. Solar energy integration in urban environment can contribute to the transition towards low-carbon energy systems with reference to the city operation and the city life, promoting on-site energy production and enhancing self-consumption, if integrated into the overall building/district energy system and coupled with electric or thermal storage.

In the first part of this chapter, the authors present the smart city approach as an integrator of smart energy solutions and associated solutions as, for example, the Positive Energy District model together with several initiatives and networks working in the development and implementation of the PED model. The second part of the chapter is dedicated to the Integrated Photovoltaic (PV) systems as a key parameter towards decarbonization using solar energy. The state of knowledge of PV integration from building level and pioneer examples are presented.

The chapter is ending with identification of the key challenges of PV integration in smart city context.

Keywords Integrated photovoltaics · Urban decarbonization · Positive Energy District

L. Aelenei (✉)
Laboratório Nacional de Energia e Geologia, Amadora, Portugal
e-mail: laura.aelenei@lneg.pt

A. Scognamiglio
ENEA Centro Ricerche Portici, Naples, Italy
e-mail: alessandra.scognamiglio@enea.it

G. C. Lazaroiu et al. (eds.), *Holistic Approach for Decision Making Towards Designing Smart Cities*, Future City 18,
https://doi.org/10.1007/978-3-030-85566-6_2

I. Smart City: The Integrator of Smart Energy Solutions

Introduction

According to Renewables in Cities 2019 Global Status Report (REN21-2019),[1] cities are responsible for around two-thirds of global final energy use, and consequently, cities account for an estimated 75% of global carbon dioxide (CO_2) emissions. Thus, the need for shifting to urban renewable energy integration and use is critical when the goal is the decarbonizing the global energy system. Due to cities' integrating framework, the energy transition in urban environment towards renewable energy can be achieved in all end-use sectors, not only in power but also in heating, cooling, and transport.

In 2015, United Nations Member States adopted the 2030 Agenda for Sustainable Development, which presents 17 Sustainable Development Goals (SDGs), with the main objective of setting a global framework and partnership towards a sustainable transformation of the society and its financial, economic and political systems. Between the proposed goals, cities play an important role (Goal 11) towards sustainability and resilience in the urban areas. On the other hand, the European Union (EU) fully recognizes the significance of sustainable development and aims to make Europe a global role model in energy transition by setting targets on greenhouse gas reduction, renewable energy production and energy efficiency for 2020 and 2030, which can pave the way to the long-term goal of a climate neutral economy by 2050.[2] With 40% of energy consumption and 36% of CO_2 emission allocated to EU buildings sector, the large-scale deployment of energy efficiency measures and renewable energy integration and use are seen as a key factor towards the climate neutral strategy of built environment transformation. Since 2010, with the publication of EPBD recast,[3] Europe has made significant progress in building level innovations as seen in the development of Net Zero Energy Buildings (NZEBs). However, in order to fulfill climate neutral targets by 2050, there is a need for scaling up solutions and strategies applied from building level to the district level and city level.

---

[1] Renewables in cities 2019 global status report, https://www.ren21.net/wp-content/uploads/2019/05/REC-2019-GSR_Full_Report_web.pdf

[2] European Commission. (2018). A European strategic long-term vision for a prosperous, modern, competitive and climate neutral economy. Communication from the Commission COM(2018) 773 final.

[3] European Union. (2010). DIRECTIVE 2010/31/EU on the Energy Performance of Buildings (recast). Official Journal L153, pp. 13–15.

Positive Energy Districts

New solutions reach visibility in the stimulation of the energy transition in urban context as, for example, the initiative on Positive Energy Blocks (PEB) in 2016[4] promoted by the European Innovation Partnership on Smart Cities and Communities (EIP-SCC). This initiative promoted the concept of PEB, a connected neighbouring building that actively manages their energy consumption and the energy flow between them and the wider energy system generating more primary energy than what they use annually.

In June 2018, the Strategic Energy Technology Plan (SET Plan) of the EC started a new Action 3.2 'Smart Cities and Communities' with the main objective of developing integrated and innovative solutions for the planning, deployment, and replication of Positive Energy Districts (PEDs).[5] This pioneering concept of PEDs is built on the paradigm of smart cities in the way that the city district is in fact an aggregator of the smart integrated solutions per sector and disruptive technologies and can be considered a systematic approach towards a sustainable urban transformation. According to SET Plan Action 3.2, the main targets of PED model implementation are (i) to bring smart energy systems capable of handling, in a smart way, buildings' energy demands with variable renewables and energy storage in order to meet the target of a smart NZEB; (ii) to develop the necessary interfaces and disruptive technologies that guarantee the connection and interaction between buildings, the urban energy system and the users; and (iii) to provide information and data on demonstration of the smart technologies and their impact on built environment energy optimization.

Initiatives on PED Urban Transformation

After the publication of the SET Plan Action 3.2, many initiatives have started working on the topic of PED with the objective of supporting the development and implementation of at least 100 PED by 2025. Although the initiatives adopted different working plans and strategies, there exist a common basic understanding about the holistic approach of this concept of PED, and this includes the implementation of energy efficiency measures and strategies in different PED sectors and the integration and use of renewable energy systems in PED environment to offset the PED energy demands. Complementary to this, the system energy flexibility should assure

---

[4] EIP-SCC. (2019, September 01). Initiative on Positive Energy Blocks. Retrieved from https://eu-smartcities.eu/initiatives/71/description

[5] SET-Plan Temporary Working Group 3.2. (2018, June). SET-Plan ACTION n°3.2 Implementation Plan: Europe to become a global role model in integrated, innovative solutions for the planning, deployment and replication of Positive Energy Districts. Retrieved from https://setis.ec.europa.eu/system/files/setplan_smartcities_implementationplan.pdf

the integration and interaction between PED systems, local grids and users. Moreover, the concept of PED should take into account inclusiveness aspects and non-technological aspect as, for example, social innovation and economic, environmental and regulatory aspects. In the following, three examples of initiatives on PED will be shortly described.

EERA (European Energy Research Alliance) JP (Joint Programme) on Smart Cities

EERA Joint Programme Smart Cities (EERA JPSC)[6] is one of the initiatives (joint programme) of the European Energy Research Alliance. Its mission is to contribute to research and innovation in smart cities – both in the development of fundamental research, innovation and co-creation with city and industry partners and in showcasing the importance of research and innovation in a field that is more and more prone to high-TRL (Technology Readiness Levels) demonstration projects. As such, during the last 10 years, EERA JPSC adapted its work plan according with European Research Needs and Politics regarding the energy in cities.

Since 2018, EERA JPSC adopted its work plan in order to support the strategy and mission of SET Plan Action 3.2 and to support the development and implementation of PEDs in Europe. The program proposed by EERA JPSC is structured in five strategic modules such as the following: Module 1 Towards European Positive Energy Cities, Module 2 PED Labs, Module 3 PED Guides and Tools, Module 4 PED Replication and Mainstreaming and Module 5 Monitoring and Evaluation.

PED-EU-NET Positive Energy District European Network

COST Action 19126[7] project was started in 2020 within the framework of the European Cooperation in Science and Technology (COST), a funding organization for the creation of research networks, called COST Actions. This COST Action on PEDs has as main objective and mission to tackle the main challenges related to PED development by calling on relevant stakeholders in academia, industry, cities and communities to actively participate in the open innovation process that transcend disciplinary, administrative and hierarchical borders. The backbone of this COST Action lies in the network of dedicated participants, who will not merely share know-hows and exchange ideas but will work together to pool resources, experiment new methods, undertake challenging missions, co-create original solutions, advance science and build capacity. The structure of this COST Action takes

---

[6] https://www.eera-sc.eu/strategic-research-agenda.html

[7] https://www.cost.eu/actions/CA19126/#tabs|Name:overview

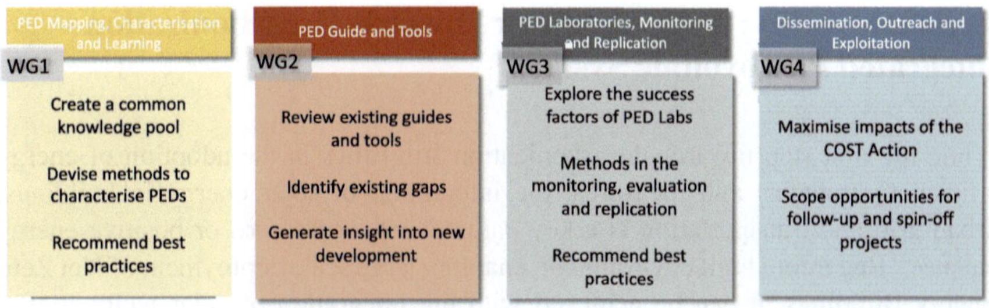

Fig. 2.1 Work plan structure, COST Action 19126

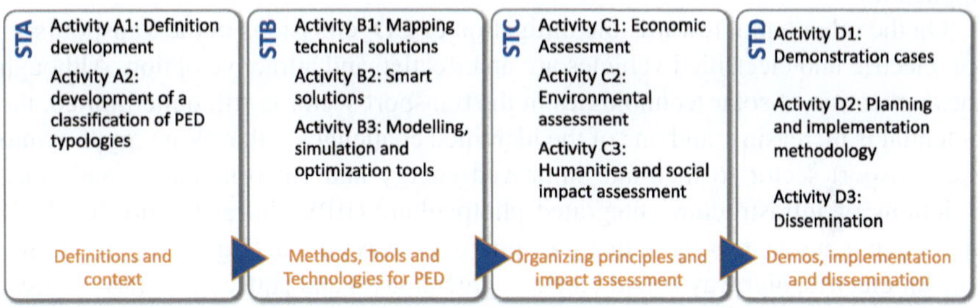

Fig. 2.2 Work plan structure of the Annex 83

reference to the 'Pathway towards PEDs' outlined in the SET Plan Action 3.2 Implementation Plan. The structure of the four-year work plan is illustrated in Fig. 2.1.

IEA-EBC-Annex 83 Positive Energy District

This project starts in 2020 as well under the framework of International Energy Agency (IEA) and Energy in Buildings and Communities Programme (EBC). The aim of Annex 83[8] is developing an in-depth definition of PED and the technologies, planning tools and planning and decision-making process related to Positive Energy Districts. Experience and data to be used in the Annex will be gained from demonstration cases. The structure of the project is illustrated in Fig. 2.2.

All these initiatives together with Horizon 2020 framework have the main scope of implementing the PED concept in European cities.

[8] https://annex83.iea-ebc.org

II. Solar Energy: Key Parameter Towards Decarbonization – Integrated Photovoltaic Systems

While the first step towards decarbonization still relies in the adoption of energy efficiency strategies and measures, the integration of solar energy in buildings, urban grid and transportation is a key parameter towards zero or positive energy balance.[9] Regarding built environment, enabling large scale deployment of Net Zero Energy Buildings, it can be achieved with the integration of solar technologies, especially PV as Building Integrated Photovoltaic (BIPV) solution offering huge opportunities for new buildings and as retrofitting solutions in particular in facade applications.

On the other hand, towards the mitigation of CO_2 emissions in the transport sector, electric and electrified vehicles are an effective and attractive option. Although the application of solar technologies in the transport sector is still quite limited, the potential is increasing, and one of the identified challenges is the PV integration into the transport sector to facilitate improved energy and environmental conditions. Solutions of infrastructure-integrated photovoltaic (IIPV) infrastructure has been applied in many sectors including transport not always urban (e.g. on top of parking lots for electric, highway sound barriers, greenhouses and aquaculture farms, street lighting).

In the following, some of these PV-integrated solutions will be addressed as the state of knowledge and also future challenges.

Built Environment Integration: Building Integrated Photovoltaics (BIPV)

State of Knowledge

Building Integrated Photovoltaic (BIPV) is already considered an emerging solar power generation technology due to its multifunctionality as integrated building component assuring promising returns in efficiently capturing solar energy, improved energy efficiency and thermal comfort in buildings, generating electricity in building footprint. After a long period of intensive studies, this technology has gained visibility over other traditional solutions, and their market acceptance has been influenced by the progress in PV technology development in which respect its performance and cost and also has gained attention of the architects and leadings real projects. Moreover, the interest in these applications and technologies have

---

[9] Aelenei, L., Petran, H., Tarrés, J., Riva, G., Ferreira, A., Camelo, S., Corrado, V., Šijanec-Zavrl, M., Stegnar, G., Gonçalves, H., Magyar, Z., Salom, J., Polychroni, E., Sfakianaki, K., (2015). New Challenge of the Public Buildings: nZEB Findings from IEE RePublic_ZEB Project, Energy Procedia, 78: 2016–2021. DOI: https://doi.org/10.1016/j.egypro.2015.11.195

been driven also by political framework and legislation, in particular by the directives Directive 2010/31/EU and the present Directive 2018/844[10] aiming to lead the construction section to build high performance new and retrofitted buildings to achieve nearly zero or net-zero energy balances.

The integration of solar energy system such as BIPV systems is a key aspect for further development of NZEBs and deployment at large scale.

During last 20 years, the building integration of solar energy systems has been studied in different design and corresponding techniques, according to geographical location and local climatic conditions and the use of buildings. In the last years, the attention is paid not only in improving solar technologies but also in the integration methods in order to improve the system efficiency and at the same time to fulfil architecture requirements. There is a common understanding that the key parameters that should be considered in the design of building integrated solar systems are integration and the type of integrated building element, the operation mode, the application and service.

Integration at Building Level

As previously referred, the studies regarding integration of photovoltaics solar systems in buildings indicate that a proper design and integration in building elements (roof and walls) can substantially improve their efficiency.

According to the publication on the building integrating issue,[11] the BIPV can be found integrated as external building envelope element such as curtain wall, sunshading and balconies or effective integration when the BIPV can assume the function of a structural element (walls or roofs). These systems can be integrated also in roofs as solar tiles or in combination with skylights.

An important design issue that should be taken into account in the parametrization of these systems and in order to improve their efficiency is the consideration of an air gap formed between the PV panels and structural building envelope element. This air gap for one hand permits that the PV panel is ventilated naturally or mechanically and so can keep its efficiency compared with the case when the PV panel is simply added to the building envelope and so can produce overheating of the panel and lowering its efficiency. This kind of system is usually identified as BIPV/T, Building Integrated Photovoltaic/Thermal, which is generating not only electricity but also heat obtained in the conversion process and transported through the system air gap inside the building and successfully recovered by natural or mechanical ventilation and used for heating interior ambient or released outside in the cooling period. The air cavity of a BIPV/T has a crucial role in the overall

[10] Directive 2018/844, https://eur-lex.europa.eu/legal-content/EN/TXT/PDF/?uri=CELEX:32018L0844&from=EN

[11] Aelenei, L., Pereira, R., Ferreira, A., Gonçalves, H., Joyce, A., (2014). Building Integrated Photovoltaic System with Integral Thermal Storage: A Case Study, Energy Procedia, 58: 172–178. DOI: https://doi.org/10.1016/j.egypro.2014.10.425

system performance, and its geometry and dimension could highly influence the air flow, heat transfer and efficiency.

Different types of integration and their respective functions have been studied according to their final application.[12] Integrated solar systems are designed to displace the traditional building components totally or partially, assuring a cross-functional role. Building Integrated Photovoltaic/Thermal was firstly designed and used in north climate architectures where the climatic conditions present a high solar radiation and high temperature gradient between indoor and outdoor temperature. A pioneer example is the ÉcoTerra residential house in Quebec, Canada[12].

ÉcoTerra has a 55 m^2 Building Integrated Photovoltaic/Thermal (BIPV/T) collector on the upper part of the roof. The outer layer of the roof consists of 21 amorphous silicon photovoltaic modules, each having an area of 2.1 m^2, a power rating of 136 W and a nominal efficiency of 6.3%. The total PV array peak capacity is 2.86 kW (direct current). This has the dual benefit of superior architectural integration and providing protection to the upper end of the modules using the roof cap. In sunny heating period, during the day time, the thermal energy released in the conversion process heated the air inside the BIPV/T roof system, and the hot air is collected and conducted to the interior of the house, by natural convention. In sunny conditions, the air temperature can increase by up to 40 °C from inlet to outlet of the collector.

Years later and due to the buildings' new legislation requirements, e.g. nearly zero energy buildings and design trends, these application starts spreading as an integrated solution in buildings located in other geographical area with different climatic conditions; in South Europe countries, the BIPV/T technology is being designed also for a good functionality in cooling period. Different aspects and challenges were addressed regarding the integration of BIPV/T in Mediterranean climate, where through experimental[13][14] or numerical[15] analysis the authors try to better design and apply these kinds of systems according to the buildings' needs and climatic conditions.

An example of such system is illustrated below which belongs to Solar XXI office nZEB building, located in Lisbon (Figs. 2.3 and 2.4).

The integration of photovoltaic systems in Solar XXI was a first priority in the building design, since the project stage. Being a demonstration building and a case

---

[12] Smith, M., Aelenei, L., (2017). BIST Classification and characterization. In: A.S. Kalogirou, ed., Building Integrated Solar Thermal Systems, Design and Applications Handbook, 1st ed. COST Action TU1205, pp. 7–21.ISBN 978-9963-697-22-9.

[13] Bot, K.; Aelenei, L.; Gonçalves, H.; Gomes, M.G.; Santos Silva, C. (2021). Performance Assessment of a Building-Integrated Photovoltaic Thermal System in a Mediterranean Climate—An Experimental Analysis Approach, Energies 14 (8), 2191, https://doi.org/10.3390/en14082191

[14] Lourenço, J.M.; Aelenei, L.; Sousa, M.; Facão, J.; Gonçalves, H. (2021). Thermal Behavior of a BIPV combined with water storage: an experimental analysis, Energies 2021, 14(9), 2545; https://doi.org/10.3390/en14092545

[15] Bot, K.; Aelenei, L.; Gomes, M.G.; Santos Silva, C. (2020). Performance Assessment of a Building Integrated Photovoltaic Thermal System in Mediterranean Climate—A Numerical Simulation Approach. Energies, 13, 2887.

Fig. 2.3 Solar XXI. (Gonçalves, H., Camelo, S., Horta, C., Graça, J. M., Aelenei, L., Oliveira Panão, M., Joyce, A., Rodrigues, C., Ramalho, A., Rocha e Silva, A., (2010). SOLAR XXI Em direção à energia zero/Towards zero energy, LNEG, Laboratório Nacional de Energia e Geologia, IBSN: 978-989-675-007-7)

Fig. 2.4 SOLARXXI building, Lisbon, photo by Helder Gonçalves

study in the field of renewable energies, 'solar photovoltaic' alongside 'solar thermal' (active and passive) was an integral and fundamental part of its design. Several challenges surrounded this option. The integration of the photovoltaic modules was done to facilitate the use of their generated thermal heat in winter, as in the case of BIPV/T designed in north climate conditions. The vertical cavity created between the photovoltaic modules and the building structure is naturally ventilated, and the heated air inside the cavity is used for space heating purpose during a sunny day in heating period or evacuated to the exterior environment to avoid overheating during cooling period. BIPV/T system installed in Solar XXI building facade include at the superior and inferior part interior and exterior ventilation vents in order to permit the air circulation with the interior environment (heating period) or exterior environment (cooling period) as it can be observed in Fig. 2.5.

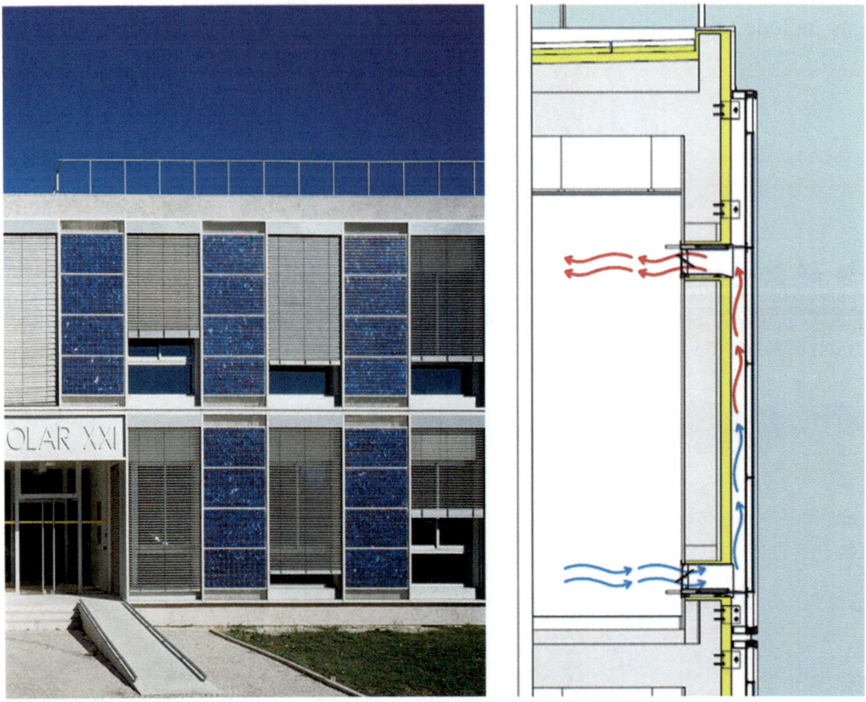

Fig. 2.5 BIPV/T Solar XXI building façade. (Gonçalves, H., Aelenei, L., Rodrigues, C., (2012). 'SOLAR XXI: A Portuguese Office Building towards Net Zero-Energy Building', REHVA Journal, ISSN: 1307-3729)

A photovoltaic system was designed to be integrated into the south facade of the building with about 100 m^2 of multicrystalline silicon modules, totalling about 12 bkWp of direct supply of electricity to the building.

Integration at Community/District Level

There already exist several projects where building photovoltaic are successfully integrated at district level. One of the first projects is the Net Zero Energy Buildings Cluster: Plus Energy Settlement, Freiburg, Germany, designed by Rolf Disch (Fig. 2.6).

The Plus Energy Settlement in Freiburg is an example where the zero energy balance is achieved in the frame of an estate. Some of the 59 built terrace houses have a positive, others a negative primary energy balance. The 59 terrace houses are made up houses with positive energy balance and others with negative energy balance, while the overall average is positive. The efficient row houses are covered with 3150 m^2 of roof top integrated PV generators. The heat is supplied by district heating. The efficiency of the houses is based on the Passive House concept, and a

Fig. 2.6 Fully roof-integrated BIPV design at the Plus Energy Settlement, Freiburg, Germany, (Aelenei, L., Waldren, D., Aelenei, D., Cúbi, E., Wittchen, K., Kim, J., (2017). Net ZEB case study buildings, measures and solution sets. In: F. Garde, J. Ayoub, L. Aelenei, D. Aelenei, and A. Scognamiglio, ed., Solution Sets for Net Zero Energy Buildings, Feedback from 30 NZEBs worldwide, 1st ed. Wiley-Ernst & Sohn. ISBN 978-3-433-03072-1). (Photos by Laura Aelenei)

consequent (urban) planning for shadow-free south orientation, position and shape of the buildings[11].

In the new context of Positive Energy Districts, the integration of renewable system plays a key role, so in the following chapter, the challenges of PV integration will be addressed.

III. Key Challenges of PV Integration

The considerations that have been done so far allow to frame the use of photovoltaics in a new smart city-related landscape. This new perspective implies that the concept of BIPV needs to be updated so as to include all the possible needs typical of a building in a smart city.

Focusing solely on the energy perspective, the wide use of renewables has changed the energy scenario, and in consequence our conventional, centralized system of energy generation is going to be replaced by a 'web' of energy generation systems, to be managed in the appropriate way to ensure a good performance of the energy grid, and the appropriate energy savings.

According to this perspective, the city can be seen as a complex system where the energy demand interacts with the energy generation from different renewable systems, which produce energy at different times and in different ways. A crucial role is played by the development of solutions and approaches that enable for matching the building energy demand and the offer from the grid.

Before going any further, it is worth to describe in a systemic, inclusive framework the hierarchy existing in the smart city environment, according to which the different exchanges of energy and information happen. The holistic approach named as smart city has been stimulating the development of new guide concepts for the city development, such as the 'smart community', that participates in the city development process through a wide use of ICT technologies and connectivity.

The novelty of the smart city approach relies in seeing from a unique perspective several aspects related to the city functioning, which were seen as separated ones until the recent past. The result is a vision of the city as a complex system of interconnected networks (transport, energy, buildings, lighting. Social relationships, water and waste, etc.).

Three different levels can be used in describing a possible smart city scheme.

The first level is the 'city government': the whole government and decision system of the city. This includes many topics: security, risks monitoring (food and environment), services for the citizens (e-Government), urban planning as well as transport planning and management. At this level, decision support systems are the most relevant methodologies of approach.

The second level can be defined as 'city operation'; it includes all the aspects for the management of the urban services in real time. It includes therefore the networks for the distribution of energy, water and waste, but also other grids like the buildings network, the public lighting network, the communication network and the security network. For each of these networks, city utilities (actors) operate to ensure the correct management and functionality of the networks. The smart city approach implies a complex city infrastructure that control and manages all the other networks. This smart urban infrastructure is made of sensor networks (related with buildings monitoring, mobility, air quality, etc.) and data transmission systems, coupled with a system of servers for collecting, elaborating and distributing (open data, cloud computing) the data and for sending them towards the vertical application of the different utilities that manage the networks. Thanks to this hybrid structure, the utilities are able to operate directly on the data on the resource demand ('resource on demand') by connecting themselves to the adjacent domains which have an influence on the demand (e.g. public lighting and traffic). This way the infrastructure costs are shared among the different utilities.

The third level can be defined as 'city life'. It is the whole system of services that are used by citizens in the everyday life. At this level the main issue is the smart community, that is, information accessibility, public involvement in the city development, participation and social inclusion. The main topics are public health, education, culture, security and privacy; the technologies used are the ones that enable advanced systems of communication between the citizen and the city itself, between the citizen and the structures of the city and between different citizens. This kind of communication can happen in the urban context (smart phones, tablets, PC, urban interaction) or in the buildings (smart homes, smart appliances, digital school, e-health, etc.).

PV can contribute to the transition towards low-carbon energy systems with reference to the city operation and the city life, promoting on-site energy production and enhancing self-consumption, if integrated into the overall building/district energy system and coupled with electric or thermal storage. Nevertheless, in order to allow the advancement of PV in the smart city direction, there is a need for developing suitable devices able to guarantee data exchanges between energy production and consumers to adequately manage received information to consequent actions decision.

In Fig. 2.7, main integration domains are illustrated, where PV plays, for multifaceted reasons, a crucial role.

From the perspective of interconnected domains of Fig. 2.7, it is possible to design an urban pattern, reducing energy consumptions due to a widespread ground use, such as transports and energy carriage. Moreover, it is possible to place renewables in the whole cluster, such as buildings' surfaces or suitable urban areas, in order to supply energy for the entire cluster and reach the net zero energy balance.

This paradigm shifts need some new research. In the approach of positive energy district or smart city, there is a general specific need for controlling the operation and maintenance of the PV system. that means the continuous monitoring and the use of predictive measures in order to match continuously the energy demand with the PV production and for guaranteeing the energy exchanges among different systems. This translates into matching on-site available renewable sources, i.e. through BIPV (choice of the most suitable surfaces; continuous monitoring, predictive measures), and to effectively use the locally produced energy in the different energy balance domains (i.e. integration of PV systems in DC networks). Moreover, considering the high number of RES-based systems that inject energy into the grid, at

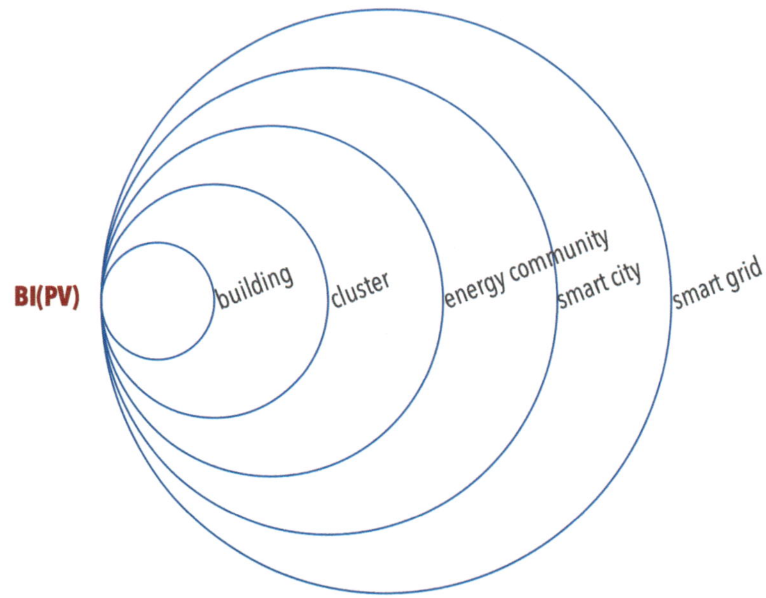

Fig. 2.7 PV in the context of enlarged smart city perspective

the smart grid level, all exchanges have to be controlled (development of forecasting tools) and managed[16,17] and new grid approaches have to be developed and adopted.

Different solutions and grid ancillary services (AS) have been proposed. For example, DC microgrids implementations are considered as promising architecture to 'embed' VERs avoiding their energy introduction in the national grid. PV and wind sources are locally managed to frequency instabilities mitigation and virtual inertia service provision.

PV can represent the most suitable technology for connecting several distributed PV generators, thanks to the scalability of such technology and to the fact that PV can guarantee to distribute intelligence at the level of the single elements of the network as well as at the cluster level. It has to be considered that deep introduction of variable energy resources (VERs) in the main AC grid will contribute to 2030 and 2050 energy targets, but some high risks and problems have to be suitably faced. In

[16]A. Scognamiglio, M. Annunziato, G. Graditi, Nearly Zero Energy Buildings, Smart Grids and Smart Cities: concepts for our cities of tomorrow, Proceedings of ZEMCH 2012 International Conference, (edited by) M. Noguchi, ZEMCH Network 2012, ISBN: 978-0-9574189-0-5, pp. 707–713.

[17]A. Scognamiglio, G. Adinolfi, G. Graditi, E. Saretta, Photovoltaics in Net Zero Energy Buildings and Clusters: Enabling the Smart City Operation, Energy Procedia, Volume 61, 2014, Pages 1171–1174, ISSN 1876-6102.

fact, lack of inertia, frequency stability and low short circuit power represent critical issue to mitigate in order to guarantee grid continuity and availability.

In this regard, PV offers the opportunity for DC-based grids, as PV generation is DC based. Considering that many sources and systems in the integrated grid are DC reliant, this suggests that DC grids coupled with storage systems can offer solutions that avoid transformation losses (local grids) which can also enhance self-consumption, avoiding problems related to the connection with the smart grid. DC microgrid implementations are considered as promising architecture to 'embed' VERs, avoiding their energy introduction in the national grid. PV and wind sources are locally managed to frequency instabilities mitigation and virtual inertia service provision[18].

Ways to improve the current generation and distribution of electricity via demand side management (DSM) and storage systems, e.g. through multi-level optimization model, which incorporates energy demand scheduler (DS) and energy storage (ES),[19] have to be developed, too.

Smart inverters provide voltage support and quality of supply at point of common coupling of the RES systems meeting the connection and operational requirements of the operators, and hybrid inverter solutions including the smartness of the system infrastructure can offer maximum use of local resources for the collective benefit of the system and the connected users (e.g. solving the mismatch due to several variable 'energy behaviors').

In a few words, PV can be a key technology for aggregating energy communities and advanced distributed controls in hierarchical set-up with the grid. Thus, a second paradigm shift in the way to consider PV, namely, BIPV, can be envisioned: from PV as a 'design factor' aimed to realize NetZEBs to PV as a 'common thread' in the smart city able to create connection and aggregation amongst buildings, clusters and energy communities so as to optimize the energy functioning of the smart city itself and its interaction with the smart grid.

In order to succeed in developing PV, there are many challenges to face; these may be related to decision support systems in the planning process or to specific PV technology features and related strategies and tools.

In general, the use of simulation tools that enable to include in the modelling environment energy generation, load matching, heating and cooling demand,

---

[18] Giovanna Adinolfi, Vincenzo Galdi, Vito Calderaro, Giorgio Graditi, Maria Valenti, DC microgrids for grid instability mitigation and virtual inetrtia ancillary service in 2030 scenarios; International Conference on Applied Energy 2020, Nov 29–Dec 02, 2020, Bangkok, Thailand (in press).

[19] Kai Zhuo Lim, Kang Hui Lim, Xian Bin Wee, Yinan Li, Xiaonan Wang, Optimal allocation of energy storage and solar photovoltaic systems with residential demand scheduling, Applied Energy, Volume 269, 2020, 115116, ISSN 0306-2619, https://doi.org/10.1016/j.apenergy.2020.115116

daylighting and economics is crucial. Moreover, in a decarbonized scenario, there will be a paradigm shift from traditional sources of energy that are easily quantified to sustainable ones that are difficult to predict as they depend on many variables, i.e. the solar availability. In the case of PV, the production, given a certain system, depends mainly on meteorological conditions, whereas in the case of BIPV, there is a higher complexity due to the fact that the BIPV and its subcomponents can have different orientations.

Storage and flexibility at all levels are of growing importance transforming demand into a key enabler for optimal use of resources bringing the end users into prime providers of flexibility.

At the level of the city government, the planning process should use decision support systems for choosing the appropriate areas/surfaces for installing PV in view of achieving the optimal energy production and to support self-consumption. Moreover, the choice of the surfaces where to install PV is relevant, as an appropriate choice of tilt and azimuth angles can orient the production in a way suitable for offsetting local demands all day long, avoiding peaks and enhancing self-consumption.[20]

Among the other challenges mainly very focused on PV technology and vision development, there is the development of control systems for grid feeding, self-consumption and local storage together with the standardization of the interoperability of such control systems (energy-matching), which is based also on digitalization, modularization and prefabrication of BIPV.

To give some insights of the PV-oriented challenges listed above, reference will be made to the EU-funded energy matching project and adaptive and adaptable envelope RES solutions for energy harvesting to optimize EU building and district load,[21] coordinated by EURAC (IT), as an effective synthesis of what was described above.[22]

The project (2017-2022) is coordinated by EURAC; it aims at maximizing the RES harvesting in the built environment by developing and demonstrating robust solutions to efficiently capture the on-site available RES through adaptive active building skin technologies and to effectively use the locally produced energy within the building and district concept, in the framework of an overall matching between energy and business vision. The expected results include versatile Click&Go substructure for different cladding systems, solar window package, modular appealing BIPV envelope solutions and RES harvesting package to heat and ventilate. Such

---

[20] G. Lobaccaro, S. Croce, C. Lindkvist, M.C. Munari Probst, A. Scognamiglio, J. Dahlberg, M. Lundgren, M. Wall, A cross-country perspective on solar energy in urban planning: Lessons learned from international case studies, Renewable and Sustainable Energy Reviews, Volume 108, 2019, Pages 209–237, ISSN 1364-0321, https://doi.org/10.1016/j.rser.2019.03.041

[21] https://www.energymatching.eu

[22] Maturi et al., 2019: Energy Matching project, Adaptive and adaptable envelope RES solutions for energy harvesting to optimize EU building and district load. Proceedings 36th EUPVSEC, doi.org/https://doi.org/10.4229/EUPVSEC20192019-6BV.4.13

solutions are integrated into energy-efficient building concepts for self-consumers connected in a local area energy network (energy LAN). This is designed to fulfill comprehensive economic rationales, through the energy harvesting business enhancer platform. Operational strategies of the energy LAN are driven by the building and district energy harvesting management system. Then, an optimization tool enables the best matching between local RES-based energy production and building load profiles and simplifies the energy demand management for the energy distributors.

Some results of this project are already available, and they are summarized in the following. A building side control system has been developed, and it is able to improve the performance for a building cluster with energy storage, electric vehicles, and considers the flexible demand shifting ability of electric vehicles. The study results show that the developed control can increase the cluster-level daily renewable self-consumption rate by 19% and meanwhile reduce the daily electricity bills by 36% compared with the conventional controls.[23]

A study has been performed that analyzes the effectiveness of one optimization approach and matches it against traditional dimensioning methods.

Three design methods are described and compared to an ideal design: the minimum capacity required by the current Italian law, the PV capacity whose annual cumulative production equals the cumulative demand of the building and an optimization technique using a constant energy demand.

The methods were all tested on a residential building located in Firenze (Italy).

The results show that the optimization approach outperforms the other methods despite the simplified input data, by leading to an improvement of NPV (Net Present Value) up to 85% compared to the other methods and can achieve >93% of the actual optimum. In countries where net billing (or net metering) incentives are still in place, the optimization technique is not so vital.[24]

The topics related to the key challenges of the use of PV in the smart city vision, due to the overlapping of different thematic issues and spatial domains, are investigated under many different calls of EU-funded projects.[6, 7, 8]

The analysis of the results of some of the running projects supports some first preliminary conclusions in identifying the existing implementation barriers, such as:

(a) The difficulty in promoting not yet mature technologies in the building sector, which is by nature very conservative

[23] Pei Huang, Marco Lovati, Xingxing Zhang, Chris Bales, A coordinated control to improve performance for a building cluster with energy storage, electric vehicles, and energy sharing considered, Applied Energy, Volume 268, 2020, 114983, ISSN 0306-2619, https://doi.org/10.1016/j.apenergy.2020.114983

[24] Marco Lovati, Mattia Dallapiccola, Jennifer Adami, Paolo Bonato, Xingxing Zhang, David Moser, Design of a residential photovoltaic system: the impact of the demand profile and the normative framework, Renewable Energy, Volume 160, 2020, Pages 1458–1467, ISSN 0960-1481.

(b) The interoperability among different digital solutions, platforms and hardware
(c) An appropriate legislative and normative national framework that allows mechanisms for the collective self-consumption (e.g. energy communities)[25]

[25] Information given to the author of the paragraph by David Moser, Coordinator of research group Photovoltaic Energy Systems, Eurac Research, Institute for Renewable Energy. Right now this research group is involved in a number of EU-funded project that well demonstrate the crucial role of PV in the smart city context. These are Build Heat, standardized approaches and products for the systemic retrofit of residential buildings, focusing on heating and cooling consumptions attenuation (www.buildheat.eu); 4RinEU; robust and reliable technology concepts and business models for triggering deep renovation of residential buildings in EU (https://4rineu.eu); cultural-E, climate and cultural-based solutions for plus energy buildings (https://www.cultural-e.eu); energy matching and adaptive and adaptable envelope RES solutions for energy harvesting to optimize EU building and district load (https://www.energymatching.eu); stardust, holistic and integrated urban model for smart cities (https://stardustproject.eu); infinite, industrialized durable building envelope retrofitting by all-in-one interconnected technology solutions (https://cordis.europa.eu/project/id/958397); Matrix, Modular Big Data Applications for Holistic Energy Services in Buildings (https://cordis.europa.eu/project/id/101000158).

Chapter 3
Smart Cities Consumers in Search of the Potential Sustainability

Manuel Villa-Arrieta and Andreas Sumper

Abstract Due to the increase in the urban population concentration, cities can be considered representative of the energy consumption and the energy sustainability of their countries, and, therefore, sustainability depends on the energy consumption behavior of the urban population. In the first urban agglomerations, primary energy resources (renewable) were transformed locally to supply a relatively low demand for energy. However, with the increase in demand, centralized energy facilities that took advantage of non-renewable energy resources were required. The use of these energy resources has caused the loss of energy sustainability. Parallel to the objective of solving environmental problems through the use of renewable energy resources in decentralized generation facilities, the smart city strategy seeks to optimize the power system and make operation more flexible by empowering the consumer. However, just as sustainability depends on the urban population's consumption behavior, the effectiveness of smart cities depends on the active participation of consumers. Faced with increased demand and the need to obtain a clean and uninterrupted energy supply, the recovery of potential sustainability depends on the consumer taking advantage of the technological deployment of smart cities and becoming a smart consumer. Based on studies on demand-side response, electrification, and energy self-consumption in cities, this chapter addresses the effectiveness of smart cities. The conclusions highlight the fact that energy sustainability is not inherent in smart cities: it depends on the consumer participation. Therefore, the effectiveness of smart cities must address the design of incentives for the participation of the population from a holistic approach and be linked to the heterogeneity of consumers and circularity in the efficient management of resources.

M. Villa-Arrieta (✉)
Escola Tècnica Superior d'Enginyeria Industrial de Barcelona (ETSEIB), Universitat Politècnica de Catalunya (UPC), Barcelona, Spain
e-mail: mvinput@protonmail.com

A. Sumper
Centre d'Innovació Tecnològica en Convertidors Estàtics i Accionaments (CITCEA-UPC), Universitat Politècnica de Catalunya (UPC), Barcelona, Spain
e-mail: andreas.sumper@upc.edu

G. C. Lazaroiu et al. (eds.), *Holistic Approach for Decision Making Towards Designing Smart Cities*, Future City 18,
https://doi.org/10.1007/978-3-030-85566-6_3

Keywords Demand side response · Smart consumers · Nearly zero energy cities

Introduction

This chapter is based on the doctoral thesis "Energy Sustainability of Smart Cities" (see Villa-Arrieta 2019) and is comprised of studies that the authors have previously published. Our intention is to use this chapter to extend the conclusions of the said thesis and contribute to the study of smart cities with a single concise piece. In this introduction, the importance of cities in the energy transition is discussed. Then, the mechanisms that these urban areas use to contribute to energy sustainability are identified, and in the following sections, these mechanisms are discussed before presenting the results and conclusions of this research.

The Importance of Cities in the Energy Transition

Cities are home to more than half of the world's population (United Nations 2018) and consume between 60% and 80% of the world's total energy production. Because energy is responsible for two-thirds of total greenhouse gas (GHG) emissions (OECD/IEA, IRENA 2017), urban areas emit 75% of global CO_2 emissions (primary GHG) (United Nations Environment Programme (UNEP) 2012). As seen in the United Nations projections for up to 2050 (projections before the COVID-19 pandemic), if the current urban population growth trend continues, global urbanization will represent 67% of the total population (United Nations 2013), and cities will demand a more significant amount of energy.

Covering this energy demand with the current model based on fossil resources will increase GHG emissions, and the consequences of global warming will put the planet's environmental sustainability at risk (IPCC 2018). Therefore, guaranteeing the security and quality of the energy supply to provide urban services using the planet's resources will pose an enormous challenge in terms of our ability to manage and restore the natural assets upon which all life depends (OCDE 2012).

The rhythm of the technological change of the systems for transforming primary energy to final energy has marked the pace of development and economic growth of cities. This final energy, mainly consumed as electricity, has improved the quality of life for humanity in the modern era but has led to the environmental consequences mentioned above. As seen in Fig. 3.1 (which includes data from after the 1973 oil crisis), electricity consumption has a more significant correlation with urban population growth rather than with the total population growth. It can be concluded, therefore, that GHG emissions are linked to urban energy consumption.

To address this problem, the world's leading economies have launched an energy transition process to move from the current economic model to one that is decarbonized and competitive. (The relationship between the COVID-19 pandemic and the

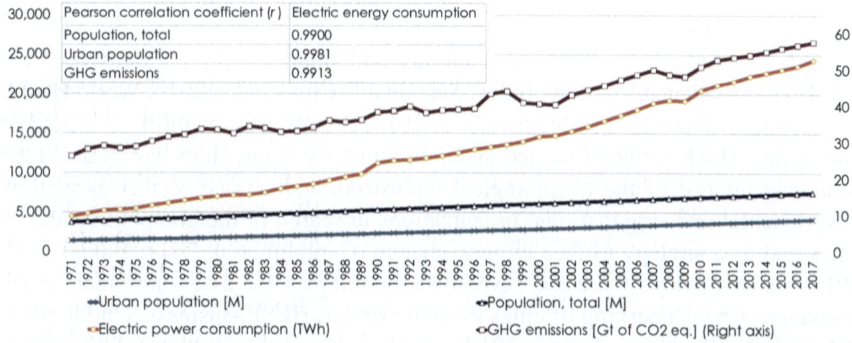

Fig. 3.1 Correlation between electric energy consumption and urban population. (Source: Adapted from Villa-Arrieta and Sumper 2019a)

energy transition has been studied by (Quitzow et al. 2021) and concludes that this crisis has deepened the gap between the countries that are leading the global energy transition and those which are progressing more slowly, which exacerbates the existing imbalances in an uneven energy transition landscape.)

The mechanisms to promote the energy transition are to decentralize and make the power system more flexible in order to save energy, increase energy efficiency, and replace the use of fossil energy with renewable energy. The flexibility of the system involves monitoring it in order to adjust the energy demand to the intermittency that characterizes renewable energies. Decentralization seeks to take advantage of distributed energy resources and bring generation closer to where energy is used.

This means that solar energy plays a remarkable role, given the uniformity of its distribution on a global scale (Check et al. 2015). Therefore, cities play a fundamental role in the energy transition process (Kammen and Sunter 2016), despite being the source of a worldwide environmental problem. Within cities, buildings use 31.43% of the total energy consumption, which is more than industry, transport, and other consumptions (International Energy Agency (OECD/IEA) 2018). In addition to focusing on energy consumption, buildings have a high margin of action to increase energy savings, energy efficiency, and the use of renewable energy through energy self-consumption. Flexibilization of the energy system also seeks to empower consumers so that through demand-side response (DSR) mechanisms, this agent can expand its participation in the power market.

Smart City and Energy Sustainability as Strategies

From a development of the society's point of view, advancing in the energy transition will lead to the sustainability of energy supply: reducing the negative impact of energy consumption on the environment will allow us to ensure the well-being of future generations. In this sense, there is a global consensus in defining the smart city concept as a critical strategy to address energy transition in cities and achieve energy sustainability (IRENA 2016).

Both the studies of energy sustainability and smart cities are extensive (EIB, UPM 2017; Sustentable Citiess Index. Sustainable Cities Index 2016; World Health Organization. The Rise of Modern Cities 2010). Energy sustainability is a field of knowledge which has been studied in depth. Similarly, smart cities has attracted the economic interest of the technological, industrial, and service sectors as well as governmental and supranational organizations in search of the competitiveness of cities and as a strategy to face climate change (Caragliu et al. 2011; Bakici et al. 2013). In this sense, there are several methodologies for assessing smart cities (rankings or benchmarking of cities as in the studies (IESE Business School 2014; JLL and The Business of Cities 2017) and the energy sustainability of cities (see International Telecommunications Union 2019).

The smart city strategy seeks to guarantee the efficacy of the energy service in cities to provide other services and efficiently manage available resources (energy and economic resources and infrastructures) (Villa-Arrieta and Sumper 2019a). The synergies between smart technologies (smart grids, distributed generation facilities, and smart meters) will make it possible to provide the energy service that citizens require as well as efficiently manage the resources needed to provide them (Lund et al. 2017). In particular, the deployment of smart meters in households and the energy self-consumption of nearly or net zero energy buildings (n/NZEB) will increase energy saving and efficiency and make the most of the local renewable energy resources (International Energy Agency (IEA) 2017a; International Energy Agency (IEA) 2017b; Ackermann et al. 2001).

In terms of sustainable development, sustainable energy is a source of the development and growth of societies and seeks a balance between economic, social, and environmental variables (Council HR and Germany 2007). According to the *World Energy Council* (WEC), energy sustainability at a country level relies on a balance between three pillars: energy security, energy equity, and environmental impact mitigation (Kim et al. 2013). The inclusion of smart technologies will allow involving those who take part in the management of the cities to find the balance between the these three pillars. Figure 3.2 shows the energy transition characteristics needed to meet the energy demands of a growing world and urban population: the smart city and energy sustainability concepts are constantly being evaluated to achieve the decarbonization targets.

Scalable Study of Energy Sustainability

The main contribution of the studies addressed in this chapter is the analysis of the scalability of urban energy sustainability up to the country level (see Fig. 3.3), as a new evaluation approach based on the hypothesis that cities tend to be representative of the country's population, energy consumption, and GHG emissions. Due to the trend of an increasing population concentration in cities, it can be argued that

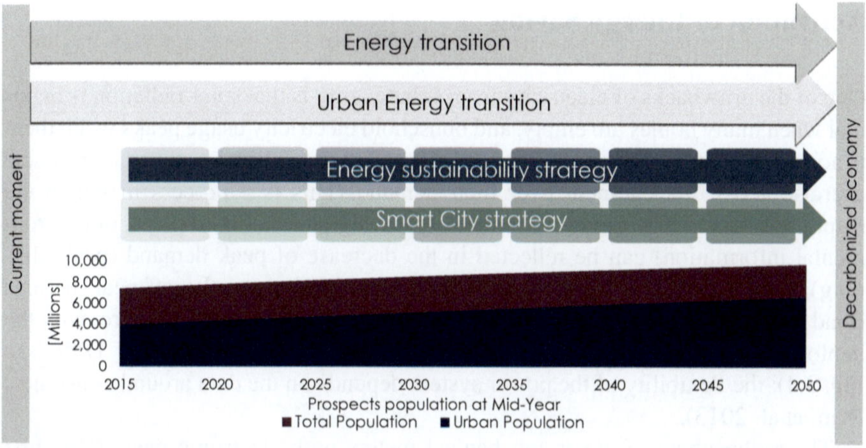

Fig. 3.2 Energy transition and population growth prospects for 2050. (Source: Adapted from Villa-Arrieta 2019)

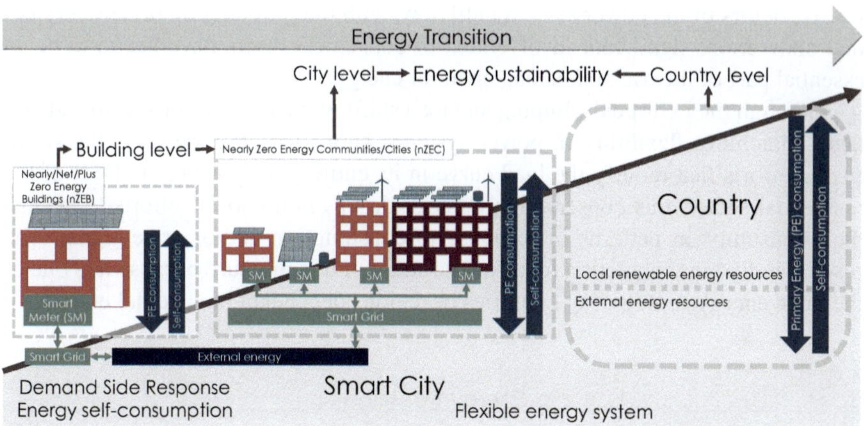

Fig. 3.3 Scalability of energy sustainability. (Source: Adapted from Villa-Arrieta 2019)

urban centers are to be representative of energy consumption and are responsible for GHG emissions. It can therefore be concluded that energy sustainability of cities is representative of the energy sustainability of their countries.

Based on a close relationship between the smart city and energy sustainability strategies, the cities' capacity to advance in energy saving and the use of solar energy resources are described below. The objective is to explain the contribution of smart technology elements to the increase of energy efficiency and urban self-consumption of cities, both of which are critical components of the energy transition to combat climate change.

Activation of Energy Saving

One of the drawbacks of electricity from solar energy is that solar radiation is maximal when many homes are empty, and household electricity usage peaks when there is no solar radiation. However, through DSR and the use of smart meters, consumers exercise active management over their demand. This is a process in which the response to energy information feedback (price signals, gamified plans, or environmental information) can be reflected in the decrease of peak demand (peak clipping), the change of consumption from the peak periods to the off-peak periods (load shifting), or in the energy-saving (strategic conservation). Therefore, with the deployment of smart meters, it is possible to activate the measures of DSR (see Fig. 3.4): the flexibility of the power system depends on the data around consumers (Kim et al. 2013).

The replacement of electromechanical meters with electronic ones is the first step in the process of empowerment, so that more meaningful and better information concerning energy consumption that leads to changes in the energy management of households can be collected (Barbu et al. 2013). Although end users were passive actors of the power system until now, with the introduction of new services that allow more significant involvement of consumers, users have begun to be an essential part of the use and management of energy.

Although the DSR peak clipping and load shifting measures offer specific advantages to achieve flexibility in power systems, strategic conservation is the set of saving efforts that modify the load curve in its entirety. Regarding the DSR of the residential sector, this conservation strategy groups households' efforts to change their consumption patterns in response to consumption and/or price signal. The results studied in this section focus on the study of this measure because it is one of the main energy strategies of countries or regions dependent on external energy.

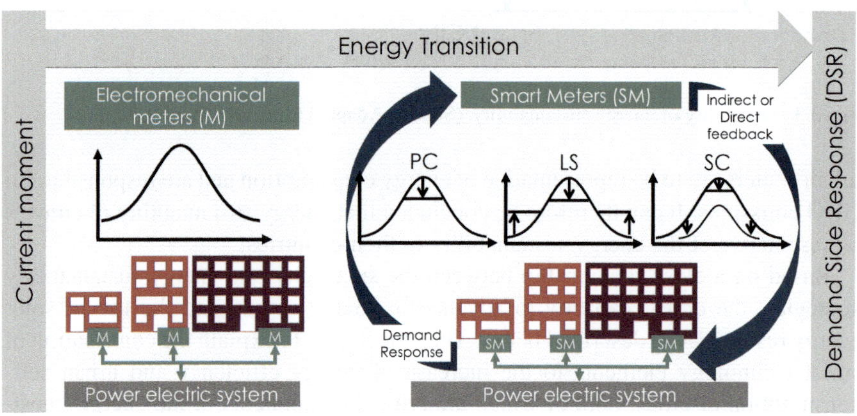

Fig. 3.4 Energy transition and demand side response (DSR). Notes: PC, peak clipping; LS, load shifting; SC, strategic conservation; M, electromechanical meters; SM, smart meters. (Source: Adapted from Villa-Arrieta 2019)

Experimental Results of the Strategic Conservation

It would seem that the use of smart meters would allow consumers to manage their demand more efficiently (Frederiks et al. 2015). However, unless users are proactive, it will not be possible for them to take advantage of the opportunities provided by new technologies. Domestic energy consumption is mainly based on routines and habits that are difficult to change. Therefore, it will be necessary to deepen the technical, psychological, social, and economic aspects that allow consumer participation and behavior adjustment (Batalla-Bejerano et al. 2020). This is precisely the field in which researchers from different disciplines seek to identify the aspects that determine consumer participation through empirical studies.

To explore this, a review of empirical works that have tried to quantify the results concerning energy savings has been carried out. This review analyzes 116 empirical studies on the strategic conservation of the DSR (to see results on peak clipping and load shifting, see (Batalla-Bejerano et al. 2020)). These studies, comprised of surveys (S), analytics (A), experiments (E), and simulations (L), looked at the indirect and direct feedback of information related to electricity consumption through smart meters. Indirect feedback includes electric bills and consumption readings, and direct feedback consists of the use of additional technologies such as smart meters (digital) that make it easier for users to check their consumption and participate in energy-saving programs. According to the results, households respond to the provision of direct and indirect information by using less electricity.

Figure 3.5 and Table 3.1 compile these results. The installation of smart meters alone in households does not guarantee the reduction of electricity consumption in the broad spectrum of their socioeconomic conditions, and it is necessary to take advantage of digital technology to ensure that households understand and react to

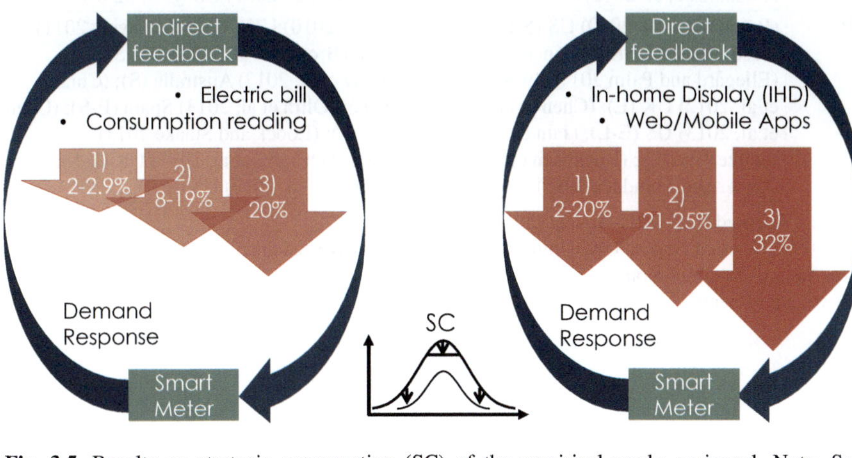

Fig. 3.5 Results on strategic conservation (SC) of the empirical works reviewed. Note: See Table 3.1. (Source: Adapted from Villa-Arrieta 2019)

Table 3.1 Strategic conservation results using indirect and direct feedback

Indirect feedback	
Figure 3.5	[Reference] Country (Methodological approach) Result
1)	(Allcott 2011) US (E-A) 2.0%; (Ayres et al. 2009) US (E) 2.1%; (Anderson and Lee 2016) US (E-L) 2.2%; (Schwartz et al. 2013) US (E-S) 2.7%; (Allcott and Rogers 2012a) US (E-L) 2.6 and 2.9%*
2)	(Gans et al. 2013) Northern Ireland (E) 11–17%; (Asensio and Delmas 2015) US (E) 8–19%
3)	(Bariss et al. 2014) Latvia (E) 20.0%; (Poznaka et al. 2015) Latvia (E-S) 23.0%
P	(Allcott and Rogers 2012b) US (E); (Ek and Söderholm 2010) Sweden (E); (Laicane et al. 2013) Latvia (S); (Lossin et al. 2016) Switzerland (E-S); (Kang et al. 2012) South Korea (S); (Rausser et al. 2018) Ireland (S); (Qingbin Wang 2016) US (S)
Direct feedback	
Figure 3.5	[Reference] Country (Methodological approach) Result
1)	(Reeves et al. 2015) US (E-S) 2.0%; (Quintal et al. 2013) Europe (E-S) 2.0%; (Fenn et al. 2012) Germany (E-L) 3.0%; (Rettie et al. 2014) UK (E-S) 3.0%; (Schleich et al. 2011) Germany and Austria (E-S) 3.7%; (Erickson et al. 2013) US (E-S) 3.7%; (Schleich et al. 2012) Austria (E-S) 4.5%; (Spagnolli et al. 2011) Europe (E) 5.0%; (Schleich et al. 2017) Austria (E) 5.0%; (Houde et al. 2012) US (S-E-S) 5.7%; (Bager and Mundaca 2017) Denmark (E) 5–7%; (Stinson et al. 2015) Scotland (E-S) 7.0%; (Shimada et al. 2014) Japan (E) 7.6%; (van Dam et al. 2010) Netherlands (S-E-S) 7.8%; (Grønhøj and Thøgersen 2011) Denmark (S-E-S) 8.1%; (Peschiera and Taylor 2012) US (E) 8.8%; (Schultz et al. 2015) US (E) 7–9%; (Jain et al. 2013a) US (E-L) 10.0%; (Nye et al. 2010) UK (E-S) 5–10%; (Chen et al. 2015) US (E) 11.0%; (Anderson et al. 2017) South Korea (E) 14.0%; (Gosnell et al. 2019) UK (E) 4–12%; (Alahmad et al. 2012) US (E-S) 12.0%; (Wood and Newborough 2003) UK (E) 15.0%; (Chen et al. 2014) US (E-S) 20.0%; (Delmas and Lessem 2014) US (E) 20.0%
2)	(Maan et al. 2011) Netherlands (E) 21.0%; (Laicane et al. 2015) Latvia (E-L) 6–23%; (Adnane Kendel 2015) France (E-S) 23.3%; (Bager and Mundaca 2017) Denmark (E) 7–25%
3)	(Costanza et al. 2012) UK (E-S) 5–32%; (Petersen et al. 2007) US (E-S) 32.0%
P	(Peschiera et al. 2010) US (S-E-S)*; (Foster et al. 2010) UK (E-S); (Strengers 2011) Australia (E-S); (Karjalainen 2011) Finland (S); (Brewer et al. 2011) US (S-E-S); (Ellegård and Palm 2011) Sweden (E-S); (Petkov et al. 2012) Australia (S); (Chiang et al. 2012) UK (E); (Chen et al. 2012) US (E-L); (Oltra et al. 2013) Spain (E-S); (Chen et al. 2013) US (E-L); (Jain et al. 2013b) US (E-L); (Loock and Staake 2013) Switzerland (E); (Buchanan et al. 2014) UK (S); (Schwartz et al. 2015) UK (E-S); (Bager and Mundaca 2015) Denmark (E); (Mogles et al. 2017) US (E-S)

Source: Adapted from Villa-Arrieta (2019)

Notes: US, United States; UK, United Kingdom; P, positive result; S, surveys; A, analytics; E, experiments; L, simulations

* Two experiments

the direct feedback of energy information. Empirical evidence shows that it is possible to reduce electricity consumption by up to 32% using such means.

Electrification and Self-Consumption in Cities

Electrification

As described previously, the deployment of smart technologies will allow cities to increase electricity generation from renewable sources to decarbonize the economy. However, according to the WEC, energy sustainability depends on the balance between energy security, energy equity and the mitigation of environmental impact. Therefore, given the intermittence of renewable energies and depending on the characteristics of each country and city (possibly similar between regions), "smart" electrification without energy storage technologies may lead to an imbalance in energy sustainability in that, although environmental benefits are possible, energy security could be at risk.

To help address this issue, the researchers studied the effect that electrification may have on countries' energy sustainability using the ETI's interactive Pathway Calculator tool. This tool can be used to determine what is necessary to improve the ranking position and understand the impact of policymaking on achieving a sustainable energy future (World Energy Council 2017). Figure 3.6 presents its elements. The procedure followed involved setting the tool variables to maximum and minimum values in order to identify the variation in the energy sustainability result of the 125 countries of the ETI 2017. This exercise was the first step to determine the photovoltaic (PV) generation capacity of cities of these countries, which will be addressed in the section below.

According to the results obtained, countries could obtain better energy sustainability by reducing energy imports and increasing the diversity of the electric generation and the diversity of the primary energy (PE) supply (see Fig. 3.7). This indicates that the use of local energy resources, such as PV electricity generation,

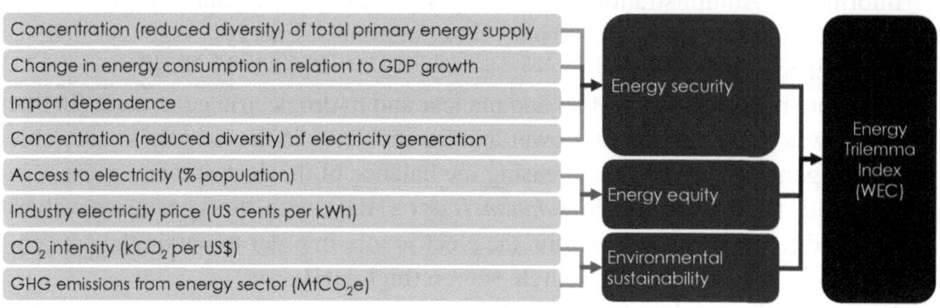

Fig. 3.6 Structure of the indicators of the Pathway Calculator of the World Energy Council (WEC). (Source: Adapted from Villa-Arrieta and Sumper 2019a)

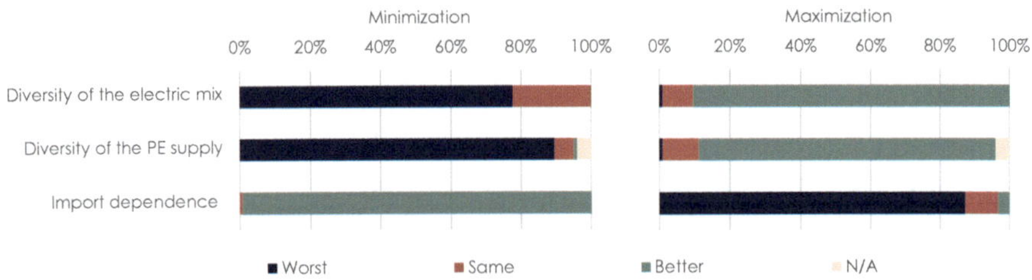

Fig. 3.7 Result of energy sustainability by variation of critical indicators. (Source: Adapted from Villa-Arrieta and Sumper 2019a)

would lead to the improvement of energy sustainability of the 125 ETI-2017 countries.

Photovoltaic Energy Self-Consumption Capacity of Cities

As mentioned above, cities are at the center of the energy transition. Given the uniformity of the solar resource, cities have the capacity to generate electricity near the same consumption points. By deploying PV systems, cities can cover part of their electricity demand and contribute to the use of local energy resources in their countries. Thus, PV generation is the leading technology to move toward the energy sustainability of cities.

To advance in this field, the contribution of urban PV generation to the energy sustainability of the ETI 2017 countries (see Villa-Arrieta and Sumper 2019a) was studied. The process was as follows:

1. Calculate the PV generation capacity of the rooftops of the most populated cities of the ETI 2017 countries (183 cities of 123 countries) taking the usable area of the Barcelona rooftops, global tilted irradiation data, and a 16% efficiency for PV panels as a reference.
2. Calculate the distribution of the electricity generation mix of the studied countries with data from the International Energy Statistics of the US Energy Information Administration (EIA) (Meyer 1991), and calculate CO_2 emissions using emission factors from International Energy Agency (IEA) (Ecometrica 2011).
3. Calculate how much fossil fuel and nuclear and hydroelectric generation (in that order) was replaced by solar power and the increases in the country's renewable energy generation without increasing the balance of its electric mix.
4. Calculate the *Herfindahl-Hirschman Index* (HHI) with the generation values obtained from each source within the electric mix in order to obtain the concentration of the generation in a single source (high HHI result).
5. Normalize the results of HHI (0–100) to compare them with the values published for each country in the 2017 ETI results.

The results of this showed that the 125 ETI 2017 countries vary the concentration (reduced diversity) of electricity generation by an average of −15.94% due to the hypothetical increase in their urban PV generation. This would have an impact on the reduction of 56.31% of the electric power generation from fossil fuels and consequently on the reduction of 64% of the CO_2 emissions of these countries. The use of the local solar resource would allow these countries to improve their sustainability due to the reduction of the dependence on energy imports. Although the diversity of the primary energy supply would be reduced in some countries (toward concentration), increased consumption of renewable resources would allow them to reduce CO_2 emissions. Table 3.2 summarizes the results obtained.

From the results, two groups of countries were identified: one group of countries that increases the diversity of electricity generation and another that reduces it. In the former, the increase in PV generation allows 87 countries to diversify the electricity mix by up to 43% concentration. On the contrary, 38 countries increase the concentration of the electricity mix to 97%. Although in both cases, following the ETI definition, as environmental protection improves, energy security can be put at risk. It should be noted that the second group of countries is made up of countries with a large proportion of hydroelectric generation in their mix, which is a technology with low intermittency.

Table 3.2 Average results of the effect of the diversification of the electric mix with urban PV generation in the ETI 2017 countries

Indicator	Initial value	Result
Number of countries analyzed	125	125
Average concentration* of electricity generation (0–100)	66.41	N/A
… after the urban PV generation (0–100)	N/A	55.16
Average variation [%]	N/A	−15.94
Number of countries that diversify electricity generation	N/A	87
Average variation of concentration* [%]	N/A	−42.44
Number of countries that concentrate electricity generation	N/A	38
Average variation of concentration* [%]	N/A	44.73
Average variation of the consumption of fossil fuels [%]	N/A	−56.31
Average variation in CO_2 emission [%]	N/A	−64.00

Source: Adapted from Villa-Arrieta and Sumper (2019a)
Note: * Reduced diversity

An Economic Evaluation of the Use of Self-Consumption Capacity

Energy self-consumption is the strategy that seeks to take advantage of the local renewable generation capacity. In the case of cities, although each of them has different characteristics, any use of local energy resources will reduce the consumption of primary external energy of fossil origin. In this sense, energy-importing countries will be able to equilibrate their energy balance with energy self-consumption in their cities. This type of distributed generation includes nearly/net zero energy buildings (n/NZEB), which are dependencies of consumers. Therefore, the advance of energy self-consumption in cities and consequently the reduction of energy imports in the countries, if it were to be the case, depend on the investment of consumers in n/NZEB.

In order to study the advancement of n/NZEB in cities from the investment point of view, the authors previously carried out technical-economic simulations with an evaluation model of self-consumption in buildings scalable to study self-consumption in cities (nearly zero energy cities, NZEC). The results of these studies are presented below.

Evaluation Model

The model, called nZEC-EATEP (see (Villa-Arrieta and Sumper 2019b)), allows evaluating the economic performance of the energy self-sufficiency process of cities. Also, it enables to jointly assess distributed generation facilities, or n/NZEB or neighborhoods with energy surpluses (positive energy neighborhoods). Figure 3.8 conceptualizes its methodology, which is based on the relationship between the energy transition, the urban energy transition, the increase in the consumption of local energy resources to the detriment of the external ones due to the advance of the n/NZEB, and the economic calculation of the future value of this process.

Case Study: Barcelona

The objective of this study was to analyze the scope of the investment in the PV self-consumption of buildings in order to promote prosumer communities. The model included 82,652 buildings and simulated 37 years (2013–2050) of energy and economic performance: in 38,700 (34.7% of 238,213 buildings with PV generation capacity data), the investment of six packages of energy rehabilitation measures and PV self-consumption (prosumer's buildings) was studied, and in 43,952 the investment in PV generation (PV extra generation) was studied. Cross-referencing with public data on the city, the PV generation capacity of the 82,652 buildings, the electricity consumption of the prosumers, and the investment cost in the energy rehabilitation measure packages were identified as well as the electricity

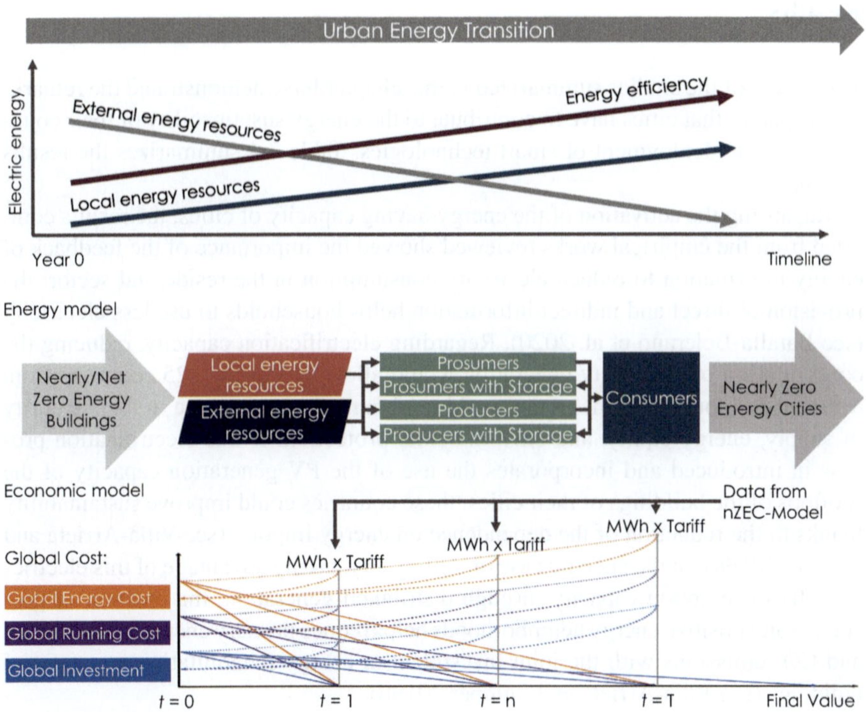

Fig. 3.8 Technical-economic evaluation of nearly zero energy cities (nZEC). (Source: Adapted from Villa-Arrieta and Sumper 2019b)

Table 3.3 Number of buildings and electric energy consumption of nZEC model for Barcelona

	No. of buildings	Electric energy consumption [GWh/Year]
Prosumers	38,700	969.363
Consumers (domestic, commercial and services, and other services	N/A	3225.927
PV_Extra	43,952	N/A
Total	82,652	4189.32

Source: Adapted from Villa-Arrieta and Sumper (2019b)

consumption of buildings extra PV generation (PV_Extra). Table 3.3 summarizes these results.

With the nZEC model studied, the city has the ability to reduce up to 9.68% of its primary energy demand, which means a reduction of up to 12.25% in energy costs and up to 11.43% in emissions of CO_2. The total investment required to achieve these savings includes the initial value and the replacement of the PV systems and the rehabilitation measures. This investment can be 1.32 times the total energy cost of the city in the same 37 years of evaluation.

Results

The results of the studies summarized in this chapter have demonstrated the remarkable capacity that cities have to contribute to the energy sustainability of their countries with the deployment of smart technologies. Table 3.4 summarizes the results obtained:

Regarding the activation of the energy-saving capacity of cities, the results compiled from the empirical works reviewed showed the importance of the feedback of energy information to reduce electricity consumption in the residential sector: the provision of direct and indirect information helps households to use less electricity (see Batalla-Bejerano et al. 2020). Regarding electrification capacity, reducing the concentration of electricity generation would allow 113 of the 125 countries from different economic regions of the world to obtain a better balance in their security of supply, energy equity, and environmental protection. If this electrification process in introduced and incorporates the use of the PV generation capacity of the rooftops of the buildings of their cities, these countries could improve sustainability thanks to the reduction of the dependence on energy imports (see Villa-Arrieta and Sumper 2019a). In the specific case of a study city, taking advantage of this electricity self-consumption capacity through distributed generation, the creation of prosumers and positive energy neighborhoods helps reduce primary energy consumption and CO_2 emissions with the joint investment in energy rehabilitation measures of buildings (see Villa-Arrieta and Sumper 2019b).

This means that energy sustainability and smart city strategies would allow for progress in the transition toward a decarbonized economy. Uniting the pillars of these strategies, it can be argued that smart technologies enable cities to be effective in the provision of urban energy service, efficiently managing available local resources to achieve a balance between energy security and equity and environmental protection. However, according to the results described above, without consumer

Table 3.4 Compilation of results

Capacity of…	Countries	Cities	Results
Energy saving	>19	>19	Reduction of between 2% and 32% of electricity consumption
Electrification	125	N/A	113 improve energy sustainability
PV self-consumption in cities	125	183	• All countries reduce on average 64% of CO_2 emissions. • 87 countries diversify the electricity mix up to 43% concentration
nZEC	1: Spain	1: Barcelona	With 34.7% of the buildings with PV generation and 16.2% of prosumer buildings: • Reduction of up to 9.68% in primary energy demand. • Reduction of up to 12.25% in energy costs. • Reduction of up to 11.43% of CO_2 emissions.

Source: Adapted from Villa-Arrieta (2019)

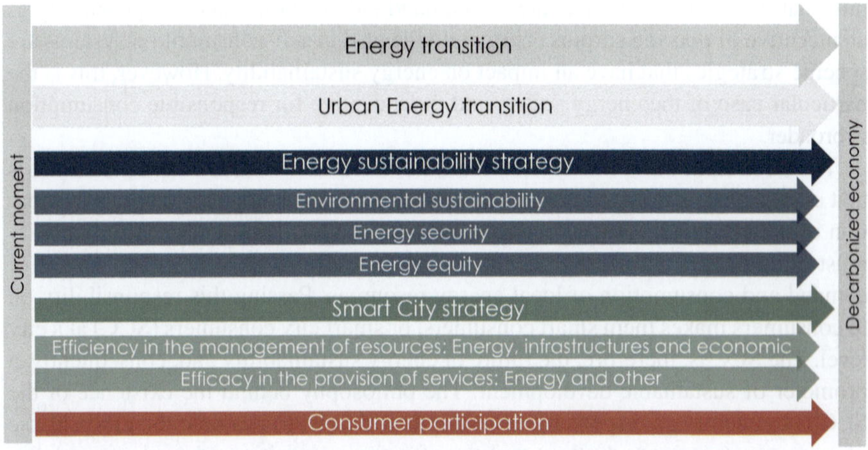

Fig. 3.9 Layers of the link between energy transition and consumer participation

participation, it is not possible to save energy or take advantage of the energy self-consumption capacity of cities through investment in buildings with almost zero energy consumption. Figure 3.9 summarizes the dependency between these layers of the energy transition.

Conclusions

The studies compiled in this chapter have been based on the study of the scalability of energy sustainability through the smart city technological strategy, from the level of buildings to that of a country. This study approach is a novel proposal to address the fact that due to the increase in the concentration of the urban population, cities tend to be representative of the energy sustainability of their countries. Smart technologies are crucial elements to keeping the balance between energy security, energy equity, and environmental sustainability of cities and their countries. The energy effectiveness of smart cities is the efficacy provision of urban energy service and the efficient management of resources around distribution systems. Smart cities allow the urban population to participate in its own sustainability, which in the context of the current energy transition process means empowering the consumer in terms of their demand to make operating the system more flexible and to optimize the efficient management of the local resources.

Of course, there are drivers that encourage consumers to participate in the achievement of global social and environmental goals, such as the sustainable development goals. However, based on specific incentive strategies for responsible consumption, consumers will exercise their participation in smart cities and in the market, helping the sustainability of local geographic frameworks that will ultimately have an impact on global sustainability. As discussed in this chapter, the information incentives to activate the demand-side response and the possibility as an incentive of pouring surplus energy into the grid in self-consumption systems are specific strategies that have an impact on energy sustainability. However, this is the particular case of the energy sector, and the incentive for responsible consumption is broader.

There are three main conclusions that can be drawn from this chapter. The first is that energy sustainability is not inherent in smart cities: it depends on the participation of the consumer to be effective in this city model. Thus, energy sustainability must be activated at an urban level to include consumers in the management of demand and consumption of local energy resources. Passing this responsibility on to consumers makes them smart consumers, or smart city consumers (SCC) at a city level. The SCC is, therefore, the future of energy sustainability and, consequently, a promoter of sustainable development. The philosophy behind the existence of the SCC is to encourage responsible consumption: that is, to promote the efficient use of resources (energy and other) and the effective acquisition of products and services with a positive impact on society and the environment.

The second conclusion is that the difficulty in reaching the SCC sustainability potential is related not only to the difference between the systems that exist in cities (different economic activities, mobility systems, types of buildings, urban characteristics, etc.) but also with the heterogeneity of consumers. Not only do cities have building owners but they also have tenants, the elderly, the student population, and short-term residents. Therefore, the decarbonization of the economy must go beyond technological innovation and address social innovation to include the heterogeneity of consumers and their different living conditions. Similarly, there must be regulatory innovation in addition to social and technological innovation in markets where there are regulatory barriers that prevent disruptive new business models within the energy transition. The objective is to obtain incentives, instruments of empowerment, and protection mechanisms for different types of consumers.

The third conclusion is related to the political implications that emerge from the conclusions above. There is a very clear need to design policies, enablers, monitoring metrics, and incentives for energy sustainability with a holistic and bottom-up approach to the economy and society, from consumers to countries. Just as sustainability is transversal to all economic activity, incentives to participate in sustainability must also be transversal to the economy, interconnecting different sectors and objectives. In this regard, two new topics must be addressed in the SCC: the decentralization of the administration and traceability of information through Blockchain technology and the inclusion of energy in the transformation toward circularity of the economy to maximize the use of resources, including the reintroduction of waste in the production chains. The design of incentive mechanisms for consumers that include the advantages of Blockchain technology to be able to cross sector borders within the framework of the circular economy will be the focus of future research by the authors.

References

Ackermann T, Andersson G, Söder L (2001) Distributed generation: a definition. Electr Pow Syst Res 57:195–204. https://doi.org/10.1016/S0378-7796(01)00101-8

Adnane Kendel NL (2015) The diffusion of smart meters in France: A discussion of the empirical evidence and the implications for smart cities. Sh Technol Res 53:194–200. https://doi.org/10.1108/JSMA-04-2015-0034

Alahmad M, Wheeler PG, Eiden J, Brumbaugh A, Alahmad MA, Schwer A (2012) A comparative study of three feedback devices for residential real-time energy monitoring. IEEE Trans Ind Electron 59:2002–2013. https://doi.org/10.1109/TIE.2011.2165456

Allcott H (2011) Social norms and energy conservation. J Public Econ 95:1082–1095. https://doi.org/10.1016/j.jpubeco.2011.03.003

Allcott H, Rogers T (2012a) How long do treatment effects last ? Persistence and durability of a descriptive norms intervention's effect on energy conservation. HKS Fac Res Work Pap Ser Harvard Univ:RWP12–RW045

Allcott H, Rogers T. The short-run and long-run effects of behavioral interventions: experimental evidence from energy conservation. 2012b

Anderson K, Lee SH (2016) An empirically grounded model for simulating normative energy use feedback interventions. Appl Energy 173:272–282. https://doi.org/10.1016/j.apenergy.2016.04.063

Anderson K, Song K, Lee SH, Krupka E, Lee H, Park M (2017) Longitudinal analysis of normative energy use feedback on dormitory occupants. Appl Energy 189:623–639. https://doi.org/10.1016/j.apenergy.2016.12.086

Asensio OI, Delmas MA (2015) Nonprice incentives and energy conservation. Proc Natl Acad Sci 112:E510–E515. https://doi.org/10.1073/pnas.1401880112

Ayres I, Raseman S, Shih A (2009) Evidence from two large field experiments that peer comparison feedback can reduce residential energy usage. Ssrn 29. https://doi.org/10.2139/ssrn.1434950

Bager S, Mundaca L (2015) How smart are electricity users with smart metering? a behavioural economics experiment. 38th Int Assoc Energy Econ Int Conf Antalya:25–27

Bager S, Mundaca L (2017) Making 'smart meters' smarter? Insights from a behavioural economics pilot field experiment in Copenhagen, Denmark. Energy Res Soc Sci 28:68–76. https://doi.org/10.1016/j.erss.2017.04.008

Bakici T, Almirall E, Wareham J (2013) A smart city initiative: the case of Barcelona. J Knowl Econ 4:135–148. https://doi.org/10.1007/s13132-012-0084-9

Barbu AD, Griffiths N, Morton G. Achieving energy efficiency through behaviour change: what does it take? Technical report No 5/2013. 2013

Bariss U, Timma L, Blumberga D (2014) Smart metering pilot project results. Energy Procedia 61:2176–2179. https://doi.org/10.1016/j.egypro.2014.12.103

Batalla-Bejerano J, Trujillo-Baute E, Villa-Arrieta M (2020) Smart meters and consumer behaviour: insights from the empirical literature. Energy Policy 144:111610. https://doi.org/10.1016/j.enpol.2020.111610

Brewer RS, Lee GE, Johnson PM (2011) The Kukui cup: a dorm energy competition focused on sustainable behavior change and energy literacy. Proc Annu Hawaii Int Conf Syst Sci:1–10. https://doi.org/10.1109/HICSS.2011.422

Buchanan K, Russo R, Anderson B (2014) Feeding back about eco-feedback: how do consumers use and respond to energy monitors? Energy Policy 73:138–146. https://doi.org/10.1016/j.enpol.2014.05.008

Caragliu A, Del Bo C, Nijkamp P (2011) Smart cities in Europe. J Urban Technol 18:65–82. https://doi.org/10.1080/10630732.2011.601117

Check R, Space-based PS, Percent K, Shingles E (2015) The future of solar energy. Mitochondrion:3–6. https://doi.org/10.1002/yd.20002

Chen J, Taylor JE, Wei HH (2012) Modeling building occupant network energy consumption decision-making: the interplay between network structure and conservation. Energ Buildings 47:515–524. https://doi.org/10.1016/j.enbuild.2011.12.026

Chen J, Jain RK, Taylor JE (2013) Block configuration modeling: a novel simulation model to emulate building occupant peer networks and their impact on building energy consumption. Appl Energy 105:358–368. https://doi.org/10.1016/j.apenergy.2012.12.036

Chen VL, Delmas MA, Kaiser WJ (2014) Real-time, appliance-level electricity use feedback system: how to engage users? Energ Buildings 70:455–462. https://doi.org/10.1016/j.enbuild.2013.11.069

Chen VL, Delmas MA, Kaiser WJ, Locke SL (2015) What can we learn from high-frequency appliance-level energy metering? Results from a field experiment. Energy Policy 77:164–175. https://doi.org/10.1016/j.enpol.2014.11.021

Chiang T, Natarajan S, Walker I (2012) A laboratory test of the efficacy of energy display interface design. Energ Buildings 55:471–480. https://doi.org/10.1016/j.enbuild.2012.07.026

Costanza E, Ramchurn SD, Jennings NR. Understanding domestic energy consumption through interactive visualisation: a field study 2012. https://doi.org/10.1145/2370216.2370251

Council HR, Germany MTO (2007) General Assembly 11759:1–38. https://doi.org/10.1093/oxfordhb/9780199560103.003.0005

Delmas MA, Lessem N (2014) Saving power to conserve your reputation? The effectiveness of private versus public information. J Environ Econ Manage 67:353–370. https://doi.org/10.1016/j.jeem.2013.12.009

Ecometrica (2011) Electricity-specific emission factors for grid electricity. Ecometrica:1–22. https://doi.org/10.13044/j.sdewes.2014.02.0030

EIB, UPM. Assessment methodology for smart city projects – application to the Mediterranean region (ASCIMER). Luxembourg: 2017

Ek K, Söderholm PP (2010) The devil is in the details: household electricity saving behavior and the role of information. Energy Policy 38:1578–1587. https://doi.org/10.1016/j.enpol.2009.11.041

Ellegård K, Palm J (2011) Visualizing energy consumption activities as a tool for making everyday life more sustainable. Appl Energy 88:1920–1926. https://doi.org/10.1016/j.apenergy.2010.11.019

Erickson T, Li M, Kim Y, Deshpande A, Sahu S, Chao T, Sukaviriya P, Naphade M (2013) The Dubuque electricity portal: evaluation of a city-scale residential electricity consumption feedback system. Proc SIGCHI Conf Hum Factors Comput Syst:1203–1212. https://doi.org/10.1145/2470654.2466155

Fenn B, Hopp O, Ahner M, Buchholz BM, Buehner V, Doss A, Hess N, Wagner W, Styczynski ZA. Advanced technologies of demand side integration by VPPs and through smart metering in households – experiences from a lighthouse project. 2012

Foster D, Lawson S, Blythe M, Cairns P (2010) Wattsup? Proc 6th Nord Conf Human-Computer Interact Extending Boundaries – Nord '10:178. https://doi.org/10.1145/1868914.1868938

Frederiks ER, Stenner K, Hobman EV (2015) Household energy use: applying behavioural economics to understand consumer decision-making and behaviour. Renew Sustain Energy Rev 41:1385–1394. https://doi.org/10.1016/j.rser.2014.09.026

Gans W, Alberini A, Longo A (2013) Smart meter devices and the effect of feedback on residential electricity consumption: evidence from a natural experiment in Northern Ireland. Energy Econ 36:729–743. https://doi.org/10.1016/j.eneco.2012.11.022

Gosnell G, Martin R, Muûls M, Coutellier Q, Strbac G, Sun M, Tindermans S (2019) Making Smart Meters Smarter the Smart Way. Cent Econ Perform

Grønhøj A, Thøgersen J (2011) Feedback on household electricity consumption: learning and social influence processes. Int J Consum Stud 35:138–145. https://doi.org/10.1111/j.1470-6431.2010.00967.x

Houde S, Todd A, Sudarshan A, Flora JA, Armel KC (2012) Real-time feedback and electricity consumption. Energy J 34:87

IESE Business School. IESE Cities in Motion Index 2014. Pamplona: 2014

International Energy Agency (IEA) (2017a) Digitalization & energy. Paris
International Energy Agency (IEA) (2017b) World Energy Outlook 2017, 2017th edn. IEA Publications, Paris
International Energy Agency (OECD/IEA) (2018) World Energy Outlook 2018. 2018th ed edn. IEA Publications, Paris
International Telecommunications Union. United for Smart Sustainable Cities (U4SSC). Geneva: 2019
IPCC. Global Warming of 1.5°C. vol. 265. 2018th ed. Geneva: 2018
IRENA (2016) Renewable Energy in Cities, International Renewable Energy Agency (IRENA), Abu Dhabi
Jain RK, Gulbinas R, Taylor JE, Culligan PJ (2013a) Can social influence drive energy savings? Detecting the impact of social influence on the energy consumption behavior of networked users exposed to normative eco-feedback. Energ Buildings 66:119–127. https://doi.org/10.1016/j.enbuild.2013.06.029
Jain RK, Gulbinas R, Taylor JE, Culligan PJ (2013b) Can social influence drive energy savings? Detecting the impact of social influence on the energy consumption behavior of networked users exposed to normative eco-feedback. Energ Buildings 66:119–127. https://doi.org/10.1016/j.enbuild.2013.06.029
JLL and The Business of Cities. The Universe of City Indices 2017 Decoding – decoding city performance. Chicago: 2017
Kammen DM, Sunter DA (2016) Urban planet: city-integrated renewable energy for urban sustainability. Science (80-) 352:922–928. https://doi.org/10.1126/science.aad9302
Kang NN, Cho SH, Kim JT (2012) The energy-saving effects of apartment residents' awareness and behavior. Energ Buildings 46:112–122. https://doi.org/10.1016/j.enbuild.2011.10.039
Karjalainen S (2011) Consumer preferences for feedback on household electricity consumption. Energ Buildings 43:458–467. https://doi.org/10.1016/j.enbuild.2010.10.010
Kim YD, Meyers K, Statham B, Ward G, Frei C. (2013) Energy Sustainability Index 2013
Laicane I, Blumberga A, Rosa M, Blumberga D (2013) Assessment of changes in households' electricity consumption. Agron Res 11:335–346
Laicane I, Blumberga D, Blumberga A, Rosa M (2015) Comparative multiple regression analysis of household electricity use in Latvia: using smart meter data to examine the effect of different household characteristics. Energy Procedia 72:49–56. https://doi.org/10.1016/j.egypro.2015.06.008
Loock C, Staake T (2013) Motivating energy-efficient behavior with green IS: an investigation of goal setting and the role of defaults. MIS Q 37:1313–1332
Lossin F, Loder A, Staake T (2016) Energy informatics for behavioral change: increasing the participation rate in an IT-based energy conservation campaign using social norms and incentives. Comput Sci – Res Dev 31:149–155. https://doi.org/10.1007/s00450-014-0295-3
Lund H, Østergaard PA, Connolly D, Mathiesen BV (2017) Smart energy and smart energy systems. Energy 137:556–565. https://doi.org/10.1016/J.ENERGY.2017.05.123
Maan S, Merkus B, Ham J, Midden C (2011) Making it not too obvious: the effect of ambient light feedback on space heating energy consumption. Energ Effic 4:175–183. https://doi.org/10.1007/s12053-010-9102-6
Meyer S. International Energy Statistics Sourcebook. CO_2 Emiss FROM FUEL Combust Highlights 1991. https://www.eia.gov/beta/international/data/browser/#/?pa=000000000000000000000000000002&c=ruvvvvvfvtvnvv1urvvvvfvvvvvvfvvvou20evvvvvvvvvnvvuvo&ct=0&tl_id=2 (accessed May 23, 2019)
Mogles N, Walker I, Ramallo-González AP, Lee JH, Natarajan S, Padget J, Gabe-Thomas E, Lovett T, Ren G, Hyniewska S, O'Neill E, Hourizi R, Coley D (2017) How smart do smart meters need to be? Build Environ 125:439–450. https://doi.org/10.1016/j.buildenv.2017.09.008
Nye M, Smith GD, Hargreaves T (2010) Burgess J. Visible energy trial: report for OFGEM
OCDE. OECD Environmental Outlook to 2050: The Consequences of Inaction 2012. https://doi.org/10.1787/9789264122246-en

OECD/IEA, IRENA. Perspectives for the energy transition: investment needs for a low-carbon energy system. 2017

Oltra C, Boso A, Espluga J, Prades A (2013) A qualitative study of users' engagement with real-time feedback from in-house energy consumption displays. Energy Policy 61:788–792. https://doi.org/10.1016/j.enpol.2013.06.127

Peschiera G, Taylor JE (2012) The impact of peer network position on electricity consumption in building occupant networks utilizing energy feedback systems. Energ Buildings 49:584–590. https://doi.org/10.1016/j.enbuild.2012.03.011

Peschiera G, Taylor JE, Siegel JA (2010) Response-relapse patterns of building occupant electricity consumption following exposure to personal, contextualized and occupant peer network utilization data. Energ Buildings 42:1329–1336. https://doi.org/10.1016/j.enbuild.2010.03.001

Petersen JE, Shunturov V, Janda K, Platt G, Weinberger K (2007) Dormitory residents reduce electricity consumption when exposed to real-time visual feedback and incentives. Int J Sustain High Educ 8:16–33. https://doi.org/10.1108/14676370710717562

Petkov P, Köbler F, Foth M, Krcmar H (2012) Motivating domestic energy conservation through comparative, community-based feedback in mobile and social media. 21. https://doi.org/10.1145/2103354.2103358

Poznaka L, Laicane I, Blumberga D, Blumberga A, Rosa M (2015) Analysis of electricity user behavior: case study based on results from extended household survey. Energy Procedia 72:79–86. https://doi.org/10.1016/j.egypro.2015.06.012

Qingbin Wang SL (2016) Are smart meters being used smartly? A case study of residential electricity customers in Vermont

Quintal F, Pereira L, Nunes N, Nisi V, Barreto M (2013) WATTSBurning: Design and evaluation of an innovative eco-feedback system. Lect. Notes Comput. Sci. (including Subser. Lect. Notes Artif. Intell. Lect. Notes Bioinformatics), vol. 8117 LNCS, Springer, Berlin, Heidelberg, p. 453–70. https://doi.org/10.1007/978-3-642-40483-2_32

Quitzow R, Bersalli G, Eicke L, Jahn J, Lilliestam J, Lira F, Marian A, Süsser D, Thapar S, Weko S, Williams S, Xue B (2021) The COVID-19 crisis deepens the gulf between leaders and laggards in the global energy transition. Energy Res Soc Sci 74:101981. https://doi.org/10.1016/j.erss.2021.101981

Rausser G, Strielkowski W, Štreimikienė D (2018) Smart meters and household electricity consumption: a case study in Ireland. Energy Environ 29:131–146. https://doi.org/10.1177/0958305X17741385

Reeves B, Cummings JJ, Scarborough JK, Yeykelis L (2015) Increasing energy efficiency with entertainment media: an experimental and field test of the influence of a social game on performance of energy behaviors. Environ Behav 47:102–115. https://doi.org/10.1177/0013916513506442

Rettie R, Burchell K, Harries T (2014) Energy consumption feedback: Engagement by design. Lect Notes Comput Sci (Including Subser Lect Notes Artif Intell Lect Notes Bioinformatics) 8519 LNCS:594–604. https://doi.org/10.1007/978-3-319-07635-5_57

Schleich J, Klobasa M, Brunner M, Gölz S, Götz K (2011) Smart metering in Germany and Austria: results of providing feedback information in a field trial

Schleich J, Klobasa M, Goelz S (2012) Effects of Feedback on Residential Electricity Demand – Results from a field trial in Austria. 2012 Int Energy Progr Eval Conf 61:1097–1106

Schleich J, Faure C, Klobasa M (2017) Persistence of the effects of providing feedback alongside smart metering devices on household electricity demand. Energy Policy 107:225–233. https://doi.org/10.1016/j.enpol.2017.05.002

Schultz PW, Estrada M, Schmitt J, Sokoloski R, Silva-Send N (2015) Using in-home displays to provide smart meter feedback about household electricity consumption: a randomized control trial comparing kilowatts, cost, and social norms. Energy 90:351–358. https://doi.org/10.1016/j.energy.2015.06.130

Schwartz D, Fischhoff B, Krishnamurti T, Sowell F (2013) The Hawthorne effect and energy awareness. Proc Natl Acad Sci 110:15242–15246. https://doi.org/10.1073/pnas.1301687110

Schwartz T, Stevens G, Jakobi T, Denef S, Ramirez L, Wulf V, Randall D (2015) What people do with consumption feedback: a long-term living lab study of a home energy management system. Interact Comput 27:551–576. https://doi.org/10.1093/iwc/iwu009

Shimada K, Ochi Y, Matsumoto T, Matsugi H, Awata T (2014) An empirical study of electric power demand control by real-time feedback of consumption levels: case of Nushima Island households. Procedia Technol 18:53–57. https://doi.org/10.1016/j.protcy.2014.11.012

Spagnolli A, Corradi N, Gamberini L, Hoggan E, Jacucci G, Katzeff C, Broms L, Jonsson L (2011) Eco-feedback on the go: motivating energy awareness. Computer (Long Beach Calif) 44:38–45. https://doi.org/10.1109/MC.2011.125

Stinson J, Willis A, Williamson JB, Currie J, Smith RS (2015) Visualising energy use for smart homes and informed users. Energy Procedia 78:579–584. https://doi.org/10.1016/j.egypro.2015.11.015

Strengers Y (2011) Negotiating everyday life: the role of energy and water consumption feedback. J Consum Cult 11:319–338. https://doi.org/10.1177/1469540511417994

Sustentable Citiess Index. Sustainable Cities Index (2016) Arcadis Glob 2016:10

United Nations: Department of Social and Economic Affairs. World population prospects: The 2012 revision, Key findings and Advance Tables. Popul Div 2013 2013:Working Paper No. ESA/P/WP.227

United Nations. World urbanization prospects: the 2018 revision, key facts. 2018. Doi:(ST/ESA/SER.A/366)

United Nations Environment Programme (UNEP). 21 Issues for the 21st Century – Results of the UNEP Foresight Process on Emerging Environmental Issues. vol. 2. Nairobi, Kenya: 2012. https://doi.org/10.1016/j.envdev.2012.03.005

van Dam SS, Bakker CA, van Hal JDM (2010) Home energy monitors: impact over the medium-term. 3218:1–26. https://doi.org/10.1080/09613218.2010.494832

Villa-Arrieta M (2019) Energy sustainability of smart cities. Universitat Politècnica de Catalunya. http://hdl.handle.net/2117/340988

Villa-Arrieta M, Sumper A (2019a) Contribution of smart cities to the energy sustainability of the binomial between city and country. Appl Sci 9:3247. https://doi.org/10.3390/app9163247

Villa-Arrieta M, Sumper A (2019b) Economic evaluation of near-zero energy cities. Appl Energy 237:404–416. https://doi.org/10.1016/j.apenergy.2018.12.082

Wood G, Newborough M (2003) Dynamic energy-consumption indicators for domestic appliances: environment, behaviour and design. Energ Buildings 35:821–841. https://doi.org/10.1016/S0378-7788(02)00241-4

World Energy Council. WEC Trilemma: pathway calculator. Pathw Calc 2017. https://trilemma.worldenergy.org/#!/pathway-calculator (accessed May 28, 2019)

World Health Organization. The Rise of Modern Cities. Hidden Cities Unmask Overcoming Heal Inequities Urban Settings 2010:3–10

Chapter 4
Security and Privacy Issues in IoT-Based Smart Grids: A Case Study in a Digital Substation

Doney Abraham, Sule Yildirim Yayilgan, Mohamed Abomhara, Alemayehu Gebremedhin, and Fisnik Dalipi

Abstract Smart Grid is one of the increasingly used critical infrastructures that is essential for the functioning of a country. This coupled with Internet of Things (IoT) has huge potentials in several areas such as remote monitoring and managing of electricity distribution, traffic signs, traffic congestion, parking spaces, road warnings, and even early detection of power influxes as a result of natural disasters, safety failures, equipment failures, or carelessness. Despite the advantages of Smart Grids, there are security threats, privacy concerns, and several open challenges related to Smart Grids. This chapter seeks to provide a review of the security and privacy perspectives inherent in IoT-enabled Smart Grids. Firstly, the chapter explores the functionalities of Smart Grids as opposed to a traditional grid. Next, the chapter provides an overview of Smart Grid architectures followed by positioning IoT concept into Smart Grid with a focus on architectures. Then, the proposed approach for identifying threats and attacks in IoT-enabled Smart Grid, namely, the security pyramid, is presented. Lastly, we work on identifying the possible threats and attacks in the digital substation use case.

Keywords Smart grid · IoT · Security · Privacy · Attacks · Threats · Cyber-physical systems

---

D. Abraham (✉) · S. Y. Yayilgan · M. Abomhara
Department of Information Security and Communication Technology, Norwegian University of Science and Technology, Trondheim, Norway
e-mail: doney.abraham@ntnu.no; sule.yildirim@ntnu.no; mohamed.abomhara@ntnu.no

A. Gebremedhin
Department of Manufacturing and Civil Engineering, Norwegian University of Science and Technology, Trondheim, Norway
e-mail: alemayehu.gebremedhin@ntnu.no

F. Dalipi
Department of Informatics, Linnaeus University, Växjö, Sweden
e-mail: fisnik.dalipi@lnu.se

G. C. Lazaroiu et al. (eds.), *Holistic Approach for Decision Making Towards Designing Smart Cities*, Future City 18,
https://doi.org/10.1007/978-3-030-85566-6_4

Introduction

The world is increasingly moving toward an Internet of Things (IoT) age, and the importance of cyber-physical systems is ever rising (Zanero 2017). IoT support numerous applications in different domains such as power grids, transportation systems, health care, water supply, oil and gas distribution, and telecommunications that are crucial for the operation of society (Sanislav and Miclea 2012). Since the critical infrastructure of a country such as energy, food, transport, telecommunications, health care, banking, and finance depends on the previously mentioned domains, its security and privacy are of utmost importance. Smart Grids based on IoT and data technologies have revolutionized the way power grids are built as opposed to traditional grids by providing features such as self-healing, bi-directional communication, feedback, and disaster recovery plans (Shabanzadeh and Moghaddam 2013). Introduction of IoT to Smart Grids has made the power grids even more reliable and can be used to monitor electricity generation, protection of transmission line, and ability by consumers to monitor usage by smart meters to name a few (Al-Turjman and Abujubbeh 2019; Fadlullah et al. 2018). Globally, IoT is estimated to have a total economic impact of around USD 11 trillion by 2025 (Jha and Sahoo 2020; Ménard 2017). Since IoT is a collection of different types of smart devices, it is also vulnerable to different types of security and privacy threats (Boroojeni et al. 2017). Some research works already focus on security and privacy challenges in IoT-integrated Smart Grids (Asghar et al. 2017; He and Yan 2016; Komninos et al. 2014; Stellios et al. 2018; Tan et al. 2017; Wang and Lu 2013). In this chapter, we provide a review of these challenges.

The remaining parts of the chapter are structured as follows: section "Overview of Smart Grid: Architecture" explains the overall architecture of a Smart Grid and shows the National Institute of Standards and Technology (NIST) conceptual model. An overview of IoT and its integration in Smart Grids with some proposed architectures are shown in section "IoT in Smart Grid". Further in section "The Proposed Approach for Use Case Analysis: The Security Pyramid", the security pyramid for analyzing use cases is proposed. Section "A Case Study: Digital Substation" focus on applying and explaining the use of security pyramid on a digital substation use case. Conclusions are drawn in section "Conclusions".

Overview of Smart Grid: Architecture

This section focuses on the overview of Smart Grid and its building blocks. It describes the architecture and shows the differences between traditional power grids and the advantages of Smart Grids.

A traditional power grid is one of the most complex critical infrastructures that have been ever built (Pagani and Aiello 2013). It consists of different parts like operations center, power generation plants, transmission towers, and power distribution centers which are physically connected by cables and wires (Saleem et al. 2019). The main functions of a power grid are electricity generation, transmission, and its distribution. Electricity is mostly generated utilizing central power plants using different energy sources and then transmitted to different load customers through high voltage lines. This in turn is distributed to consumers using distribution centers at a lower voltage (Ali 2013). Transmission and distribution of electricity are owned by power companies which make profit from the consumers. The electricity and information flow in a traditional power grid are unidirectional (Saleem et al. 2019) which results in lack of flexibility, lack of information sharing to customers, and control mechanisms that respond slowly in the event of power failures or attacks to the power grid. Traditional grids also lack self-healing and self-restoring capability in case of a down time (Fang et al. 2011). Additionally, due to the high usage of electronic devices, traditional power grids have a large amount of wastage of resources due to inefficient distribution of electricity, lack of monitoring and communication, and inadequate methods to store energy. All these coupled together has led to the introduction of Smart Grids.

Smart Grids enable the integration of both cyber and physical systems in the sense that Information and Communications Technology (ICT) is integrated with power networks to enable generation, transmission, and distribution of electricity in a more efficient manner (Yu and Xue 2016). The following are some key features and characteristics of a Smart Grid (Jain 2016; Paul et al. 2014):

1. Information and electricity flow are bidirectional.
2. Robust and uninterrupted power supply as Smart Grid has self-healing capabilities that enable real-time state monitoring to analyze faults and respond to them.
3. Integrates modern advanced sensor technology, measurement technology, communication technology, information technology, computing technology, and control technology.
4. Optimizes asset utilization and operates efficiently by reducing the cost of operations and investments. This is done by aptly managing power loss and improving power efficiency.
5. Operates resiliently against attacks and natural disasters.
6. Interoperable as it enables logical grouping of standards among diverse components of the Smart Grid.
7. Enables active participation by customers, new products, services, and markets.

The NIST Conceptual Model of Smart Grid

The main domains in NIST conceptual model as described by (Cunjiang et al. 2012) are as follows:

1. **Power Generation**: This is the domain where energy is generated in large quantities from different sources like wind, hydro, solar, biomass, etc., and converted

to electricity. These are normally linked straight to transmission systems that in turn offer applications smart in nature.
2. **Transmission**: Electricity is transmitted from the sites of bulk generation to long distances to the substations of areas where electricity demand is higher.
3. **Distribution**: This is the domain where electricity is distributed to customers.
4. **Consumer**: This is the domain where the electricity is consumed, managed, and generated in some cases (e.g., smart houses with solar panels). Consumer domain is subcategorized into individual houses, commercial/building, and industrial complexes with varying energy needs for each of these categories.
5. **Service Provider**: This provides services to business processes of power system producers, customers, and distributors. These are utility services such as electricity billing, managing customer accounts, management of energy use in homes, etc.
6. **Operations**: Operations domain ensures continuous functioning of the power system. The typical applications of operations domain are network operation monitoring including breaker and switch states, fault management (identification of faults, informing customers, coordination), maintenance and construction (inspection, cleaning, and adjustment of equipment), customer support for purchase, provision and troubleshooting of power system services, etc.
7. **Markets**: This is the domain where power grid assets are traded. The supply/demand and prices are exchanged in this domain.

The following section focuses on the integration of IoT in Smart Grids.

IoT in Smart Grid

IoT in broad terms are all devices that are interconnected and communicate over the Internet (Hanes et al. 2017). IoT devices are broadly scattered with low storage capability, processing capabilities that can improve performance, security, and reliability. Some examples of IoT devices are smart mobile phones, smart fridge, smart meters to measure power consumption, automobiles, and smart buildings, to name a few (Abdul-Qawy et al. 2015; Hanes et al. 2017). Figure 4.1 shows a high level overview of IoT features (Xia et al. 2012). It shows a communication dimension that can be maintained by anyone irrespective of the location of the person.

IoT technology has an important role in building Smart Grid infrastructure. Some of its usages are as follows (Ghasempour 2019):

1. Monitoring electricity generation in power plants that operate with coal, wind, solar, and biomass. It can help calculate the energy demands of customers and store energy accordingly. This facilitates efficient use of energy at a later stage when demand is higher.
2. Monitoring and protecting power transmission lines, controlling the devices (e.g., temperature, current, voltage sensors) used for transmission and assessing the power consumption.

Fig. 4.1 IoT features

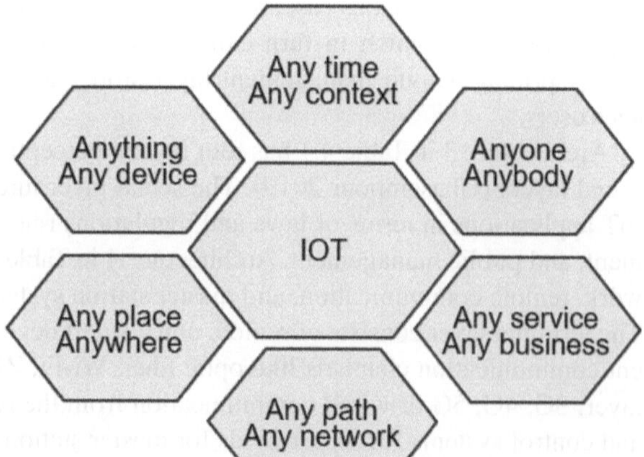

Table 4.1 IoT architectures in Smart Grid proposals (Ghasempour 2019)

Layers/ architectures	Architecture 1	Architecture 2	Architecture 3	Architecture 4
Layer 4		Application	Social	Master station system
Layer 3	Application	Cloud management	Application	Remote communication
Layer 2	Network	Network	Network	Field network
Layer 1	Perception	Perception	Perception	Terminal

3. From a consumer point, IoT have various uses like smart meters to monitor power usage, control charging of electric vehicles, schedule the energy use among IoT devices in a household, and ensure the continuous connectivity across networks.

Table 4.1 shows some of the IoT architectures utilized in Smart Grids which vary in layers they are built of. The proposed architectures are either three or four layered. Architecture 1 has three layers, namely, perception, network, and application layer (Ghasempour 2019; Mauri et al. 2016; Ou et al. 2012). Perception layer collects data using various sensors, tags, and readers. Network layer maps data gathered by perception layer to different communication protocols using wired or industry standard wireless networks. The industry standards include 3G, 4G, 5G, broadband, Zigbee, or Wi-Fi, and these further transmit data to application layer that can monitor the IoT devices in real time. Application layer contains an application structure that can compute and process information and enable interfacing and integration.

Architecture 2 has four layers, namely, perception, network, cloud management, and application layer (Ghasempour 2019; Viswanath et al. 2016). Here, the perception layer consists of two additional layers that include a thing layer comprised of different IoT sensors, tags, readers to sense, control, collect data, a gateway layer comprised of microcontrollers, and displays that control elements connecting to the thing layer. As in previous case, the network layer transmits data from perception to application layer which in turn can provide services to consumers like manage energy pricing. Cloud management layer stores and analyzes data and also manages users.

Architecture 3 in Table 4.1 has four layers: perception, network, application, and social layers (Ghasempour 2019). The social layer integrates and regulates various IoT applications in terms of laws and regulations relevant for IoT devices, government, and public management. Architecture 4 in Table 4.1 has a terminal, field network, remote communication, and master station system layer (Ghasempour 2019). The terminal layer consists of remote units, smart devices, and smart meters; different communication channels like optic fiber, Wi-Fi, Zigbee, etc., for field network layer; 3G, 4G, 5G or wired communication from the remote communication layer; and control systems for Smart Grids for master station layer.

Even though IoT technologies are being adopted in Smart Grids, there are security and privacy challenges which need to be addressed. Communication in IoT-enabled Smart Grids is conducted over the open Internet which is already vulnerable to cyber attacks (Aloul et al. 2012). Various approaches, measures, and recommendations are proposed in the literature to tackle these challenges (Dalipi and Yayilgan 2016; Parra et al. 2019; Sadiku et al. 2017). Besides, the diversity of devices and applications in Smart Grids adds up to the complexity of handling these challenges. In order to leverage the full capability of IoT integration, there is a need to clearly understand and locate these challenges in IoT-enabled Smart Grid. Based on the literature and in cooperation with partners of digital substation (DS) projects, we have identified several challenges (Ahmed et al. 2019; Andrea et al. 2015; Bekara 2014; Chen et al. 2017; Dalipi and Yayilgan 2016; Gunduz and Das 2020; He et al. 2017; Karimipour and Dinavahi 2017; Khan and Salah 2018; Li et al. 2019; Ni et al. 2017; Qureshi et al. 2020; Rice and AlMajali 2014; Rodriguez-Calvo et al. 2018; Salpekar 2018; Tian et al. 2018; Valea et al. 2019; Whitman and Mattord 2011; Yamada et al. 2018; Zhang et al. 2019).

The Proposed Approach for Use Case Analysis: The Security Pyramid

In this chapter, we propose the security pyramid (Fig. 4.2) as an approach in identifying and analyzing threats and attacks in case studies of IoT-enabled Smart Grid. First and foremost, such an identification and analysis require setting the security objectives that stakeholders of an IoT-enabled Smart Grid e.g., digital substation, prioritize to comply with (peak of the pyramid). Next, in the middle layer of the pyramid, threats and attacks categories which particularly fall under the set objectives are identified. As the last step, in the bottom layer of the pyramid, actual threats and attacks which are identified for each attack and threat category are defined for the case study in hand.

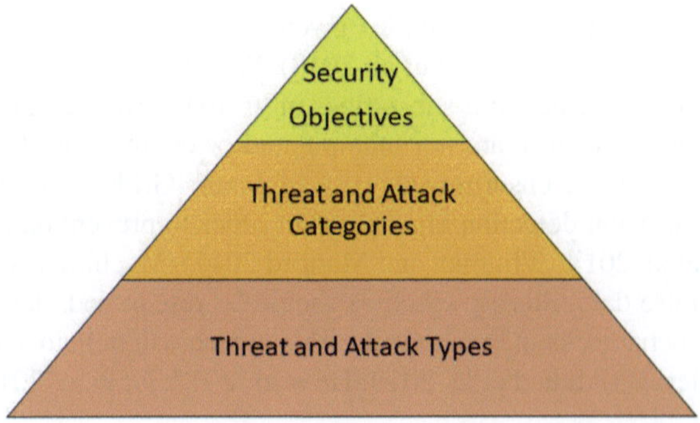

Fig. 4.2 The proposed security pyramid

Security Objectives

In a power grid, the effective security objects and their descriptions are provided in this section. We assume the same security objectives hold for the IoT-enabled Smart Grid. NIST has defined certain criteria to maintain the security and privacy of Smart Grids. These are confidentiality, integrity, and availability (CIA) principles (Harvey et al. 2014). Confidentiality is information protection from access without a proper authorization. Integrity is the assurance of information not being modified without authorization. Availability guarantees that data, applications and resources are available to authorized users whenever they need them.

Threat and Attack Categories in IoT-Enabled Smart Grids

In the previous sections, we have referred to the importance of ensuring security and privacy of Smart Grids. Besides, IoT devices are particularly vulnerable to security and privacy threats. In this section, we provide a summary of the threats and attacks that are likely to occur in Smart Grids, in IoT-integrated Smart Grid architectures, and in stand-alone IoT architectures and the IoT devices in them.

Data Manipulation

These are the types of data integrity threats where unauthorized users aim to mislead the Smart Grids toward making wrong actions. For example, an adversary can illegally manipulate a smart meter in order to modify energy consumption to decrease cost during the peak demands (Bekara 2014). Another example is when an attacker accesses an unprotected substation and uses his/her own device to influence the communication among substation assets. This way, attacker is able to inject false data into the communication channels.

Solution In the examples given above, data integrity is affected (Karimipour and Dinavahi 2017; Tian et al. 2018). We need to ensure that the communication channels, whether in a substation, in an IoT infrastructure, or in IoT devices (such as smart meters), are not manipulated or compromised by unauthorized parties. One method of ensuring integrity of Smart Grids is implementing different kinds of intrusion detection algorithms in order to prevent data manipulation attacks (Chen et al. 2017; Whitman and Mattord 2011). Machine learning based models and other false data filtering schemes should be researched, developed, and applied to detect such attacks in Smart Grids. This in turn will help to achieve sufficient data integrity levels (Ahmed et al. 2019; He et al. 2017; Ni et al. 2017).

Unauthorized Access of Data

It is important that data is available only to authorized users throughout the entire communication process in Smart Grids. It is very critical that data collected via IoT sensors will not be used to uncover information to anyone except for the Smart Grid operators. Otherwise, the confidentiality of data is lost.

Solution For the purpose of achieving confidentiality of data, data acquisition frameworks which are confidentiality assured, secure key management, access control, and trust management mechanisms are proposed (Zhang et al. 2019). Providing integrity and encryption of the data from IoT devices used by consumers ensures confidentiality (Valea et al. 2019).

Attacks Due to Unauthorized Access of Users or Devices

Attacks Due to Unauthorized Access of Devices to IoT-Enabled Smart Grid

It is important that only authorized devices and applications can be connected to IoT. For example, all the laptops located in a digital substation should be authorized to connect to digital substation, whereas any other laptops which are in the digital substation accidentally or on malicious intentions should not connect to the digital substation. The process of authentication ensures that information distributed in the Smart Grid network is legitimate.

Solution Authenticity of devices is an integrity problem. The identity of any IoT device being used in context should be timely authenticated in order to escape any potential manipulation of the system. Nevertheless, in the time-sensitive and traffic-intensive nature of IoT-based Smart Grid communications, authentication of data and object (e.g., smart meter) is an intricate mission, and more research is needed in this direction (Rice and AlMajali 2014; Salpekar 2018). Still, if the authenticity of the devices is guaranteed but the user accesses more than what he/she is authorized for, then the confidentiality is compromised (elevation of privilege). Elevation of privilege is a big threat of "insider users."

Attacks Due to Unauthorized Access of Users to Devices

Data and network objects, such as smart meters, transformers, and cables are available only for authorized users and services upon request. Authorized access can occur due to many facts such as failure of a device, user getting access because of elevation of privilege, errors in the access control mechanisms and policies, loss of device or use of fake device in the system, etc.

Solution Authorization and access control is a solution here. Granting access privileges to Smart Grid devices and functionalities can significantly reduce the probability of unauthorized and malicious access to network devices. As indicated in the literature, restricting access to objects through, for example, role-based access control (RBAC) (Sandhu 1998) or/and attribute-based access control (ABAC) (Hu et al. 2013) can enhance the system reliability by eliminating potential cyber threats (Dalipi and Yayilgan 2016). Since the controlling of IoT-enabled Smart Grid is performed remotely over the open Internet, access control mechanisms are instrumental in order to prevent and restrain users or devices access in the network.

In a complex and interaction-intensive IoT-enabled Smart Grids, it is vital to enforce security and privacy objectives among different types of objects, layers, and applications (Andrea et al. 2015). Trust is an assured reliance on a claimed identity, and an assured reliance comes in the form of authentication/authorization. Authentication/authorization can in turn be used to prevent unauthorized access. Providing trust management to millions of IoT devices and ensuring trusted governance of IoT device ownership remain as one of the open issues for future research to be addressed.

Threats on the Privacy of Customers

Smart Grid network resources contain private information of millions of users, including energy consumption and consumption patterns of users. Such information not only can be used for marketing purposes but also could provide clear indications to intruders about the whereabouts of the consumers.

Solution To ensure privacy and protection of personal information from unauthorized access or misuse, securing access to data for IoT-enabled Smart Grids should be enforced by constantly adapting encryption mechanisms and privacy preservation schemes.

Attack and Threat Types

Depending on the use case, the actual attacks and threats are identified, and they are linked to the attack and threat categories in the middle layer. Particularly, a close cooperation is required with the stakeholders of the case study in order to be able to identify the assets and the attacks and threats that can occur along these assets.

A Case Study: Digital Substation

In this section, we examine digital substation as a case study for identifying possible threats and attacks in this particular critical infrastructure. A digital substation is both a physical and soft infrastructure. Below are the three steps followed in the case study.

Step 1

All the three security objectives, confidentiality, integrity, and availability, are recognized and acknowledged by the stakeholder of a digital substation. Even though a digital substation composes of a generic infrastructure, we particularly focus on the digital substations in Norway within the scope of the ECoDiS (Engineering and Condition monitoring in Digital Substations) Project.[1] There are three pilot sites in the project from Norway: Furuset substation owned by Statnett SF-Statelig Foretak, Hafslund, and Skagerak substations.

Step 2

All the attack categories from the previous section are valid for examination for digital substation, particularly for our pilots, namely:

1. Data manipulation
2. Unauthorized access of data
3. Attacks due to unauthorized access of users or devices
4. Attacks on the privacy of customers

Step 3

In our previous work (Khodabakhsh et al. 2020a; Khodabakhsh et al. 2020b), we have identified some attacks and developed a generic map of these attacks in digital substation pilots as shown in Fig. 4.3. The identified assets in this digital substation pilots are SCADA (Supervisory Control and Data Acquisition), Gateway, IED (Intelligent Electronic Device), HMI (Human Machine Interface), Communication Network, MU (Merging Unit), Switch and NCIT-CT&VT (Non-conventional Instrument Transformers-Current and Voltage Transformer).

Now we will link each of the identified attacks in the map into categories in the middle layer of the security pyramid, by listing each asset and clarifying how each

[1] https://www.sintef.no/en/projects/2019/ecodis/

asset is identified under one or more of the attack and threat categories of the middle layer. These are elaborated in the following section.

Smart Grid Threat and Attack Categories vs. Attacks on DS

1. **SCADA:** Malware software can be designed to steal/copy data from devices, and it does not only manipulate the data. In this case, if a malware managed to access data for unauthorized access, then the privacy of the users is violated. If users gain an unauthorized access to a device, then they might take a control of the system, e.g., taking control of SCADA system. Manipulated software will produce manipulated data. Changing control data can stop functionality in the system. Access to data storage can lead to data manipulation. For example, if you manage to access data in servers, you can change anything there. The type of sensory data in a SCADA system can be temperature values from transformers, humidity of the devices, etc.
2. **Gateway:** Attackers can use DoS attack on gateways to make system resource and system functionalities unavailable. We relate this to unauthorized access of users to devices.
3. **IED:** One of the ways an attacker can gain unauthorized access to IEDs is using brute force attack. An attacker can then access the data in RAM of the IED, and

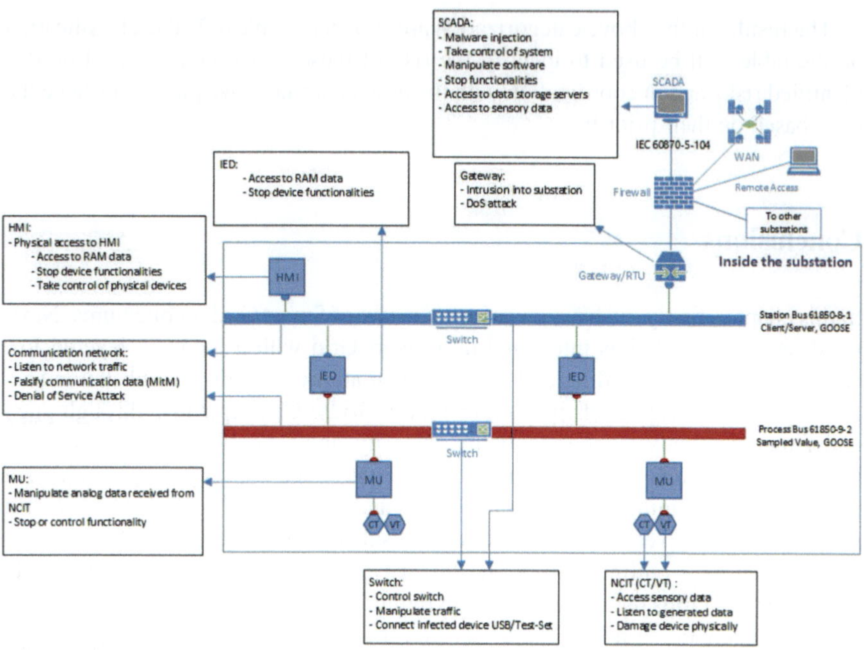

Fig. 4.3 Revised version of the map of possible attacks in digital substation from (Khodabakhsh et al. 2020b)

depending on type of data in the RAM, privacy may be influenced. The attacker can also stop the device functionalities by manipulating different data parameters. For example, modifying the messages from IEDs can mislead the operators about the actual state of DS.

4. **HMI:** By gaining access to HMI, an attacker can take control of the physical devices, stop the device functionalities, and also gain access to RAM data. Depending on the type of data in RAM, it can also influence privacy.
5. **Communication Network:** Depending on the type of data listened to in the network traffic, data privacy may be influenced. This can be caused by Man-in-the-Middle (MitM) attack. A DoS attack here can make system resource and system functionalities unavailable. We relate this to unauthorized access of users to devices.
6. **MU:** By gaining access to MU, an attacker can control different functionalities in the MU. Data received from NCIT can also be manipulated with this attack.
7. **Switch:** The switches on the station bus and/or process bus can be controlled with this attack. If the attacker manages to manipulate the traffic in these switches, he might gain unauthorized access to data. The attacker can then, for example, redirect traffic to the server.
8. **NCIT-CT&VT:** The attacker here can gain access to different data from the sensors like current or voltage measurements and can physically damage the devices. Once the access is gained, then different values in the sensors can be manipulated causing disruptions to services and sending incorrect values further to other devices in the substation.

The results of the above categorization are shown in Table 4.2. The classification on the table will be used to identify the risk of these attack goals. Based on this identified risks and in coordination with the project partners, we plan to mitigate the risks based on their priorities.

Conclusions

In this chapter, firstly we have given an overview of Smart Grid architectures. Next, we discussed how IoT is integrated into Smart Grid with a focus on layered IoT architectures. Understanding and locating such attacks in a digital substation as a case study are essential in protecting Smart Grids. Consequently, through such understanding, analysis, confidentiality, integrity, and availability of data in Smart Grids can be maintained. For that reason, we introduce the security pyramid as a proposal for examining and analyzing attacks and threats in IoT-enabled Smart Grid use cases. For a given use case, pilot digital substations of the ECoDiS project in Norway, we have outlined attack and threat types that are likely to occur in IoT-enabled Smart Grids using the security pyramid. Further works will involve (but not limited to) generating normal data based on the DS architecture as in Fig. 4.3 and then simulating various DS attacks listed in Table 4.1. As part of future work, there will also be studies conducted on the different methods to detect these attacks.

Table 4.2 Classification of attack and threat types from digital substation into attack and threat categories

			Smart grid threat and attack categories					
					Unauthorized access of devices or users			
DS Assets	Asset Access Type	Attack/Threat Types	Data manipulation attacks	Unauthorized access of data	Unauthorized access of devices to Smart Grid	Unauthorized access of users	Attacks on privacy	Disruption attacks
SCADA	Gaining access to SCADA	Malware injection	X	X	X	X	X	
		Take control of system			X	X		X
		Manipulate software	X	X		X		
		Stop functionality	X			X		X
		Access to data storage servers	X	X			X	
		Access to sensory data		X				
Gateway	Access to gateway	Intrusion into substation		X		X		
		Denial of service attack						X
IED	Access to IED	Access to RAM data	X	X		X	X	
		Stop device functionalities	X			X		X
HMI	Physical access to HMI	Access to RAM data		X		X		
		Stop device functionalities	X		X	X		X
		Take control of physical devices	X	X		X	X	X
Communication network	Access to communication network	Listen to network traffic		X			X	
		Falsify communication data (MitM)	X	X				
		Denial of service attack						X

(continued)

Table 4.2 (continued)

DS Assets	Asset Access Type	Attack/Threat Types	Smart grid threat and attack categories					
					Unauthorized access of devices or users			
			Data manipulation attacks	Unauthorized access of data	Unauthorized access of devices to Smart Grid	Unauthorized access of users	Attacks on privacy	Disruption attacks
MU	Access to MU	Manipulate analog data received from NCIT	X	X		X		
		Stop or control functionality	X			X		X
Switch	Access to switch	Control the switch				X		
		manipulate Traffic	X	X	X		X	X
		Connect infected device (e.g., USB)	X	X		X	X	
NCIT (CT\VT)	Access to NCIT (CT/VT)	Access to sensory data		X				
		Listen to generated data		X			X	
		Damage device physically				X		

The following abbreviations are used in this manuscript:

CIA	Confidentiality, integrity, and availability
CT	Current transformer
DMS	Distribution Management Systems
DoS	Denial of service
DS	Digital substation
ECoDiS	Engineering and Condition monitoring in Digital Substations
EMS	Energy management systems
HMI	Human Machine Interface
ICT	Information and Communications Technology
IED	Intelligent electronic device
IoT	Internet of Things
MAC	Mandatory access control
MitM	Man-in-the-Middle
MU	Merging Unit
NCIT	Non-conventional Instrument Transformers
NIST	National Institute of Standards and Technology
RAM	Random Access Memory
RBAC	Role-Based Access Control
SCADA	Supervisory Control and Data Acquisition
VT	Voltage Transformer

Acknowledgments This work is partly carried out in the Engineering and Condition monitoring in Digital Substation (ECoDiS) which is a project funded by Research Council of Norway (NFR), Innovation Project for the Industrial Sector – ENERGIX program, project number 296550, coordinated by Statnett RD group. Statnett is the system operator of the Norwegian power system. The project aims in exploiting the full potential of a digital station to increase security of supply and observability and reduce costs in a changing energy system.

References

Abdul-Qawy AS, Pramod P, Magesh E, Srinivasulu T (2015) The internet of things (IoT): an overview. Int J Eng Res Appl 1(5):71–82

Ahmed S, Lee Y, Hyun S, Koo I (2019) Unsupervised machine learning-based detection of covert data integrity assault in smart grid networks utilizing isolation forest. IEEE Tran Inf Forensics Secur 14(10):2765–2777

Al-Turjman F, Abujubbeh M (2019) IoT-enabled smart grid via SM: an overview. Futur Gener Comput Syst 96:579–590

Ali AS (2013) Smart grids: opportunities, developments, and trends. Springer, London

Aloul F, Al-Ali A, Al-Dalky R, Al-Mardini M, El-Hajj W (2012) Smart grid security: threats, vulnerabilities and solutions. Int J Smart Grid Clean Energy 1(1):1–6

Andrea I, Chrysostomou C, Hadjichristofi G (2015) Internet of things: security vulnerabilities and challenges. In: 2015 IEEE symposium on computers and communication (ISCC). IEEE, Piscataway, pp 180–187

Asghar MR, D'an G, Miorandi D, Chlamtac I (2017) Smart meter data privacy: A survey. IEEE Commun Surv Tutorials 19(4):2820–2835

Bekara C (2014) Security issues and challenges for the IoT-based smart grid. In: FNC/MobiSPC, pp 532–537

Boroojeni KG, Amini MH, Iyengar S (2017) Overview of the security and privacy issues in smart grids. In: Smart grids: security and privacy issues. Springer, Cham, pp 1–16

Chen DD, Li ZQ, Tian L, Teng MX, Feng X (2017) Big data based intrusion detection of smart meters. DEStech Trans Comput Sci Eng. https://doi.org/10.12783/dtcse/cnsce2017/8884

Cunjiang Y, Huaxun Z, Lei Z (2012) Architecture design for smart grid. Energy Procedia 17:1524–1528. https://doi.org/10.1016/j.egypro.2012.02.276

Dalipi F, Yayilgan SY (2016) Security and privacy considerations for IoT application on smart grids: survey and research challenges. In: 2016 IEEE 4th International conference on future internet of things and cloud workshops (FiCloudW). IEEE, Piscataway, pp 63–68

Fadlullah ZM, Pathan ASK, Singh K (2018) Smart grid internet of things. Mob Netw Appl 23(4):879–880

Fang X, Misra S, Xue G, Yang D (2011) Smart grid—the new and improved power grid: a survey. IEEE Commun Surv Tutor 14(4):944–980

Ghasempour A (2019) Internet of things in smart grid: architecture, applications, services, key technologies, and challenges. Inventions 4(1):22

Gunduz MZ, Das R (2020) Cyber-security on smart grid: Threats and potential solutions. Comput Netw 169:107094

Hanes D, Salgueiro G, Grossetete P, Barton R, Henry J (2017) IoT fundamentals: networking technologies, protocols, and use cases for the internet of things. Cisco Press, Indianapolis

Harvey M, Long D, Reinhard K (2014) Visualizing nistir 7628, guidelines for smart grid cyber security. In: 2014 Power and Energy Conference at Illinois (PECI). IEEE, Piscataway, pp 1–8

He H, Yan J (2016) Cyber-physical attacks and defences in the smart grid: a survey. IET Cyber Phys Syst Theory Appl 1(1):13–27

He Y, Mendis GJ, Wei J (2017) Real-time detection of false data injection attacks in smart grid: a deep learning-based intelligent mechanism. IEEE Tran Smart Grid 8(5):2505–2516

Hu VC, Ferraiolo D, Kuhn R, Friedman AR, Lang AJ, Cogdell MM, Schnitzer A, Sandlin K, Miller R, Scarfone K et al (2013) Guide to attribute based access control (ABAC) definition and considerations (draft). NIST special publication 800(162):4–14

Jain A (2016) Changes and challenges in smart grid towards smarter grid, December. https://doi.org/10.1109/ICEPES.2016.7915907

Jha RS, Sahoo PR (2020) Internet of things (IoT) – enabler for connecting world. In: ICT for competitive strategies: Proceedings of 4th international conference on information and communication technology for competitive strategies (ICTCS 2019), December 13th–14, 2019, Udaipur, India. CRC Press, Boca Raton, p 1

Karimipour H, Dinavahi V (2017) On false data injection attack against dynamic state estimation on smart power grids. In: 2017 IEEE international conference on smart energy grid engineering (SEGE). IEEE, Piscataway, pp 388–393

Khan MA, Salah K (2018) IoT security: review, blockchain solutions, and open challenges. Futur Gener Comput Syst 82:395–411

Khodabakhsh A, Yayilgan SY, Abomhara M, Istad M, Hurzuk N (2020a) Cyber-risk identification for a digital substation. ACM. https://doi.org/10.1145/3407023.3409227

Khodabakhsh A, Yildirim Yayilgan S, Houmb SH, Hurzuk N, Foros J, Istad MK (2020b) Cybersecurity gaps in a digital substation: from sensors to SCADA. https://doi.org/10.1109/MECO49872.2020.9134350

Komninos N, Philippou E, Pitsillides A (2014) Survey in smart grid and smart home security: issues, challenges and countermeasures. IEEE Commun Surv Tutor 16(4):1933–1954

Li S, Song H, Iqbal M (2019) Privacy and security for resource-constrained IoT devices and networks: research challenges and opportunities. Sensors 19(8):1935

Mauri JL, Tomás J, Canovas A, Parra L (2016) An integrated IoT architecture for smart metering. IEEE Commun Mag 54:50–57

Ménard A (2017) How can we recognize the real power of the internet of things? McKinsey. https://www.mckinsey.com/business-functions/mckinsey-digital/our-insights/how-can-we-recognize-the-real-power-of-the-internet-of-things#. November. Accessed on 18 Oct 2020

Ni J, Zhang K, Alharbi K, Lin X, Zhang N, Shen XS (2017) Differentially private smart metering with fault tolerance and range-based filtering. IEEE Trans Smart Grid 8(5):2483–2493

Ou Q, Zhen Y, Li X, Zhang Y, Zeng L (2012): Application of internet of things in smart grid power transmission, June, pp 96–100. https://doi.org/10.1109/MUSIC.2012.24

Pagani GA, Aiello M (2013) The power grid as a complex network: a survey. Physica A 392(11):2688–2700

Parra GDLT, Rad P, Choo KKR (2019) Implementation of deep packet inspection in smart grids and industrial internet of things: challenges and opportunities. J Netw Comput Appl 135:32–46

Paul S, Rabbani MS, Kundu RK, Zaman SMR (2014) A review of smart technology (smart grid) and its features. In: 2014 1st international conference on non conventional energy (ICONCE 2014). IEEE, Piscataway, pp 200–203

Qureshi KN, Hussain R, Jeon G (2020) A distributed software defined networking model to improve the scalability and quality of services for flexible green energy internet for smart grid systems. Comput Electr Eng 84:106634

Rice EB, AlMajali A (2014) Mitigating the risk of cyber attack on smart grid systems. In: CSER, pp 575–582

Rodriguez-Calvo A, Cossent R, Frías P (2018) Scalability and replicability analysis of large-scale smart grid implementations: approaches and proposals in Europe. Renew Sust Energ Rev 93:1–15

Sadiku MN, Tembely M, Musa SM (2017) Home area networks: A primer. Int J Adv Res Comput Sci Softw Eng 7(5):634–635

Saleem Y, Crespi N, Rehmani MH, Copeland R (2019) Internet of things-aided smart grid: technologies, architectures, applications, prototypes, and future research directions. IEEE Access 7:62962–63003

Salpekar M (2018) Protecting smart grid and advanced metering infrastructure. In: 2018 2nd international conference on I-SMAC (IoT in social, mobile, analytics and cloud) (I-SMAC). IEEE, Piscataway, pp 22–26

Sandhu RS (1998) Role-based access control. In: Advances in computers, vol 46. Elsevier, London, pp 237–286

Sanislav T, Miclea L (2012) Cyber-physical systems-concept, challenges and research areas. J Control Eng App Inf 14(2):28–33

Shabanzadeh M, Moghaddam M (2013) What is the smart grid? Definitions, perspectives, and ultimate goals, November. https://doi.org/10.13140/2.1.2826.7525

Stellios I, Kotzanikolaou P, Psarakis M, Alcaraz C, Lopez J (2018) A survey of IoT-enabled cyber-attacks: assessing attack paths to critical infrastructures and services. IEEE Commun Surv Tutor 20(4):3453–3495

Tan S, De D, Song WZ, Yang J, Das SK (2017) Survey of security advances in smart grid: a data driven approach. IEEE Commun Surv Tutor 19(1):397–422

Tian J, Wang B, Li X (2018) Data-driven and low-sparsity false data injection attacks in smart grid. Secur Commun Netw 2018:1–11

Valea E, Da Silva M, Flottes ML, Di Natale G, Dupuis S, Rouzeyre B (2019) Providing confidentiality and integrity in ultra low power IoT devices. In: 2019 14th international conference on design & technology of integrated systems in nanoscale era (DTIS). IEEE, Piscataway, pp 1–6

Viswanath SK, Yuen C, Tushar W, Li WT, Wen CK, Hu K, Chen C, Liu X (2016) System design of internet-of-things for residential smart grid. IEEE Wirel Commun 23(5):90–99

Wang W, Lu Z (2013) Cyber security in the smart grid: survey and challenges. Comput Netw 57(5):1344–1371

Whitman ME, Mattord HJ (2011) Principles of information security. Cengage Learning, Boston

Xia F, Yang LT, Wang L, Vinel A (2012) Internet of things. Int J Commun Syst 25(9):1101–1102. https://doi.org/10.1002/dac.2417

Yamada T, Suzuki K, Ninagawa C (2018) Scalability analysis on the smart grid fast automated demand response aggregation web services for widely distributed building facilities. Electr Eng Jpn 205(1):17–25

Yu X, Xue Y (2016) Smart grids: a cyber–physical systems perspective. Proc IEEE 104(5):1058–1070

Zanero S (2017) Cyber-physical systems. Computer 50(4):14–16. https://doi.org/10.1109/MC.2017.105

Zhang Y, He Q, Chen G, Zhang X, Xiang Y (2019) A low-overhead, confidentiality assured, and authenticated data acquisition framework for IoT. IEEE Trans Ind Inf 16(12):7566–7578

Chapter 5
Smart Lighting Systems for Smart Cities

Georges Zissis, Pascal Dupuis, Laurent Canale, and Nazim Pigenet

Abstract Artificial light absorbs 13–14% of the world's electricity annual production. Today, we are witnessing a transition from the conventional "analogue" lighting technologies toward "digital" lighting. Smart lighting will become the backbone for smart cities and homes. The objective is switching to smart human-centric lighting driven by both "efficiency" and "quality of light." Next-generation street lighting systems should provide the "Right Light" with the best efficiency and quality, when and where it is needed. This chapter will highlight all the above-mentioned issues and will focus on the future of lighting systems and their contributions to the sustainable development of smart cities.

Keywords Smart lighting · Lighting systems · Light-emitting diodes · Solid-state lighting · Visual light communication · Social life · Vulnerabilities · Urban space

Artificial Lighting Evolutions and Revolutions

Humans cannot live normally without the possibility to light-up habitat and its surroundings. Our poor night vision and primitive fear of the night translate into an imperative need to use artificial light to illuminate our environment. It is impossible to imagine the evolution of our society and, more generally, of our social life without the possibility of enlightening ourselves "at will." Therefore, from the very dawn of human civilizations, it has been understood that fire and heated or

G. Zissis (✉) · P. Dupuis · L. Canale
Univesrité Toulouse 3, Toulouse, France
e-mail: georges.zissis@laplace.univ-tlse.fr; pascal.dupuis@laplace.univ-tlse.fr; canale@laplace.univ-tlse.fr

N. Pigenet
Kawantech SAS, Toulouse, France
e-mail: nazim.pigenet@kawantech.com

G. C. Lazaroiu et al. (eds.), *Holistic Approach for Decision Making Towards Designing Smart Cities*, Future City 18,
https://doi.org/10.1007/978-3-030-85566-6_5

incandescent objects emit light that can be used for lighting purposes; artificial lighting has been discovered. Since then, to satisfy human needs, the energy demand for producing artificial lighting has witnessed a tremendous increase reaching up to 2% of our annual primary energy use. More precisely, following Zissis (2021) in 2019, artificial light production absorbed around 2900 TWh of electricity, corresponding to 13–14% of the world's electricity annual production.

Further, since the control of fire by humans, lighting technology has co-evolved for millennia with our species; Fig. 5.1 summarizes this co-evolution and shows the way forward to the next decades.

The first technological revolution occurred toward the end of the nineteenth century with the invention of "electric light." This allowed humans satiating their need to "light-up" on demand, by replacing technologies based on the consumption of substances that were more hazardous. Since then, artificial lighting has been the subject of a continuous and fascinating technological evolution; twentieth-century scientists and development engineers worldwide created such a wide range of light sources and lighting solutions for every application.

The advent of the first oil crisis in the early 1970s gave the opportunity to governments to realize that lighting is responsible for huge energy consumption and that a non-negligible part of that was just squandered. The concepts of energy efficient lighting and of rational use of energy were born. Compact fluorescent and halogen lamps appeared, and for the first time, sensors (developed independently a

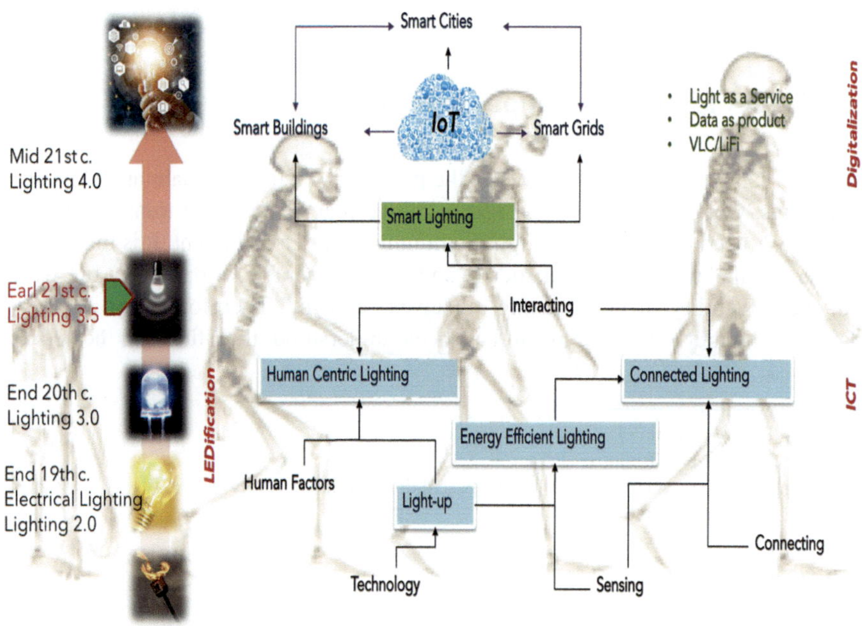

Fig. 5.1 Evolution of artificial lighting technology through ages and the way forward to lighting 4.0 era

few decades earlier) were integrated with more or less success to the lighting systems.

The third revolution in the domain of lighting came at the end of the twentieth century with the invention of the blue light-emitting diode which paved the way for the production of electric light by solid sources of light (SSLs) and more precisely conventional light-emitting diodes (LEDs) and organic light-emitting diodes (OLEDs). The arrival of SSLs, with efficiencies that today far exceed those of the old technologies, has had a direct effect, the decrease in electricity demand for lighting (19% in 2000 decreased to 15% in 2015 to less than 14% today). However, this revolution has a "secondary" effect because it has also paved the way for the era of "digital light" which will allow the quantity and quality of light to be adapted according to its use and the needs of its user without compromising its well-being and health. On the one hand, "human-centric" lighting can be seen as part of this digitalization. From the other hand, "connected lighting" is rapidly developing and consists on networking light sources with sensors to better serve human needs without squandering energy and harming the environment. This is the point that we are today.

Lighting 4.0 consists on integrating smart human-centric lighting into a larger network known as the "Internet of Things" allowing interactions not only between technological components but also with the end user who can express its needs. In this concept, the light will become the product that offers services, and the connected lighting system will be a major component of an intelligent system where the user is an actor. This is a radical transformation that will imply a fundamental paradigm change of the lighting industry and its economic model. This revolution is imminent.

City Lighting: The Present

Nowadays, artificial lighting is literally "submerging" the urban environment. As shown in Fig. 5.2, artificial light in the cities has different usages ranging from street lighting to architectural/monument lighting and from transport lighting to signage, without omitting different "non-lighting" applications like the lighting of urban horticultural units for plant growth, water, air purification/sterilization appliances, etc.

In fact, the diversity of usages can be explained by the fact that artificial lighting in urban spaces fulfills distinct functions:

- It allows the urban space users (pedestrians or motorized) **to see** correctly the environment and that way **to locate** and **to identify** as soon as possible objects, other persons, animals, etc., with less incertitude or error.[1]

[1] Usually, in daylight conditions, humans can identify a shape within 100 ms and recognize an object/person within 400 ms.

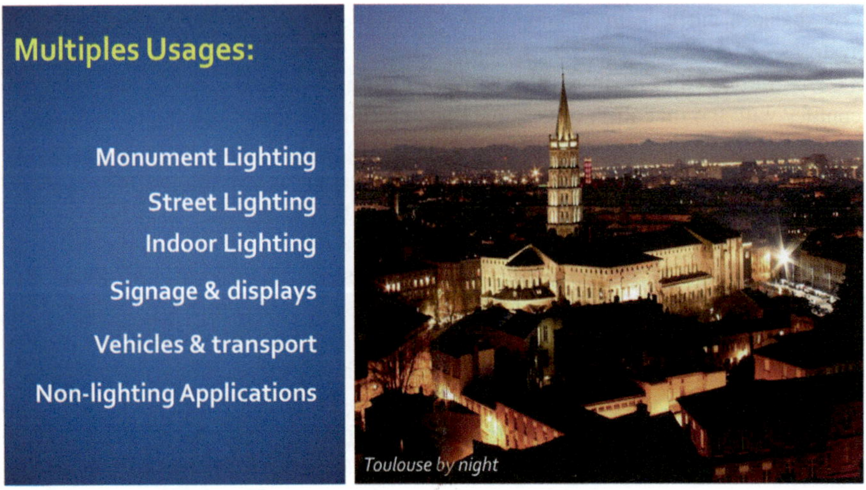

Fig. 5.2 Artificial lighting cities has many different usages

- It allows **being seen** and, eventually, be identified by other urban space users or/and surveillance systems.
- It contributes **to secure** traveling by any means across the urban space because it allows locating potential obstacles and other passive or active hazards in the vicinity of the user.
- It allows urban space users **to feel secure**. In a well-lit environment, humans feel secure because they can (or at least they think so) recognize rapidly objects, persons, obstacles, and potential dangers or aggressive behaviors. Another way round, evolving in dark space (know or worse unknown) induces anxiety and stress. All in all, nobody is really ready to go down a dark alley in an unknown gruesome neighborhood.
- It incites users **to communicate** between them. This can be seen as a direct consequence of the fact that when someone feels secure, it is inclined to exchange with unknown people that seem friendly around him/her.
- It contributes **to appreciate** the space and that way (1) **to find the way** and **to be oriented** by locating and recognizing "cardinal points" and (2) **to appreciate beauty** and architectural style.
- It has the power **to enhance impressions** and to make the city more attractive and fancier. That way, the city attracts people at the right moment and at the right place. A Franco-Suisse architect, Le Corbusier,[2] said, "…Architecture is the skillful, correct and magnificent play of volumes assembled under the light…".
- Last, but not the least, it contributes **to create ambiance**, or "mood," in the urban spaces. This creates emotionally stimulating environments and appealing atmospheres that attract and satisfies the expectations of modern city users.

[2] Charles-Édouard Le Corbusier, 1887–1965

Worldwide, cities absorb more than 70% of the world's energy supply today, and this quantity is growing over time. Further, beyond the fact that the human population is increasing with an average annual rate of 1.09%,[3] cities attract more and more people. United Nations (2018) estimates that the populations of the world's cities with 500,000 inhabitants or more grew at an average annual rate of 2.4%. The projections show that by 2050, two-thirds of the world's population will be condensed in urban areas.

Considered since the post-war II period as a necessity, a means of security, and attraction or enhancement, we witness a phenomenal rise of public lighting. There are currently 317 million total streetlights in the world. Following Northeast Group LLC (2017), this number will grow to 363 million total streetlights by 2027. That said, city lighting is responsible for roughly 250 TWh of electricity per annum and the release of more than 100 million metric tons of greenhouse gas in our atmosphere. Without coordinated action undertaken rapidly by national governments, this quantity will grow not only due to the population increase but also by the fact that city authorities and also private establishments (or even individuals) are using more and more light to "be seen" and attract people. Today, as shown in Fig. 5.3, more and more cities are "shining in all colors" for a large part of the night, and cities are visible from the space; this is not the best thing to do.

This growth is no longer sustainable, and a drastic reduction in energy consumption is required.

Beyond the above cited facts, street lighting impacts seriously industry as well as public finances. Concerning industry income, based on recent data, Statista reported in 2019, following van Gelder (2020), the value of the global outdoor lighting

Fig. 5.3 Our cities are "shining in all colors" for a large part of the night (Photo G. Zissis, 2019)

[3] This average has been calculated by using the following data from www.worldometers.info: world population growth per year in 2020, 1,05%; 2019, 1,08%; 2018, 1,10%; and 2017, 1,12%.

market amounted to about US$ 10.7 billion and forecasted to rise to US$ 23.8 billion by 2030. Concerning impact on public finances, let's take as an example what happens today in France: nearly ten million lamps light-up the French cities for an installed power for lighting of 1.26 GW. Following the French Illuminating Engineering Society AFE (2019), beyond large cities, lighting an average rural municipality in France accounts for about 50% of its energy bill that represents up to 25% of the municipal budget.

Hence, beyond the economic and energy resources it absorbs, the artificial lighting of urban spaces today overflows its objective by polluting earthly nights to the point that the stars "extinguish" the night sky of our modern megacities and impacts seriously the fauna and flora with serious consequences on our ecosystem.

Smart City Lighting: The Future

As has been explained in the above paragraph, this growth is no longer sustainable, and a drastic reduction in energy consumption is required. However, even if there is space for achieving important energy savings, urban/street lighting should continue fulfilling several additional functions like traffic and pedestrian security as well as on the end user's well-being, without omitting that artificial "architectural/scenic" lighting plays a very important role to the city beautification and "invites" people to visit the city and its stores/restaurants/bars/theaters contributing to the local economy and social life.

During the last decade, solid-state lighting challenges conventional technologies. In particular, LED has turned into a game changer beating conventional technologies in all aspects. Decreasing the cost of LED lights and increasing reliability are some of the major factors driving their adoption by cities. In the short term, the largest LED street lighting markets will be North America, Europe, and South Asia, with adoption being driven by cities looking to reduce high energy and labor costs.

That way, following Griffiths (2017), in 2012, 10% of new public streetlights were LED-based, but this figure rises to 80% in 2020. The Northeast Group LLC (2017), estimates that across 125 countries, 264 million LED streetlights will be added over the next ten years, reaching a penetration rate of 89% by 2027. Further, Northeast Group LLC (2016), the same analyst group has identified more than 1000 unique LED streetlight projects in over 90 countries.

However, switching to LED technology alone will not be enough to meet cities' energy-demand abatement targets contributing to be in-line with international engagements as expressed by Paris agreements. Adaptive, interoperable lighting solutions are needed to bring savings to the next level, facilitated by connecting LED lamps with a central management system over the Internet. These networked street lighting systems allow operators to monitor and regulate light levels in unprecedented ways, resulting in increased energy savings and lower operational costs. As Griffiths (2017) predicted, the 50% energy savings that are realized by switching to LEDs increase to 80% when connectivity and a central management system are added. All in all, experts believe unanimously that "smart lighting" technologies will help to achieve those ambitious targets and even going beyond. However, even if the path toward the future seems clear, several points have to be clarified.

The first one to be clarified is the definition itself of "**smart lighting**." Even if this seems surprising, today it is very difficult to find in the literature a compact and coherent definition. For this reason, the authors decided to propose the following definition:

A **smart lighting system** principal function is to produce, at any moment, the right light: where it is needed and when it is necessary. It should adapt the quantity and quality of light to enhance visual performance in agreement with the type of executed tasks. It must guarantee well-being, health, and safety of the end users. It should not squander passively the resources of our planet and limit actively the effects of light pollution on the biotope, or any other impacts on the environment. Ideally, the system could offer additional services (geo-localization, data connectivity, etc.) to the end users through different communication protocols.

That said, such smart-lighting system must fulfill a number of requirements:

- Optimizing visual performance by adapting dynamically the light quantity, light distribution in space, and light quality (CCT, IRC, spectrum, etc.)
- Avoiding any visual disturbance (glare, light flicker, strobe effects, shadows, etc.) that can compromise end user's security and well-being at any moment.
- Reducing always energy consumption of the installation incorporating, when possible, renewable energy sources without compromising the above conditions.
- Respect normative requirements of any kind without compromising the above conditions.
- Limiting actively the effects of light pollution on the ecosystem and biotope and respecting that new sky-protection legislations or standards.
- If necessary, providing additional services to end users and relaying information via visual communication or other channels.

Such a system is designed to create emotionally stimulating environments and appealing atmospheres. On the one hand, visual cues in the design of a space aid the interactions with surroundings. On the other hand, the use of the right artificial light to illuminate objects in a space while minimizing its negative effects such as glare and low contrast is always beneficial. Figure 5.4 illustrates the ideal smart lighting system configuration that integrates all the above requirements.

The intelligence degree of a smart lighting system can be measured in terms of collected, analyzed, and then used and/or transmitted information. To make the system intelligent, it is first necessary to equip it with sensors capable of acquiring the relevant information and then process it rapidly. For instance, at a first glance, it vital is to detect street occupancy.

There are many ways to detect street occupancy, which are trade-offs between sensitivity and costs. One of the more ubiquitous consists in capturing pictures of the pathway in the vicinity of the lighting pole and, by image processing, to detect moving targets. In that case, the input sensor is a wide-angle low-resolution monochrome camera operated from a USB link. However, image analysis requires a

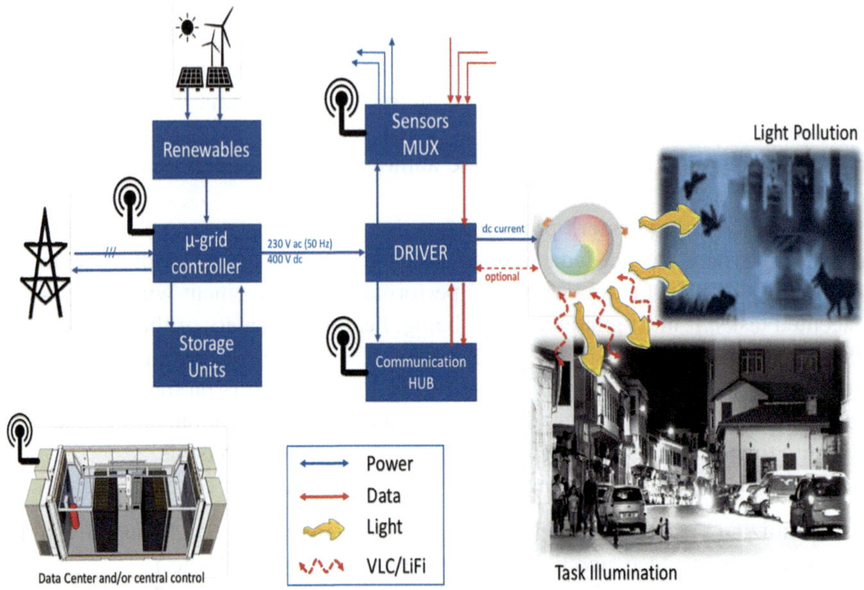

Fig. 5.4 The ideal configuration of a smart streetlighting system, Zissis (2021)

significant amount of computing resources. With respect to the constraints of simplicity and economy, one option is to use embedded calculators based upon a system-on-chip architecture, built in a similar way to Raspberry Pi devices. This approach is called "edge computing": the analysis is performed locally, at the edges of the network, without the need to concentrate all the video streams to a central analysis platform. The input images are segmented according to contrast. The algorithm detects the most frequent pathways, which permits over time to differentiate between sidewalks and road. Then it is able to isolate a reduced number of BLOBs (binary large objects) and tries to associate a category with each BLOB: pedestrian, bicycle, motorcycles, cars, and animal. The pedestrian category can be further refined to detect people with a stroller, person using a wheelchair, person with strange trajectory, etc. However, the system is strictly respecting privacy; thus, it does not have the capacity to perform facial recognition or mineralogical plate resolution. An important aspect is tracking BLOBs' motion: cars are expected to stay for a short duration in the field of view, bicycles and motorcycles are slower, and of course pedestrians are the slowest. Animals have a more erratic behavior. However, pedestrians have also the particularity to be able to appear or disappear "out of nowhere," i.e., coming from or entering a nearby building. The algorithm should also discard parasitic information like tree branches moving in a periodic fashion under heavy winds; night butterflies and mosquitos orbiting around a lamp, or event rain drops moving downward. Further, the camera can estimate luminance level, and that way it acts as a twilight detector when the sky is overcast during the daylight period.

The next level is being able to use that information to control the local lighting conditions in standalone mode. To this end, the system contains specific interfaces to change the light level on the street/pedestrian area. The most used protocols are analog 1-10 V and open access DALI.[4] The objective is to adjust the light at any situation. Even if cars are moving faster than pedestrians, their motion can be anticipated, and the final result will be a light bubble centered on the target and moving at the same speed, whereas other lighting poles stay in off or standby position. Of course, the system has to be resilient to ensure that the street will be always lit when necessary. Thus, when the control system detects fault conditions in the sensor or in the data treatment, it is setting the associated lamp to operate at its full power during the programmed working hours (an astronomical clock can be used). This way, in case of hardware or software issues, the basic functionality of the lamp, which is to illuminate the street, is fulfilled. In addition, flash memory cards can store the operating system status and the operating logs, and that information can be used for maintenance issues.

The occupancy detection precision and reactivity of the system can be largely enhanced by exchanging information between adjacent sensors; thus, information transmission capacities are necessary: short-range high-speed transmitters can be used for inter-device data exchanges complemented eventually by long-range low-speed transmitters to transmit street occupancy statistics to a central station. Data transmitted to the central station can be used in a deep learning process to create and adapt usage scenarios that can be used to operate the system in a predictive mode that increase the end-user satisfaction. For example, when a bus approaching a given bus-stop at a given time, the flow of pedestrians in the surrounding streets can be anticipated, and the system can illuminate the paths accordingly. The possibilities are endless.

Further, lighting infrastructure can be used, among others, as a basis for (a) environment monitoring and protection of biotope; (b) electric vehicle charging; (c) Wi-Fi or LiFi provider for Internet nomad devices connectivity; (d) digital signage, public communication, and alerts broadcasting; (e) localization services especially in underground premises and tunnels; and (f) remote meter reading and analytics at the edge.

By using a central or, better, a cloud-based management system, it is possible to implement auxiliary functions which will be detailed later. All in all, communication capacity is a cornerstone for a smart lighting system. For instance, a smart lighting system can be used for connecting communicating objects thanks to visual light communication (VLC) and LiFi functionalities. Objects can communicate between them or/and a central system. VLC/LiFi offers a large band-pass to allow large data flows.

Smart lighting can be used in public areas to increase the sense of security of people. In fact, as explained in the first part of this chapter, safety in everyday life is

[4] For drivers supporting the new DALI-2 protocol, bidirectional access is provided, meaning the controller can access the electrical data (power and current) from the lamp driver.

improved by overcoming barriers such as dark areas at night without being in obligation to squander energy for illuminating empty spaces, for instance, elder people, which can enhance their participation in public life. Connected smart lighting can reduce vulnerabilities as it can recognize social alarms, hazardous citations, and accidents and broadcast information to vulnerable public space users (i.e., elder or disable people). More, anectodical, smart lighting poles can be used to fight against urban crimes: A large lighting company signed an agreement with STT to incorporate the gunfire shot sensors into their LED street lighting systems. That way, Griffiths (2017) said that the system can broadcast in real-time alerts to emergency and dispatch centers, patrol cars, etc.

A communicating lighting system also permits to regulate traffic using VLC/LiFi which can be used for communicating from car to car or/and from city infrastructure to car. This information can be used to adjust trajectories, reduce speed, adjust traffic lights phase and duration, and prevent collisions. Transfer signals information to cars may include traffic lights situation in the neighborhood, speed limits, congestion, closed roads, etc. It can reduce car roaming for parking. Locating a parking spot in a city is a time-consuming and polluting action. Smart lighting poles can locate available parking spots and broadcast information to driver's mobile phones or directly to smart cars.

That way, in smart cities, the so-called smart street lighting will become a category of "smart connected urban objects" forming the basis of a dense mesh that can be used for various applications and services. Indeed, both traditional lighting poles and more advanced smart lighting installations have the potential to act as a smart city platform, enabling a range of other smart city applications through the integration of data collection devices such as sensors, cameras, and detectors of any kind.

Looking closer to all these transformations implies that illumination will become a sub-function in a more complex ICT (Information Communication Technology) system, which will be a sub-system in the Internet-of-Things (IoT) global system. Light will be the vector of new services and the carrier of dataflows that will spawn additional products. This is a fundamental paradigm change that will push the lighting industry to recast its economic models, and it will contribute to the rise of the ICT industry. However, this radical change of paradigm could hide some dangers: A fully smart system able to recognize situations/actions and/or people, that can collect and transmit data, could become a privileged target of hackers who are looking for various types of information that can compromise individual privacy or who are searching a "get way" to more sensitive city systems for serving any type of criminal actions. For sure, smart lighting emmergence is fully depending on cybersecurity development to defeat any kind of attacks.

Harnessing Energy Demand Through Smart Lighting Systems

As has been explained in the above paragraphs, highly efficient street lighting systems are one route to help governments reduce energy demand in line with the international agreements for the environment. A way to achieve that is dimming-down light when streets are less crowded or empty during late night time following the analysis of the information gleaned by the various sensors. Further, it has been demonstrated that photopic luminance does not correlate well with perceived brightness and by suggesting to adopt the quantity of light, and its spectrum at any moment's demand was pioneering the smart lighting concept (Zissis 2006). Today, with the use of light-emitting diodes incorporated to smart lighting systems, this saving potential can be multiplied by factor of 5 to 7 as shown in the next paragraphs.

To illustrate the benefits of implementing smart lighting systems in our cities, we intentionally chose to discuss, through an example, the reduction of energy demand as a function of the "smartness degree" of the system.

In this example, we are evaluating the energy savings and environmental impact of large-scale lighting systems through various usage scenarios. This example is applicable for lighting systems for cities with more than 10,000 inhabitants. Taking into account that in urban configuration we have in average one lighting pole per 10 inhabitants we conclude that our simulations should be carried out for a system with N = 1000 lighting points.

However, if 1000 lighting points are necessary for getting the desired illumination conditions with legacy technologies, we expect that it could be possible, by adjusting the geometry and photometry using LEDs, to get similar lighting conditions with some less or few more lighting poles. Thus, we decided to run our simulations with a variable number of lighting poles ranging from 600 up to 1200.

Both reference and smart lighting systems are equipped with astronomical clock in order to switch on and off the system following the sunset and sunrise time. In our model, the night duration is calculated by the sunset and sunrise times as obtained, using the codes developed by McClain (1998) for NOAA[5] based on astronomical algorithms proposed by Jean Meeus, as function of the geographical position in the globe. It should be noticed that in any case the full annual operating duration never exceeds 4300 h.

In our example, the reference system is set by using classic high-pressure sodium (HPS) technology, with either (1) ferromagnetic ballast without dimming nor sensors or (2) electronic ballast with dimming capacities but without sensors. The characteristics of the reference system are given in Table 5.1:

The considered smart lighting system is based on LEDs coupled with electronic ballast/driver[6] and sensors. We can obtain the average nominal power of the LEDs lights assuming:

---

[5] NOAA the US National Oceanic and Atmospheric Administration

[6] Following CIE lighting glossary, the term ballast is usable for both discharge and LED technologies; however, the term driver is also acceptable for LEDs only

Table 5.1 Reference lighting system configurations

Basic configuration	Type of lamp: High-pressure sodium (luminous flux 16,000 lm). Fixture with DLOR 60%. Type of ballast: Ferromagnetic. Dimming capacity: No. Lamp nominal power: 150 W. Ballast losses: 11,5% of the lamp's nominal power (17 W).
Enhanced configuration	Type of lamp: High-pressure sodium (luminous flux 16,000 lm). Fixture with DLOR 65%. Type of ballast: Ferromagnetic. Dimming capacity: Yes, maximum level 50% of nominal power. Lamp nominal power: 150 W. Ballast losses: Variable as function of consumed power as shown in Fig. 5.5.

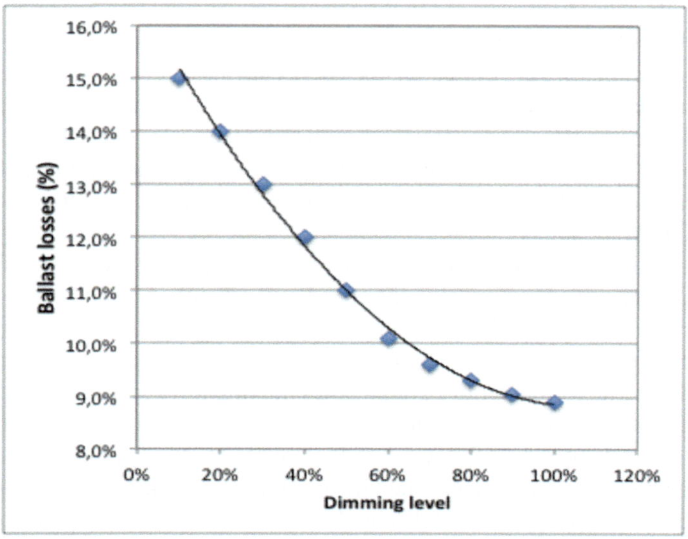

Fig. 5.5 150 W Electronic ballast losses as function of the dimming level (100% = nominal)

1. Luminous flux directed toward the street should be the similar to that obtained by the reference system (16,000 lm x 0.65 DLOR).
2. LED fixture orients all useful flux toward the street (ULOR = 0%) as requested by regulations for the reduction of pollution of the skies.
3. High-quality optics for the fixture (global fixture efficacy of 90%).
4. An average lamp-fixture luminous efficacy of 120 lm/W (following IEA-4E-SSL Annex (2019), this corresponds to the 15% top-products found in the market in 2020). Under these assumptions, the average LED power is found to be equal to 95 W.

The system has dimming capacities, and the lower dimming level can be as low as 10% of the nominal power. Total system extinction is also possible if necessary, during a predetermined time slot. In order to better analyze system capacities, we decided to decouple various system capacities (the "smartness degree") and study them independently. This assumption leads us to elaborate four different scenarios.

Scenario 1

The smart lighting system operates all night without extinction periods, no dimming, and no sensing capacities. This is the most pessimistic scenario corresponding to a business-as-usual case (BAU). The annual energy consumed is given by Eq. (5.1):

$$E = N\left(P_n + P_b\right) T_n \tag{5.1}$$

where N, the number of poles in the lighting scheme; P_n, the nominal lamp power, in W; P_b, the nominal ballast loss, in W; T_n, the night duration, in h.

Scenario 2

The smart lighting system operates a part night. The extinction period is fixed every night from 01:00 till 05:00.[7] No dimming and no sensing capacities. The annual energy consumed is given by eq. (5.2):

$$E = N\left(P_n + P_b\right)\left(T_n - T_{ex}\right) \tag{5.2}$$

where T_{ex} is the annual extinction duration, in h.

This configuration is well adapted for rural or suburban lighting systems when the use of road/pathways is very seldom.

Scenario 3

The smart lighting system operates all night, but it can be dimmed at various levels and various time slots. Extinction is possible for a given time slot, and no sensors are coupled with the system. We defined the following sub-scenarios (Table 5.2):

The annual energy consumed is given by Eq. (5.3):

[7] Extinction must be allowed by national regulations and/or standards.

Table 5.2 Extinction/dimming sub-scenarios implemented

Scenario 3a	Single level dimming. Dimming from 01:00 to 05:00. Dimming level: 25% of the nominal power. No extinction.
Scenario 3b	Single level dimming. Dimming from 01:00 to 05:00. Dimming level: 50% of the nominal power. No extinction.
Scenario 3c	Single level dimming. Dimming from 01:00 to 05:00. Dimming level: 75% of the nominal power. No extinction.
Scenario 3d	2-level dimming Dimming 50% at from 11:00 to 01:00 and from 05:00 to 06:00. Dimming 75% at from 01:00 to 04:00 and from 05:00 to 06:00. No extinction.
Scenario 3e	2-level dimming and extinction Dimming 50% at from 11:00 to 01:00 and from 05:00 to 06:00. Dimming 75% at from 01:00 to 02:00 and from 04:00 to 05:00. Extinction from 01:00 to 04:00.

$$E = N\sum_{i}\left\{\frac{D_i}{100}P_n + P_b\left(P_i\right)\right\}T_{d;i} \tag{5.3}$$

where $T_{d;i}$ is the operation duration under the ith dimming level, in h; D_i is the dimming level, in %; P_i is the input power under the ith dimming level ($P_i = P_n D_i/100$), in W; $P_b(P_i)$ are the ballast losses under the ith dimming level, in W. It is worth to note here that dimming must be operated with care in order to avoid impact on the illuminance horizontal uniformity on the macadam level which is potentially undesirable.

Scenario 4

The smart lighting system operates all night in standby condition. It is punctually activated, pole-by-pole, when an event is detected by the sensors (e.g., pedestrian, vehicle). Each lighting pole is equipped with a sensor (camera) with a power consumption of 3 W and transmitters consuming an additional 1 W.

Each event is characterized by the event duration, τ_{ev}; during that period, the system is operating at full power (it can be done successively pole-by-pole if necessary). In our simulations, we consider that a typical event duration is approximately 3 minutes. The number of events per hour, n_{ev}, is the main input parameter (this can be fixed or variable).

Table 5.3 Extinction/dimming sub-scenarios implemented

Scenario 4a	Extinction between two successive events.
Scenario 4b	Dimming at 10% of the nominal between two successive events.

We distinguish two cases: (1) the system is switched off between two successive events, or (2) the system is maintained at a minimum standby level between two successive events. This leads to the following two sub-scenarios (Table 5.3). The system reaches a saturation level if the product $n_{ev}\tau_{ev}$ is higher than 1 hour; then the system operates at full power continuously.

The annual energy consumed is given by Eq. (5.4):

$$E = N\left\{(P_n + P_b)T_f + \left[\frac{D_{sb}}{100}P_n + P_b(P_{sb})\right]T_d + T_n P_{sc}\right\} \tag{5.4}$$

where D_{sb}, standby dimming level, in %; P_{sb}, the standby state consumption ($P_{sb} = P_n D_{sb}/100$); $P_b(P_{sb})$, ballast losses under standby dimming level, in W (this quantity is set to zero in the case of extinction); T_f, operation duration under full power with $T_f = n_{ev}\ \tau_{ev}$, in h; T_d, operation duration under non-zero dimming with $T_d = T_n\text{-}T_f$, in h; P_{sc}, sensors and transmitter power consumption, in W (operating all night, switched off in daytime).

Results and Discussion

This section shows the main results obtained by these simulations. Figure 5.6 shows all simulations together.

Figure 5.7 shows the additional annual energy savings (in %) compared to the reference system (basic configuration) for the different scenarios presented in the previous section. Here, the number of lighting points is set equal to 1000.

It can be seen that even under the most pessimistic BAU scenario (n°1) the system can achieve more than 35% energy savings. This is due to the fact that installed LED power per pole is much lower than standard HPS lamps.

Scenarios that allow dimming (scenarios 3) can achieve important additional energy savings; however, some of them are less efficient than simple extinction during the off-peak hours in the city (scenario n°2).

In all cases, lighting systems equipped with sensors are the most efficient (scenarios 4a and 4b) provided that saturation conditions are not reached at any time. In case of saturation, scenarios 4 are less efficient than scenario 1 because the power consumption of sensors and transmitters has to be added to the lamp consumption.

Figure 5.8 shows the influence of the number of lighting poles under the different scenarios. It can be seen that in the case that a higher number of poles is necessary for achieving desired lighting conditions, it is possible to maintain a high level of

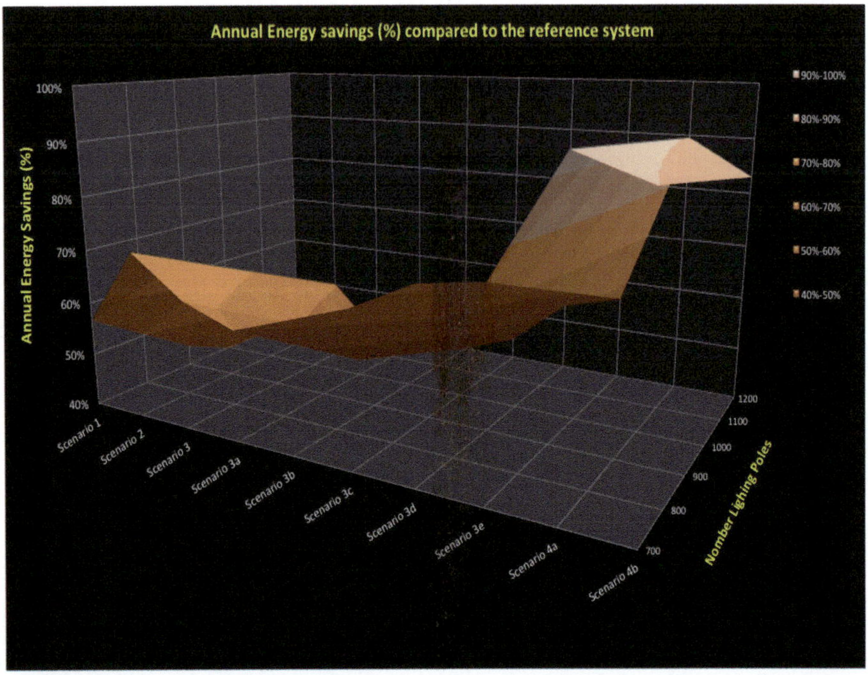

Fig. 5.6 Energy savings for all mentioned scenarios as function of the number of lighting points

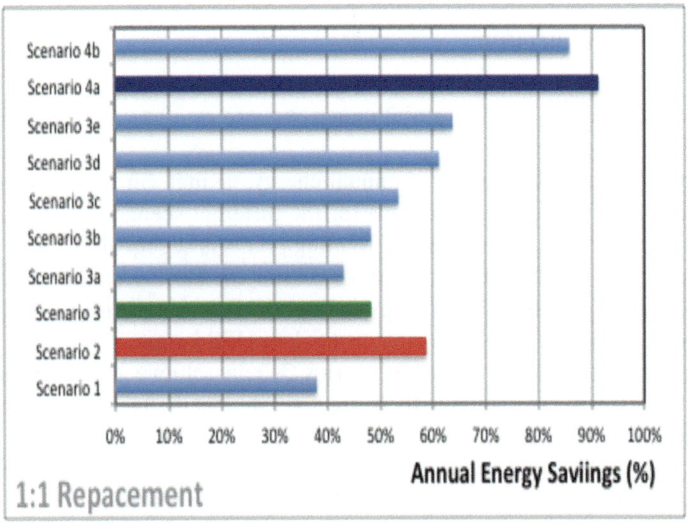

Fig. 5.7 Annual energy savings, in %, compared to the reference system (basic configuration) [N = 1000 poles]

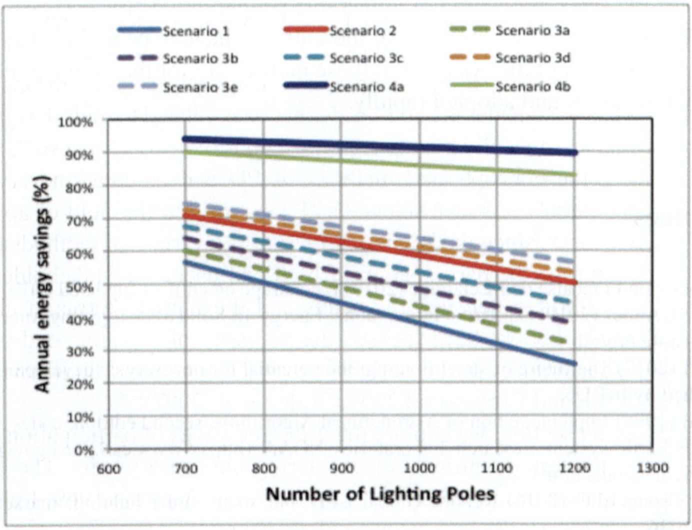

Fig. 5.8 Additional energy savings, in %, compared to the reference system (basic configuration) as function of the number of lighting points

energy savings by using smart lighting system capacities (dimming and sensors). We can also observe that for scenarios 4 (use of sensors), the energy consumption is almost independent on the number of lighting poles.

The most important conclusion that can be drawn from this exercise is that smart lighting systems with high smartness degrees can achieve more than 90% energy savings when replacing legacy street lighting in cities. However, in some cases, it is preferable adjusting the smartness degree following the usage of the system, otherwise the energy savings will vanish and in the worst case overconsumption can occur. Scenarios that allow dimming can achieve important additional energy savings; however, some of them are less more efficient than simple extinction during the off-peak hours in the city.

Further, an equilibrium between initial installation cost increases rapidly with smartness degree and achieves cost savings on the city's energy invoice. Overestimating the smartness degree of the installed system can lead to an important return on the investment periods that cannot always be affordable for a given city.

It should be noticed here that the above model excludes the power consumption linked to data harvesting and associated data centers. This can be a critical issue for the next years, and it should be included in the simulations, but for the moment, it is difficult to do so due to missing information and experience. On the opposite, coupling smart lighting systems with renewable sources and using the right business model can reduce energy costs for the city even if the initial investment is heavier.

Finally, it should be kept in mind that all the above discussion is based on a single parameter: the energy savings and the associated ownership cost models. In reality, a smart lighting system has many additional advantages including reduction of the pollution of the skies, protection of vulnerable people in the city, and possibility to offer additional services to citizens. All these effects have to be included to a global efficiency evaluation of the system, but these metrics are for the moment missing; it should be developed and adopted rapidly.

References

AFE Association Française de l'Eclairage (2019) L'éclairage en chiffres (updated 2019)

IEA-4E-SSL Annex (2019) Data from International Energy 4E Solid State Lighting Annex, https://www.iea-4e.org/ssl

Griffiths H (2017) The future of street lighting: the potential for new service development, report published by IoT UK

McClain B (1998) Implementation of Astronomical Algorithms, second edition, codes accessible through Earth System Research Laboratories NOAA (https://www.esrl.noaa.gov/gmd/grad/solcalc/calcdetails.html)

Northeast Group LLC (2016) Report, Global LED and smart street lighting: market forecast 2016–2026

Northeast Group LLC (2017) Excerpt of study for Telensa, Global LED and Smart Street Lighting: Market Forecast 2017–2027

United Nations (2018) The World's cities in 2018, Data Booklet

Van Gelder K (2020) Global outdoor lighting market size from 2019 to 2030, available via Statista, https://www.statista.com/statistics/1132346/outdoor-lighting-market-value-worldwide. Accessed March 10th, 2021

Zissis G, Mucklejhon S (2006) Standardising Mesopic Vision Conditions and Incidence on Light Sources Science and Technology, International Standardization as a Strategic Tool, IEC Centenary Challenge, 1st Prize paper, pp. 119–130, editions IEC Geneva, ISBN 2-8318-8867-0 (2006)

Zissis G, Bertoldi P, Serrenho T (2021) Update on the Status of LED-Lighting world market since 2018. JRC Techinical report n°JRC122760, Paolo Bertoldi (ed), Publications Office of the European Union, Luxemburg, https://doi.org/10.2760/759859

Part II
Smart Infrastructures for Smart Cities

Chapter 6
The Role of Geoinformation in the Development of Smart Cities

Dariusz Gotlib and Robert Olszewski

Abstract The issue of smart cities is multifaceted. It includes humanistic, social, natural, and technical issues. It touches upon problems of philosophical, sociological, economic, political, management, and technological nature. The complexity of this issue precludes us from discussing or analysing all these aspects. Therefore, we will only focus on the role of geoinformation and ICT in managing a smart city. However, when considering the role of geoinformation technologies in the development of smart cities, we do not forget that technology is supposed to serve the inhabitants, and not vice versa, to determine their way of life or to decide where they should go for a walk or where to eat or sleep. In our opinion, it is not possible to achieve a smart city level without access to comprehensive, accurate, and up-to-date information about space and tools enabling the transformation of 'raw' data into spatial information and knowledge. Most of the decisions made in the city have a spatial context, just as most modern information systems need access to spatial data.

Today, open data are widely available (especially for large cities), which is a stimulus for developing innovative geoinformation services. The tools used today to process big data – e.g. data commonly available in social media, cloud computing – allow for obtaining useful real-time information about the needs and expectations of inhabitants, their migration within the city, or potential threats.

Appropriate use of geoinformation and innovative geoinformation technologies in city management will allow to obtain highly satisfactory results and gain an advantage over cities unable to use this potential. Appropriate actions in this respect may contribute (as one of many components) to increasing the 'intelligence' of an urban organism.

Keywords Geoinformation · ICT · Smart cities · City management

D. Gotlib (✉) · R. Olszewski
Faculty of Geodesy and Cartography, Warsaw University of Technology, Warszawa, Poland
e-mail: dariusz.gotlib@pw.edu.pl; robert.olszewski@pw.edu.pl

G. C. Lazaroiu et al. (eds.), *Holistic Approach for Decision Making Towards Designing Smart Cities*, Future City 18,
https://doi.org/10.1007/978-3-030-85566-6_6

Introduction

The creation, development, and operation of smart cities is a multi-faceted issue covering humanistic, social, natural, and technical aspects and involving problems of philosophical, sociological, economic, political, managerial, and technological nature. However, what plays a unique part in modern cities' functioning are the technologies; it is currently impossible to achieve the goals established for smart cities in the areas mentioned above without them. A *smart city* is a city that skilfully uses available technologies as a driving force. According to the BSI PAS 180:2014 specification, a smart city implies 'the effective integration of physical, digital and human systems in the built environment to deliver a sustainable, prosperous and inclusive future for its citizens'. In Manville et al. (2014), a smart city is defined as 'a city seeking to address public issues via ICT-based solutions on the basis of a multi-stakeholder, municipally based partnership'.

The scope of technologies essential for a smart city is enormous: from materials, energy, transport, construction, and environmental technologies to the information and communications technologies mentioned above. This chapter focuses only on one technological area, namely, geoinformation. In the authors' opinion, it is a crucial area due to its horizontal and integrative nature. Geoinformation is of particular importance for the operation of a smart city. Most of the data and city resources are directly or indirectly related to the Earth's surface (georeference). The location of various buildings or the entire city plays a crucial role in its functioning; it can stimulate or hinder its development. A city's spatial relations with its surroundings influence economic and social processes. They condition the movement of people, develop local communities of inhabitants, and influence the sense of comfort and the city's image. The location of a city encourages (or discourages) investors and tourists. The spatial relations within the city are just as important: topography, communication network, spatial arrangement of the water and sewage, telecommunications, gas, and electricity infrastructures; the distribution of buildings and green areas; the distribution of educational, cultural, sports, and healthcare facilities; and the vicinity of residential and industrial areas, all affect the residents' quality of life as well as the city's economy (Gotlib and Olszewski 2016a, b).

It means that geoinformation may play an integrative role in the functioning of city systems, and the role of geoinformation technologies is not only to provide information on the location of objects and phenomena in space (e.g. in the form of cartographic visualisation) but also spatial data mining and acquiring spatial knowledge essential for the effective operation and development of the city.

Thanks to a 2D, 3D, or 4D map (understood as a model of space), it possible to discern relationships that are difficult to discover in any other way. A map can be an excellent interface to a large amount of data. It also helps ordinary residents as well as decision-makers to improve their understanding of phenomena in the city. Moreover, a map becomes the basis for the operation of many systems and robots; its recipients or users are no longer just people. Without a precise model of space, it is hard to imagine the development of autonomous systems, e.g. driverless cars or drones autonomously carrying out many essential tasks. Obviously, in the end, the beneficiary is always a person.

One should emphasise that creating a smart city is an 'organic' process (as the development of the city itself). The construction and development of a sustainable urban ecosystem and the implementation of innovative e-services based on new technologies should all be considered imponderables in the field of smart city development. In a particular way, perfecting the methods of acquiring geoinformation related to the development of the Internet of Things (IoT) and developing methods of machine learning and spatial data mining predestine geoinformation technologies play a vital role in shaping smart cities.

The concept of creating a smart city based on access to geoinformation enables effective delivery of technological, organisational, and management models as well as new legal regulations in order to launch, support, and continually develop the city in relation to:

- Material urban infrastructure (transport, information, communication, energy, water, etc.) and the environment
- Efficiency in the consumption of material and organisational resources
- Social activities and interactions based on a sustainable model of life, work, education, mobility, security, accessible and open administration, etc.
- Partnership-based social capital

It is impossible to implement such tasks without access to comprehensive, accurate, and up-to-date information about space or tools that enable transforming 'raw' data into spatial knowledge or information.

Still, many cities do not fully exploit the potential of geoinformation technologies. There may be several reasons behind this, but these two seem the most significant:

- A lack of appropriate (up-to-date, reliable, and available) resources of spatial data
- Siloed (vertical) management of urban resources, including spatial data

The first case is becoming rarer with the recent rising popularity of spatial data acquisition technologies, including satellites, aeroplanes, drones, networks of sensors in a city (IoT), cars, and even people using smartphones. Increasingly, often, these data are made available under open data licenses. Access to the data stimulates the development of innovative geoinformation services, in line with the idea of 'citizens as sensors' (Goodchild 2007). Citizens themselves create some of these data as volunteered geographic information (VGI). Some are obtained by processing large data sets commonly available on social media. Thanks to sharing the data using cloud storage, it is easily possible to provide the data to the interested people, institutions, and companies as well as obtain valuable real-time information on residents' needs and expectations or potential threats.

Siloed management of urban resources is the second factor harming the efficiency of using geoinformation. Various organisational units create and use official spatial data without commonly accepted strategies for shared data use or coherent plans for creating systems based on open and universal platforms. Such plans may serve as the basis for much more effective use of geoinformation through their integration with other data and systems. Changes are already occurring in this respect, but only the cities that introduce them faster are likely to win the race to become smart.

The recent (2014) work of such organisations as the British Standards Institution (BSI) and ISO led to creating several documents meant to guide or organise creating smart city-type solutions, for example, BSI PAS3 180, BSI PAS 181, BSI PAS 182, PD 8100, PD 8101, ISO 37101, and ISO 37120 (Bielawski and Gotlib 2016).

When defining the role of spatial data and geoinformation technologies in the creation and development of a smart city, ISO geographic information standards (19,100 series) and Open Geospatial Consortium (OGC) studies are essential. Regional guidelines and good practices are also emerging, such as those connected to the implementation of the INSPIRE Directive in Europe. Although it is worth noting that no universally recognised standards in using geoinformation in a smart city have been established, many modern cities take extensive measures to use geoinformation effectively.

Spatial information analysis, for instance, provided the basis for the plans to create two currently developed smart cities: Songdo in South Korea and Masdar City in the United Arab Emirates. Advanced spatial data analyses are essential for improving the existing urban agglomerations, generally regarded as smart cities' flagship solutions: London, New York, Amsterdam, Barcelona, Singapore, and many other metropolises, e.g. Warsaw or Poznań.

Using spatially localised information to create, operate, and develop smart cities in various aspects and areas has been the subject of many scientific publications. Thompson (2016), Bhattachary (2017), Laurinia and Favetta (2017), Popov et al. (2017), Azri (2018), and Inajit et al. (2018) analysed the issue of spatial planning and the acquisition of spatial knowledge based on the analysis of data collected by various social organisations and institutions. The publications by Wang (2014), Homainejad (2015), Mahavir and Bedi (2017), and Olszewski et al. (2019) addressed the issues of residents' mobility, developing geoinformation societies in smart cities, modelling social behaviour, preserving cultural heritage, and acquiring spatial knowledge based on data from stationary and mobile sensors that are part of the so-called Internet of Things. Spatial information also provides the foundation for many analyses connected with the planning and monitoring of energy networks in smart cities, as exemplified by the research presented in (Kim et al. 2012).

Appropriate use of geoinformation and innovative geoinformation technologies in city management facilitates obtaining highly satisfactory results and gaining an advantage over cities that cannot use this potential (Gotlib and Olszewski 2016a, b). Appropriate measures in this area may be one of many components that contribute to augmenting the 'intelligence' of the urban organism.

Singapore's example should be particularly interesting in this respect. Through consistent implementation of the long-term regional development policy supported by modern information and communications technology and geoinformation analysis, this city-state with an area of 716 km^2 and a population of up to six million achieved the status of a metropolis that fulfils all the requirements for a smart city. Singapore has been implementing the 'Smart Nation' project since 2014, and its assumptions have exceeded the mere optimisation of the city transport through an IoT network of sensors. Most of all, the project aims at providing comprehensive geoinformation services for residents. The use of up-to-date geoinformation, a network of stationary and mobile sensors, edge computing technology, and machine learning methods supports both the health protection of residents (especially seniors) and the development of modern business services that employ spatial data.

The Impact of ICT on the Development of Smart Cities

Although it is crucial to view smart cities' development holistically and, thus, the perceived quality of life is of the utmost importance, the degree to which advanced technologies are used unquestionably influences smart cities' development. The functioning of a smart city has rested on changing the 'systemic' paradigm, which dominated the second half of the twentieth century, in favour of the 'network' approach prevailing in the early twenty-first century. 'Network' in this sense covers not only strictly technological solutions but also social ones. Data transmission in networks and universal communication both imply developing an open information society, which simultaneously stimulates smart cities' development. In the technological sense, the basic ICT technique used to develop smart cities is indeed a network, both fixed and wireless. The expectation for both types of networks is to have a high bit rate, i.e. the ability to transmit a large number of bits per second (bit/s) (Muraszkiewicz 2016). Not only does employing the network approach enable real-time data transmission, but it also improves city management and the operation of service-providing businesses, and it increases social participation. Collaboration in the ABC triangle (Administration-Business-Citizens) supported by information and communications technologies (using geoinformation) facilitates the improvement of information flow and the development of a local community that shapes a smart city.

This collaboration also requires technological support to enable the collection and analysis of urban spatial big data. A set of methods and tools collectively called big data is one of the latest ICT techniques with particularly great potential from smart cities' perspective. The volumes of such sets, often multimedia, are measured in petabytes and exabytes (Muraszkiewicz 2016).

Such data are generated by telecommunications networks systems (also spatially localised), medical equipment systems (e.g. MRI), meteorological systems, commercial network information systems, or geoinformation systems. According to the report by McKinsey Global Institute, big data techniques encompass a new generation of technologies and architectures designed to extract informational values from very large volumes of various data through rapid data acquisition (Fig. 6.1), knowledge discovery, or analysis. This approach explicitly presents the main purpose of maintaining and processing huge data resources: analysing and exploring data (data mining) and transforming them into knowledge (knowledge discovery).

Fig. 6.1 Visualisation of a large set of spatial data (point cloud) obtained in the laser scanning process LIDAR. (Source: OPEGIEKA Company)

A developing network of city sensors that are part of the Internet of Things leads to a significant increase in the volume of urban data. Analysing them and creating variants of forecasting models of the city's operation and development requires using traditional data processing methods and advanced artificial intelligence methods, including machine learning algorithms (mainly deep-learning neural networks). Huge sets of 'teaching' examples collected by individual cities facilitate developing traffic control systems or optimising waste management that uses artificial intelligence methods. Having access to powerful analytical tools that operate on a huge amount of constantly generated data enables decision-makers to understand the city's phenomena and, thus, optimise various processes, aided by modern data visualisation methods using advanced 3D graphics and animation technologies. The outcome of access to data, their processing, and visualisation are the so-called digital twins, i.e. digital maps of cities on which one may carry out complex simulations (Fig. 6.2).

Presenting the surrounding reality in a virtual world by building a 3D model of a city creates unique possibilities for analysis, simulation, and visualisation. The vaster the part of intangible and invisible data surrounding us every day in the real world, e.g. air pollution level or noise propagation in the 'urban jungle', the more valuable these possibilities are (K. Leszek 2016).

Three conceptual and technological trends for smart cities' development and operation seem particularly noteworthy: the aforementioned digital twins, the Internet of Things, and processing or accessing of data in the cloud. The possibilities opened up by modern technology for collecting huge resources of information

Fig. 6.2 Three-dimensional vector model of the city. (Source: Fotokart company)

by IoT city sensors enable effective traffic control or optimisation of city transport, planning mass events, counteracting a pandemic, and warning residents about increased air pollution, flood risk in cities, etc. These solutions are at the stage of wide implementation (especially in the Far East, in Shanghai, Singapore, or Tokyo), or at the stage of subsequent studies Zanella et al. (2014), Dustdar et al. (2017), O'Dwyer et al. (2020), and Al Solami (2021).

The idea of digital twins and their use as decision-support tools in creating and developing a smart city emerged at the beginning of the twenty-first century. At the heart of this approach lies creating a digital copy of an existing or a planned city in the form of an advanced model (Fig. 6.2). These models use both detailed spatial data and information on the operation of urban processes: public transport, garbage disposal, ways of spending free time by residents, etc.

Analyses done on digital twins enable testing many alternative city management variants and predictive 'WHAT-IF?' analyses. However, conducting such analyses requires building extremely detailed spatial data models and keeping them up-to-date; it also requires the collaboration between specialists and the integration of social, economic, planning, and geospatial knowledge. There are many examples of effective implementation of solutions using the idea of a digital twin in smart cities' development. Dembski et al. (2020) described an example of using a digital twin to model a smart city's operation on the example of the city of Herrenberg in Germany. O'Dwyer et al. (2020), White et al. (2021), and Raes et al. (2021) also conducted similar studies.

Another crucial in information and communications technology is storing and sharing data in the cloud. It enables a large number of users to have access to shared data resources at the same time, significantly decreases data redundancy and thus facilitates data analysis, ensures data security, and facilitates the flexible use of the computing power of hundreds or thousands of computers, depending on the needs (high scalability). More than ever, in terms of organisation, it enables the collaboration between various stakeholders in the city. On the one hand, such solutions provide new opportunities and, in certain important aspects, increase data security; on the other hand, if poorly implemented and managed, they may pose new threats. Especially in smart and hyper-connected 'digital cities,' cybersecurity issues must be given priority.

Other important technological solutions for smart cities' operation are universal access to smartphones with fast wireless data transmission, universal Internet access points, miniaturised GNSS receivers and evolving indoor positioning systems, edge computing, and access to large sets of geospatial data in the cloud. Artificial intelligence techniques are also growing in importance.

How 'Smart' Are Smart Cities

Classifying or assessing smart cities stems from formulating the concept of a smart city. One of the first rankings defining the 'smartness' of modern cities was the so-called Vienna Ranking that considered six basic factors. The nature of currently developed classifications is rather advisory, indicating the appropriate developmental direction for a given city that is significant for a specific agglomeration. When assessing the level of implementation of a smart city concept, one should consider the method and scope of applying ICT and geoinformation. Such are the criteria presented, e.g. in the paper ('Smart Governance in the Capital City of Warsaw with the use of ICT and Geoinformation Technologies'). The authors of the research regarded the criteria of 'the "smartness" of the IT system' and 'the use of geoinformation' as particularly important in analysing existing solutions. When evaluating 'the "smartness" of the IT system,' two specific criteria have been proposed:

1. The degree of connection with other systems:
   - Systems logically linked to other systems and sharing data
   - Independent systems or systems minimally dependent on other systems
2. The degree of the functional and algorithmic complexity of a system:
   - Complex systems equipped with advanced applications and algorithms (e.g. artificial intelligence) that enable the implementation of at least the basic characteristics of the 'smartness' mentioned above
   - Systems that perform tasks without any 'smart' characteristics

In the opinion of the authors of the research, 'the use of geoinformation' criterion is also essential for evaluating the systems' potential because the majority of city systems have, or may have, a spatial reference. Thus, geoinformational identification may serve as the basis for creating an integrated platform for target ICT systems comprising the starting point of a smart city.

Spatial Data Sets and Geospatial Analyses

City dwellers have ever-higher requirements for the quality of life. Growing needs coincide with the development of ICT technologies and the emergence of modern digital services, and with increased vehicular traffic, air pollution, increasingly frequent adverse weather phenomena, and the intensification of various threats, including terrorist ones. Performing advanced spatial analyses and making appropriate administrative and business decisions require knowledge of the location of objects and phenomena, both those located on the ground and above/below the ground.

City dwellers expect, among others, the relevant services to react quickly and efficiently to threats such as accidents, floods, damage caused by storms, vandalism or terrorism, convenient and nearby public transport, and public services within reach. It is no longer enough to plan a city's development intuitively or spontaneously and have ad hoc reactions to events. Achieving these goals in a complex urban organism requires using precise spatial models of cities together with a wide range of specialist data.

This type of analysis entails collecting spatial data in multi-year cycles (e.g. flood risk of varying severity) and real-time analytical operations on data from sensors, video monitoring, or citizens' reports, e.g. 112 calls. The most important types of spatial data needed in urban information systems include:

1. Records of towns, streets, and addresses
2. City corporate limits and the limits of other territorial divisions, e.g. tax offices, fire departments, statistical regions or census areas, and constituencies

3. Geographical names, e.g. housing estates, parks, rivers, and other physiographic objects
4. The course and parameters of roads, railway tracks, cycle lanes, and sidewalks
5. Data on the organisation of road traffic (road signs and the resulting graph describing the permissible traffic)
6. Location of public transport facilities, e.g. bus and train stations, airports, airports, harbours and marinas, public transport stops
7. Location and parameters of buildings
8. Location and parameters of other engineering structures, e.g. bridges, viaducts, embankments, masts
9. Location and parameters of underground, above-ground, and overhead technical infrastructure, e.g. a power grid, a water and sewage network, a gas network
10. Information on land cover, e.g. the range of water, tree and shrubbery cover, buildings, grasslands, rocky areas
11. Information on soils and land use
12. Ranges of geological divisions
13. Topography (information on the topographical relief)
14. Ranges of divisions of local land-use development plans
15. Ranges of protected natural areas
16. The location of tourist attractions and national heritage sites
17. Location of facilities important for residents and visitors to the city, e.g. headquarters of companies and institutions, restaurants, shops, supermarkets, wholesalers, industrial plants
18. Location of objects important from the point of view of the city's economy, e.g. ranges of economic zones, landfills, points of raw materials extraction or reloading
19. Location of medical facilities
20. Location of educational institutions
21. The location of sensors and video monitoring cameras and individual sensors' specification (e.g. air pollution, water contamination, etc.)

The study Gotlib and Olszewski (2016a, b) includes a list of examples of selected applications for this type of geospatial data:

1. Searching for and presenting the best areas for development
2. Supporting spatial planning, social geoparticipation, the process of landscape protection and consolidation that considers landscape values, etc.
3. Supporting the processes of project settlement in investment processes between various companies managing municipal infrastructure
4. Creating effective management systems for municipal real estate, including the support of photovoltaic panels installations and other elements of renewable energy sources
5. Supporting the processes of crisis management, emergency services, public safety and security services, and intelligence agencies
6. Supporting tourism management
7. Assisting public transport passengers
8. Supporting video monitoring systems

The analysis of collected information is the basic goal of building urban spatial information systems, enabling the comprehension of the regularities in the spatial distribution of phenomena and their interdependencies, accelerating the computational process, facilitating decision-making support, etc. One should emphasise that the integration or (at the minimum) cooperation of individual urban systems using geoinformation is essential. Performing complex spatial analyses and using, e.g. algorithms and machine learning methods are only effective when the authorities implementing the concept of a smart city depart from the traditional siloed approach, favouring horizontal solutions that enable synergy.

Most modern geoinformation applications allow for the construction of hierarchical spatial analyses and the creation of complex analytical diagrams and multi-criteria modelling. To analyse big data stored in municipal offices' organisational units, it is also possible to use artificial intelligence methods (e.g. neural networks, decision trees, or genetic algorithms) to construct and evaluate complex development models.

Below are examples of spatial analyses performed in modern cities:

1. Analyses of flows of public transport passengers, pedestrians, and cars to devise an optimal road network development plan, change timetables, and avoid threats.
2. Calculation of visibility ranges from selected positions, lines or areas that may be useful, for example, in spatial planning or the operation of services such as the police (including anti-terrorist units) (Fig. 6.3a, b)
3. Shade analysis to plan the location of buildings and selection of places for photovoltaic installations
4. Calculation of the travel time to the selected object from any other point in the city
5. Creation of distribution maps of various phenomena based on source data obtained at measuring points
6. Socio-demographic analyses, e.g. selecting areas inhabited by people with the highest purchasing potential for a specific type of service or production activity
7. Detection of areas at risk of landslides
8. Calculation of air flows to determine the city's ventilation corridors
9. Calculation of ranges of a potential flood wave (Fig. 6.4)

Another example of employing geoinformation may be the Integrated Traffic Management System implemented in Warsaw, which combines traffic lights (556 adaptive traffic control signals) at intersections and traffic control. Based on a network of cameras and sensors, the system detects the traffic volume and adjusts the traffic light phases accordingly.

Virtually, every department of the city hall, municipal police, police, and municipal companies either uses or should use this type of analysis. Cities pursuing the title of a smart city strive for more complex informational connections between public bodies, commercial entities, or non-governmental organisations. Thus, it is even more important to have effective access to harmonised spatial data sets.

It is worth noting that most of these analyses should use 3D models. The era of 2D models paved the way for the era of 3D and 4D data that make it possible to

Fig. 6.3 a, b Analysis of area visibility on the model of the city. (Source: geoportal.gov.pl)

generate the elevation information and the temporal variability of a given phenomenon. 3D spatial databases allow designers to create realistic visualisations of reality, making it easier to assess, for example, the development potential of a given urban concept (Fig. 6.5).

Fig. 6.4 Map of flood risk for a part of Warsaw city. (Source: geoportal.gov.pl)

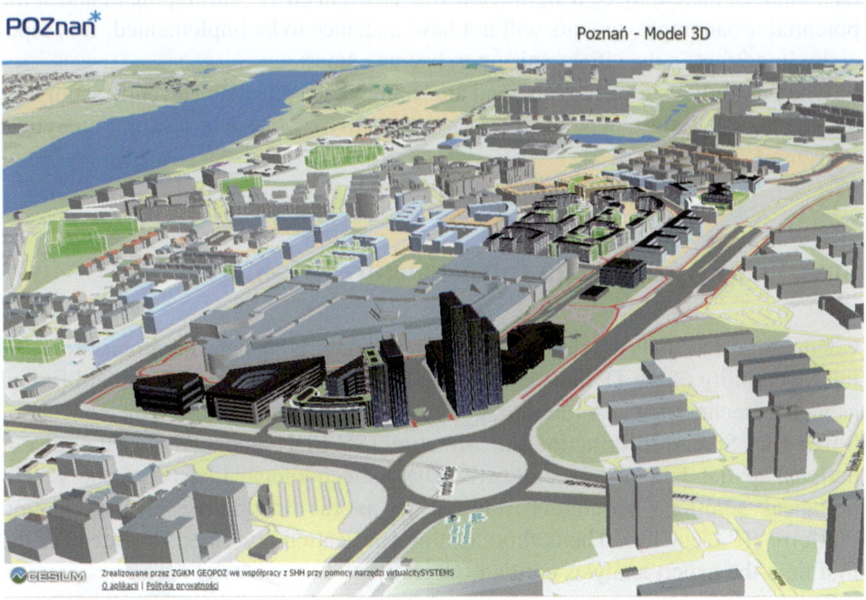

Fig. 6.5 Digital twin of Poznań, Poland. (Source: Board of Geodesy and City Cadastre GEOPOZ, http://sip.poznan.pl/model3d/#/)

A 3D model is not just a modern mock-up of a city; it constitutes a city's spatial information system. It facilitates interactive simulations that are understandable to both engineers and ordinary residents (Katarzyna Leszek 2016). Using a 3D model may accelerate decision-making processes in the city and, thus, counteract threats more effectively.

Land management is another process in the city that requires 3D models. It becomes a huge challenge in modern and developing cities, especially when there is a shortage of space and when implementing large infrastructure projects under and above the ground (subway, viaducts, construction of additional storeys). The possibility of dividing the existing cadastral parcels' space vertically and assigning property rights to them is one of the basic benefits of introducing the 3D cadastre system (Karabin 2016). Many global research and implementation projects on 3D cadastre focus on issues related to the visualisation of spatial objects; this tendency is visible when analysing the topics of the papers at periodically organised workshops on the 3D cadastre (Delft 2001), (Delft 2011), (Schenzhen 2012), (Dubai 2014). The International Federation of Surveyors (FIG) has already developed a 3D cadastre conceptual model: it is a Land Administration Domain Model (LADM, ISO 19152: 2012) (Karabin 2016). Without introducing a 3D cadastre based on advanced spatial data models, there may be a significant slowdown in cities' development, and some potential urban improvements will not have a chance to be implemented. It is especially important in the cities aspiring to become smart.

The possibilities for accessing and analysing geospatial data currently seem limitless, and the results depend only on the quality of the data and the analytical tools used.

Smart City Spatial Data Infrastructure (SC-SDI)

Conducting the analyses described above requires both data harmonisation and interoperability of the IT tools used. What is needed are organisational and formal as well as technological solutions. A solution may be implementing the concept of Smart City Spatial Information Infrastructure (SC-SDI) modelled after the idea of national spatial data infrastructures (NSDI). Its technological basis may be a technological solution, hereinafter referred to as the Integrative Geoinformation Platform (IP-GeoCity). The authors of this publication presented this idea for the first time the expert analysis entitled 'Objectives of the SMART CITY program in the capital city of Warsaw' commissioned by the Digitisation Bureau of the Warsaw City Hall (Comtegra 2017) and mentioned in the publication ('Smart Governance in the Capital City of Warsaw with the Use of ICT and Geoinformation Technologies').

In this paper, the Integrative Geospatial Platform is understood as the modern system that uses advanced geoinformation methods and technologies. Such a system may play an integrative role for domain-specific IT systems, presenting the information contained inside against a 3D city model and many other (official and social) data on space. The developed geospatial platform should be integrated with the cadastral system, the geodetic record of the utility network, land-use planning, data from the flood protection system, road data banks, traffic monitoring systems, and public transport systems, as well as data on the natural environment, and location of commercial facilities, healthcare facilities, tourist attractions, security-related events, etc.

The primary functional tasks for IP-GeoCity are as follows:

1) Displaying information on the city's space from many data sources: the geodetic and cartographic resources, public institutions resources, resources of commercial companies, and social media resources.
2) Providing an interface for accessing spatial data about the city, in particular, their display and joint analysis by providing programming interfaces, e.g. API.
3) Making it possible to implement the so-called spatial services, e.g. geocoding, searching for the optimal route to drive or walk to a specific object, making a cross-section of the terrain, calculating the number of people in a given area, determining the visibility of the terrain and objects from a given place, calculating the distance and area, and calculating the time availability of given places. Figure 6.6 shows the general idea of integrating different systems.

As mentioned above, the concept of SC-SDI implementation cannot be limited only to technological solutions. The success of this type of solution depends on the simultaneous introduction of relevant organisational and formal regulations. The National Spatial Data Infrastructure (NSDI) is a good example of this type of solution in various countries; legally established, it ensures the harmonised development of spatial data resources. In Europe, INSPIRE exemplifies such infrastructure. Some of the solutions proposed in NSDI may apply to SC-SDI implementation; furthermore, both infrastructures should be connected and coherent. Below are the primary organisational and formal assumptions proposed in this regard:

1. City spatial data resources are part of the NSDI. It is necessary to define a set of georeferenced data and their relation to the georeferenced data at the national level. In this context, georeferenced data are primarily topographic data, land and property register data (cadastral data), data from the register of borders, geodetic records of utility networks, records of towns, streets and addresses, and geographic nomenclature. The fundamental georeferenced data resource should be a 3D topographic database with models of buildings at LoD3 (Level of Detail) and key public buildings at LoD4. A digital terrain model (topography), a digital land cover model, and collections of orthophotomaps from aerial and satellite photographs should constitute the database's integral elements. In the future, a 3D cadastre system may become an extremely significant element.
2. In the city's organisational structure, it is important to indicate the 'leading bodies' responsible for supplying IP-GeoCity with various detailed and specialised data on the city. The leading bodies' role should be ensuring appropriate geocoding of the data at their disposal, ensuring the correct use of shared reference data resources, and harmonising their data with other data administrators.

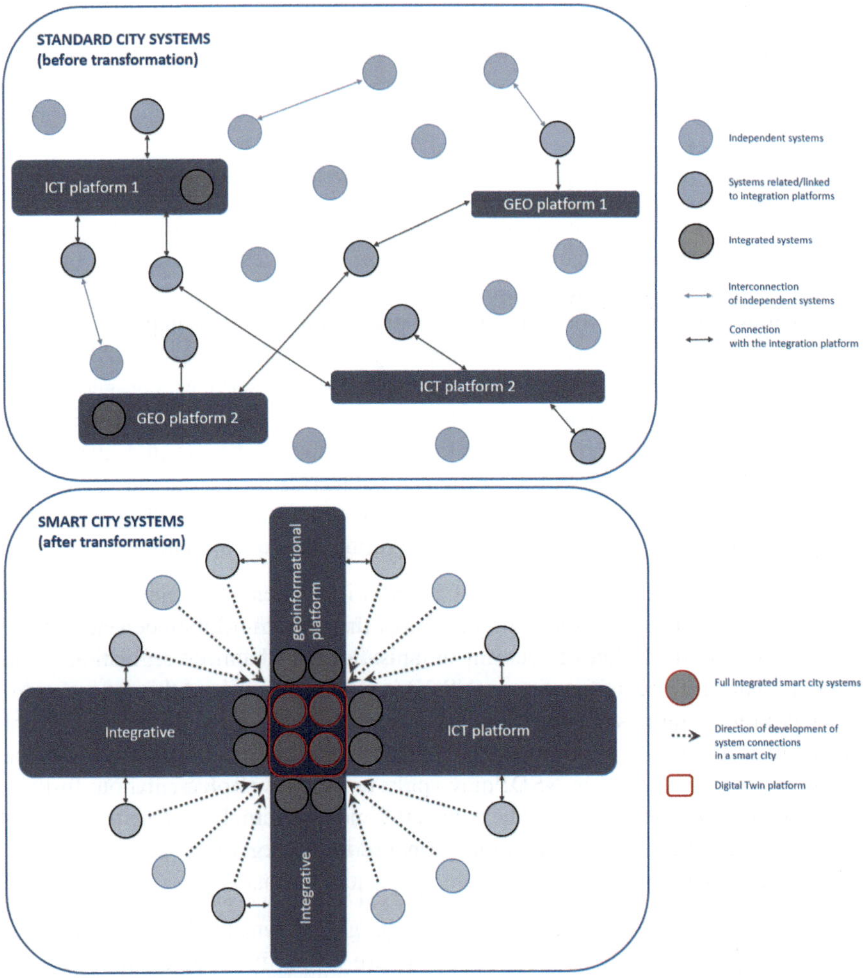

Fig. 6.6 The process of integrating various systems: from autonomous systems to systems connected through integrative platforms

All the data mentioned above must be visible to users in the form of one coherent **city map of the digital twin 3D model;** they must match, i.e. maintain their spatial relations and remain available for homogeneous spatial analyses. However, this does not mean that one common database managed by one entity retains all the data. These resources may have a high degree of autonomy provided the application of common standards for data sharing and exchange.

3. It is necessary to adopt common technological standards for sharing spatial data and require the systems to develop accordingly.
4. It is necessary to define the scope of open data available to all interested parties free of charge and without restrictions on use.

The aim of the proposed concept of the Integrative Geoinformation Platform is to enable the interaction of many information systems in a city. Simultaneously, this type of solution may enable the creation of many innovative products, greater standardisation of systems, and reduction of their operating costs.

It is important to collect and process data from many registers and open data resources to implement the smart city idea. The essence of a smart city, however, lies not only in the integration of existing IT systems, registers, and databases but also in the creation of innovative solutions that, by achieving synergy, will enable a distinct and continuing improvement of urban infrastructure and social interactions (Gotlib and Olszewski 2016a, b).

Geoinformation and Social Aspects

According to the authors, the development of technology and implementation of modern IT solutions, such as geoinformation infrastructure, big data, cloud computing, social media, or geoinformation services, should, above all, simultaneously stimulate social development (Olszewski and Wieszaczewska 2016).

Using appropriate and diverse communication and consultation techniques that include as many groups of interested parties as possible may shift the centre of gravity and reframe a 'smart city' into a 'learning city.' The term 'learning city' indicates its processual nature that changes depending on the inhabitants' internal needs, in a dialogue between all parties interested in developing the metropolis (Olszewski and Wieszaczewska 2016).

It is noteworthy that the process of even the most complex data analysis carried out in the city hall is currently limited to the processing of 'hard' data, e.g. official data or data from the network of city sensors. Social geoparticipation and the resulting cooperation, co-decision-making and co-management, require employing active forms, such as dedicated geo-surveys – surveys with the capability of indicating places in space. Using modern geoinformation technologies and ICT tools to create geo-surveys and advanced 3D modelling methods, at times with augmented reality elements, is a necessary but insufficient component of engaging the residents in a smart city. What is crucial to change passive and atomised individuals into an open information society, creating a vision for the development of the capital in the process of social participation, is the liberation of social energy.

Summary

Whether it is a digital spatial database or an analogue form, a map is an excellent platform for presenting a city's past, present, and future; thus, it is a useful decision-making support tool. Most of the key decisions in a city are related to the location of objects and phenomena in space. A map is not just a spatial model of the city;

thanks to appropriately applied geoinformation technologies, it can be an extremely effective interface for accessing various urban data.

Building a digital twin model of a city and conducting advanced, multi-variant analyses on its basis enables performing a detailed and reliable analysis of many alternative solutions and selecting the optimal one. It is important to emphasise that multi-variant analyses of urban spatial big data require adequate IT infrastructure and artificial intelligence methods to acquire spatial knowledge from data, which is particularly significant due to the growing importance of the Internet of Things and the expansion of the IoT network of city sensors.

Building an integrative geospatial platform (as one of the elements of a city's complex information system) may greatly facilitate the collaboration between many organisational units of the city and citizens, the implementation of the idea of participatory support for investment planning, and solutions enabling the residents to determine participatory budget allocation, support the process of making economic decisions, and increase the level of security.

In the authors' opinion, it is vital to pay attention to the difference between a modern city (in the colloquial sense) and a smart city. A 'modern' city is a city that implements various new organisational, technical, and financial solutions in a traditional manner. In contrast, in a smart city, various solutions are built and implemented in an integrated manner, achieving synergy by integrating interoperable systems and services that compose a smart city's ecosystem.

Extensive use of geoinformation, which may constitute a platform integrating city systems, is also important for implementing the idea of a smart city. Likewise, the collaboration between local authorities, residents, and companies providing modern information services is of key importance. Thanks to the potential of urban spatial big data, clearly defined modern geoinformation services may be the driving force behind smart cities' development and may constitute a development tool helping a local community to progress towards a geoinformation society.

When considering the role of geoinformation technologies in the development of smart cities, we do not forget that technology is supposed to serve the inhabitants, and not vice versa, to determine their way of life or to decide where they should go for a walk or where to eat or sleep. In our opinion, it is not possible to carry out the above-mentioned tasks set out for a city striving to become smart without access to comprehensive, accurate, and up-to-date information about space and the tools enabling the transformation of 'raw' data into spatial information and knowledge. Most of the decisions made in the city have a spatial context, just as most modern information systems need access to spatial data.

Today, open data are widely available (especially for large cities), which is a stimulus for the development of innovative geoinformation services. The tools used today to process big data – e.g. data commonly available in social media or cloud computing – allow for obtaining useful real-time information about the needs and expectations of inhabitants, their migration within the city, or potential threats.

In the authors' opinion, the implementation of the smart city idea does not consist only in the implementation of advanced technological solutions but also in the intelligent use of information resources and infrastructure in all spheres of city activity. The essence of a smart city is to create innovative solutions that will allow achieving a lasting improvement of urban infrastructure and social interaction through obtaining the synergy effect.

Acknowledgements We want to express our gratitude to Comtegra S.A. company for the invitation to implement the project, the development of a detailed inventory of the systems in Warsaw, and effective cooperation, and to the Warsaw City Hall for their substantive cooperation and granting us access to the materials.

Works Cited

Al Solami E (2021) Replication-aware secure resource administration scheme for Internet of Things-smart city applications. Trans Emerg Telecomm Technol 32(3):e4200. https://doi.org/10.1002/ett.4200

Bielawski B, Gotlib D (2016) Techniki komunikacyjne i informacyjne dla inteligentnych miast. In: Gotlib D, Olszewski R (eds) Smart city. Informacja przestrzenna w zarządzaniu inteligentnym miastem

Dembski F, Wössner U, Letzgus M, Ruddat M, Yamu C (2020) Urban digital twins for smart cities and citizens: the case study of Herrenberg, Germany. Sustainability 12(6):2307. https://doi.org/10.3390/su12062307

Dustdar S, Nastić S, Sćekić O (2017) Smart cities. Springer, Cham

Goodchild MF (2007) Citizens as sensors: the world of volunteered geography. GeoJournal 69:211–221. https://doi.org/10.1007/s10708-007-9111-y

Gotlib D, Olszewski R (2016a) Możliwości wykorzystania informacji przestrzennej i technologii geoinformacyjnych w zarządzaniu nowoczesnym miastem. In: Gotlib D, Olszewski R (eds) Smart city. Informacja przestrzenna w zarządzaniu inteligentnym miastem

Gotlib D, Olszewski R (2016b) Wprowadzenie. In: Gotlib D, Olszewski R (eds) Smart city. Informacja przestrzenna w zarządzaniu inteligentnym miastem

Homainejad AS (2015) With geospatial in path of smart city. ISPRS Ann Photogram Rem Sens Spat Inf Sci XL-7/W3:1381–1387

Inajit A, Ploeger HD, Van Loenen B, Van Oosterom PJM, Van Oosterom P, Dubbeling D (2018) Designing open spatial information infrastructure to support 3D urban planning in Jakarta smart city. Proceedings of the 6th International FIG 3D Cadastre Workshop

Kim SA, Shin D, Choe Y, Seibert T, Walz SP (2012) Integrated energy monitoring and visualization system for smart green city development: designing a spatial information integrated energy monitoring model in the context of massive data management on a web based platform. Autom Constr 22:51–59

Laurinia R, Favetta F (2017) About external geographic information and knowledge in smart cities. ISPRS Ann Photogram Rem Sens Spat Inf Sci XLII-4-W3:51–58

Leszek K (2016) Technologie i koncepcje tworzenia modeli 3D miast. In: Gotlib D, Olszewski R (eds) Smart city. Informacja przestrzenna w zarządzaniu inteligentnym miastem

Mahavir, Bedi P (2017) Integration of existing information systems as a spatial eDemocracy tool for smart cities in India. In: Vinod KT (ed) E-democracy for smart cities. Advances in 21st century human settlements. Springer, Singapore, pp 369–390

Manville C et al (2014) Mapping smart cities in the EU. European Parliament, Directorate-General for Internal Policies, Policy Department A: Economic and Scientific Policy (Brussels). http://www.europarl.europa.eu/RegData/etudes/etudes/join/2014/507480/IPOL-ITRE_ET%282014%29507480_EN.pdf

Muraszkiewicz M (2016) Wybrane standardy i normy dotyczące tworzenia systemów inteligentnych miast. In: Gotlib D, Olszewski R (eds) Smart city. Informacja przestrzenna w zarządzaniu inteligentnym miastem

O'Dwyer E, Pan I, Charlesworth R, Butler S, Shah N (2020) Integration of an energy management tool and digital twin for coordination and control of multi-vector smart energy systems. Sustain Cities Soc 62:102412

Olszewski R, Wieszaczewska A (2016) Geopartycypacja społeczna z wykorzystaniem nowoczesnych technologii geoinformacyjnych. In: Gotlib D, Olszewski R (eds) Smart city. Informacja przestrzenna w zarządzaniu inteligentnym miastem

Olszewski R, Pałka P, Turek A, Kietlińska B, Płatkowski T, Borkowski M (2019) Spatiotemporal modeling of the smart city residents' activity with multi-agent systems. Appl Sci 9

Popov M, Fedorovsky O, Stankevich S, Filipovich V, Khyzhniak A, Piestova I, Lubskyi M, Svideniuk M (2017) Remote sensing technologies and geospatial modelling hierarchy for smart city support. ISPRS Ann Photogram Rem Sens Spat Inf Sci IV-5/W1:51–56

Raes L, Michiels P, Dalianis T, Adolphi T, Tampère C, McAleer S, Kogut P (2021) DUET: a framework for building secure and trusted digital twins of smart cities. IEEE Internet Comput. https://doi.org/10.1109/MIC.2021.3060962

White G, Zink A, Codecá L, Clarke S (2021) A digital twin smart city for citizen feedback. Cities 110:103064

Zanella A, Bui N, Castellani A, Vangelista L, Zorzi M (2014) Internet of things for smart cities. IEEE Internet Things J 1(1):22–32

Chapter 7
Tools for Citizen Engagement in Urban Planning

Elie Daher, Mohaddeseh Maktabifard, Sylvain Kubicki, Régis Decorme, Burak Pak, and Rachel Desmaris

Abstract Engaging citizens in urban planning has the potential to generate effective ideas to reinvent our cities. Particularly, designing easy to grasp and effective tools for co-creating meaningful urban spaces remains a significant challenge and an emerging need. Such tools that can involve the community in an intelligent manner are in strong demand. It is, however, required to design, develop, and implement well-executed engagement tools that open the horizons of evaluating and responding to urban-related problems while involving city councils, architects, ICT developers, and urban planners. In this chapter, a framework aiming to investigate future forms of citizen involvement within urban planning activities is prefigured. It addresses the research aim to present and discuss technology-driven civic engagement in the planning process via a toolset and the outcomes of a market study for this tool. Furthermore, it identifies and discusses the gaps between four main existing groups of urban planning software; physical planning tools, physical/civic engagement tools, civic scenario planner tools, and data analysis methods and how to bridge these gaps. In addition to these, it highlights the potential of such urban planning participatory tools to generate possible socioeconomic impacts and concludes by assessing (1) the degree to which the technology creates an inclusive environment to exchange and implement urban planning related ideas and (2) the extent to which such tools could lead to an integrated and coherent engaging method for citizen engagement.

E. Daher (✉) · S. Kubicki
Luxembourg Institute of Science and Technology, Esch-sur-Alzette, Luxembourg
e-mail: elie.daher@list.lu; sylvain.kubicki@list.lu

M. Maktabifard · R. Decorme · R. Desmaris
2R2M Solution, Paris, France
e-mail: mohaddeseh.maktabifard@r2msolution.com; regis.decorme@r2msolution.com; rachel.desmaris@r2msolution.com

B. Pak
KU Leuven, Leuven, Belgium
e-mail: burak.pak@kuleuven.be

G. C. Lazaroiu et al. (eds.), *Holistic Approach for Decision Making Towards Designing Smart Cities*, Future City 18,
https://doi.org/10.1007/978-3-030-85566-6_7

Keywords Urban planning · Digital tools · Citizen engagement · Co-creation · Technology

Participation in Urban Planning

The word "participation" gained popularity and attention during the late 1960s (Pateman 1976). Participation was not only popular in the political context; it also gained a wide importance in the design of products. "Community design" involves different aspects of collective activities such as "community planning and architecture, social architecture, and community participation" (Sanoff 2000; Toker 2007; Sanders and Stappers 2008). This approach has emerged from the fact that the mismanagement of the physical environment is a major factor contributing to social ills (Barcellini et al. 2015). In participatory design, the roles of different stakeholders and profiles involved in the design are different. The end-users who play a passive role in the conventional and traditional design process are given the opportunity to express their preferences, share knowledge and feedback, and, therefore, play an important role in concept development (Sanders and Stappers 2008).

In Participatory Design (PD), the process is responsive, which means that the designers collect and consider the end-user's feedback related to their design output, but the end-users are not fully empowered or in control of this process (Cimerman 2000). However, the concepts and methods of PD and co-creation vary widely among different studies in social science and according to the field of application (politics, design, etc.). Since the beginning of 1960s, a variety of methods has been established to improve the participation of both end-users and stakeholders in the design process to bridge between the needs and the output (Abras et al. 2004). Various research and books have stressed the necessity of engaging end-users and other relevant stakeholders in the design of their designed and built environment (Horelli n.d.; Kanji and Greenwood 2001; Butler et al. 2007).

In the following section, we explore the different waves of participation in urban planning in the western countries' context during the last 50 years. The extended waves of participation are based on different research (Arnstein 1969; Kelly 2001; Claridge 2004; Miessen 2017a; Meeus and Pak 2018).

1968–1990s Emergence of Participation

In the early 1960s, the role of communities in the built environment gained a considerable importance with some architects and urban planners. In Great Britain, the idea that the public should participate was first raised in the early 1960s with the British architect John Turner. His ideas were developed through a number of articles and publications where he was putting emphasis on the role of households in the architectural design process (Turner 1963). Turner developed an influence approach related to the self-help housing and role of the community in the design process. John Habraken was another writer and Dutch architect who developed his ideas about participation and the role of the user in the design process in the early 60s. His idea for a flexible design approach to housing was published in a book so-called Supports and demonstrated in practice in the mid-1960s. A different component in the community engagement in the built environment was related to the planning approach in the urban spaces (Davidoff 1965). This approach was based on Community Design Centers assisting citizens on a number of issues related to architecture and urban planning. Sherry Arnstein, based on the experimentations of this approach, has created a ladder of participation which is still used as reference for community participation nowadays (Arnstein 1969).

In the 1970s, the Scandinavian countries started to research the role of the user in the design process and the development of projects. The "collective resource approach" idea developed new innovative strategies for workers to influence the design and the use of new technologies in the workplace. It attempted to empower "trade unions and workers at the local level" (Kraft and Bansler 1994).

In the 1980s, there was a new wave to change and find alternatives to the top-down design approaches. Several activities, particularly driven by non-governmental organizations, were leading the change to a bottom-up approach (Kelly 2001; Claridge 2004). The 1980s witnessed as well flourishing of activities, particularly among non-government organizations (NGOs) in seeking alternatives to top-down outsider-driven development. The emphasis was on participatory appraisal and analysis in rural communities. In the 1990s, the participation approach continued to be an important element in the design process, and it became synonymous with sustainable development.

Early 2000s Emergence of e-Participation and Crowdsourcing

In the recent two decades, the use of information and communication technology (ICT)-based participation tools and methodologies for urban planning has gained traction. In order to provide urban stakeholders with platforms for involvement, many technology-enabled participatory tools, approaches, and applications have been developed (Gün et al. 2019). Today, the employment of ICT-based participation platforms for addressing urban issues is gaining traction as governmental authorities consider innovative ways to provide novel, open, and democratic communication channels.

To start with, the e-participation era is an important period in the short history of this field, which emerged in the 2000s. E-participation involves ICTs to engage non-designer groups with diverse backgrounds, expertise, ambitions, and positions in diverse collaborative design activities throughout the design process (Sanders et al. 2010). This method enables non-expert end-users and lay-persons who are not traditionally involved in urban planning, research, and design to participate in design processes. ICT-based engagement platforms, in this sense, are digital arenas for retrieving, analyzing, visualizing, and sharing information, expertise, and alternatives that serve to social causes and policy challenges (Desouza and Bhagwatwar 2014). Particularly, 3D modelling software and games offer new opportunities for civic engagement in urban planning. These complex and interactive software solutions allow spatial configurations and data to be visualized, accessed, and interacted with representation formats convenient for novice participants without any planning or spatial design expertise.

Among the ICT-enabled participation paradigms, crowdsourcing is a recent approach emerged in the 2010s which dominated the field in the last decade. According to Saxton et al. (2013), crowdsourcing refers to a certain form of outsourcing tasks to a specific mass of users: it is a sourcing model where organizations use Internet technologies to track and employ the efforts of a virtual crowd to carry out designated tasks. By using online platforms to harvest the *wisdom of the crowd* and make use of the collective intelligence of multitudes of users, crowdsourcing opens up a whole slew of possibilities for urban planning and government and non-profit applications (Brabham 2010).

Since the emergence of crowdsourcing as a dominant ICT-enabled participation paradigm in the 2010s, starting with the English-speaking countries, a variety of platforms have been put in action to address urban issues in different parts of the world. Among those, the frequently referenced platforms were *FixmyStreet, OpenPlans, CitySourced, and Neighborland*. The empowerment capacity of these crowdsourcing platforms are found to be quite limited since these enable limited information exchange and idea sharing, located in the lower rungs of the ladder of citizen participation (Pak and Verbeke 2013). From the perspective of bottom-up spatial design and planning, these platforms lack the necessary tools and capabilities to enable a wide range of stakeholders to co-create urban plans or designs since they mainly aim at collecting feedback from the users (Pak and Verbeke 2013).

An in-depth study of 25 crowdsourcing platforms in Europe, chosen from a total of 106 platforms (Gun et al. 2020), showed that the vast majority (77%) of the platforms strive for one-sided information sharing, the lowest degree of empowerment on the participation scale introduced by Senbel and Church (2011). Hardly 12% of systems, including Maptionnaire, MinStad, and Smarticipate, aimed to allow end-users to build their own goals and ideas. According to this study, *BetriReykjavik, BlockByBlock, FindingPlaces, Quakit, ZO!City,* and *Unlimited Cities* were among the few cases that allowed users to engage in the co-creation of urban plans and ideas.

The limitations and challenges of the use of crowdsourcing platforms in urban contexts have been covered in several studies focusing on real-world cases, and these were not only restricted to lower levels of design empowerment. Two early analytic studies found that crowdsourcing platforms mainly benefited people with a greater level of education and wealth (Helsper 2008). Later research revealed that crowdsourcing platforms can be influenced by existing social and spatial inequalities such as the digital divide, the inability of disadvantaged citizens to access the Internet to utilize platforms, lack of skills such as map and plan interpretation required to use platforms effectively, and variations across interest groups in terms of age, social position, expectations, ethnicity, and economic level (Evans-Cowley and Hollander 2010; Desouza and Bhagwatwar 2014; Vicente and Novo 2014; Pak et al. 2017).

The overrepresentation of the privileged parts of the (Western) society on crowdsourcing platforms is prominent and a major problem according to Bryson et al. (2013, p. 29). These particular users can be profiled as male, middle-aged, political *techies*, and sensitive to urban and neighborhood issues. Overrepresentation can potentially happen when exclusive groups produce more content, reports, ideas, and feedback, echoing the well-document exclusion issues in face-to-face civic planning practices; however, in crowdsourcing, this can happen on a far larger scale (Bryson et al. 2013).

These findings imply that crowdsourcing might increase inequality in terms of both representation and cause unjust socio-spatial effects (Pak et al. 2017). When used uncarefully, crowdsourcing techniques may unavoidably exacerbate socio-spatial inequality in our cities. Therefore, it is of utmost importance for crowdsourcing practices to make sure the concerns of disadvantaged groups (digital illiteracy, language hurdles, and immigrant origins) are representatively addressed. Communities where the majority of the population is immigrants and undereducated, for example, are unable to adequately report concerns in their neighborhoods, putting them at danger of degradation, whereas well-educated high-income citizens profit from the crowdsourcing platform. This can lead to large socio-spatial disparities over time which serves against the democratic and egalitarian principles behind participatory planning (Pak et al. 2017).

2010s Emergence of Critical Spatial Practice as an Alternative to Participation

In a critique on Jane Rendell's approach (Gallo and Pellitteri 2018) to critical spatial practice limiting it to the expansion of the spatial to the artistic frontier, Miessen argues for the staging of a reorganized relationship, combining a re-thinking of existing disciplines with the production of a new body of recognizable work (Miessen 2017b). Critical spatial design aims to go beyond architecture as a physical construction and explore the construction of alternate realities, criticizing existing protocols, and generating new protocols for this venture "reflective practicum in designing" (Schön 1987). In parallel to this line of thought, Awan, Tatjana Schneider, and Jeremy Till (Awan et al. 2013) describe their three major criticisms toward the notion of "architectural." The first argument condemns the architectural obsession with buildings and objects as its primary location. An object-oriented and a temporal regard to architecture neglects the occupation of the building, its temporality, and the relation to society and nature. This resonates with Manzini's (2015) argument that design – in general – needs to be redefined as he notices a shift away from the tangible object toward services, experience design, and organizational structure.

Breaking free from prevailing modes of urban-architectural design requires the establishment of a working practice exploiting the productive encounters between different disciplines (De Smet et al. 2019). While being about a network, our critical spatial practice represents itself as a networked practice.

This involves developing the participants' capacity of reflecting-in-action and reflecting-on-action as a crossbencher and going off autopilot actions; in other words, instead of blindly following and repeating what is learned in the past, questioning the existing protocols within the field of architecture with the will to develop new critical practices is set-forth. The "crossbench practitioner" (Miessen 2017b) is describing "a participator who is not limited by existing protocols, and who enters the arena with the will to generate change." This requires participants to get out of the comfortable boundaries of traditional expertise in architecture, toward the unknown, the intentional and skillful mastering of incompetence in the ocean of practices. Critical spatial practice in this context is a form of participatory action research. The participants combine their role of agent of change with one of researcher and therefore commit themselves to reflexivity, paying attention to the process of action and reflection as they unfold.

Critical spatial practice provides a testing ground for phenomena, methods, and tools as elements of a transdisciplinary framework. This is also the staging of a reorganized relationship, combining a re-thinking of existing disciplines with the production of a new body of recognizable work. At the same time, research and design consortiums have developed similar characteristics: new emerging alliances, defining new partnerships, focusing on new ways of transdisciplinary thinking and setting up new kinds of joint professional or academic projects.

Benefits and Importance of Participation in Urban Planning

Community design engagement is about involving the citizens and the civic organizations in the design of a project. It does not only consist of sharing information and telling people what is done and what is decided, but it has to be a two-way information sharing (activity, task, process). In general, this process is about gathering citizens' and civic organizations' preferences, viewpoints on what needs to be achieved, priorities, and significance and including them in the design process.

The importance of participation in the design process has been the subject of different research studies and is well established in the literature. Al-Kodmany (1999) recognized different benefits from participation, stating that participation gives a stronger sense of commitment and increases the users' satisfaction regarding the design solution, by providing realistic expectations of outcomes (Al-Kodmany 1999).

Additional viewpoints enlarge the choices of possible design solution spaces and improve the value of the ultimate decision (Kelly and Van Vlaenderen 1995; Kelly 2001). The more opinions collected during the process of design and while planning, the more probable the end product will address the concerns of the community

(Bamberger 1991). Sharing additional viewpoints and ideas leads to better facilitation of actions and activities with a higher potential for generating ecological and sustainable solutions (Price and Mylius 1991).

Experts and decision-makers recognize that the community, using the designed and planned spaces, has a close link with this space, while they do not often have this relationship with the planned spaces. This close relationship provides knowledge that should be considered leading to a better decision.

Participation is recognized to encourage the sense of responsibility and concern to solve the issues and problems in planning (White 1981). It also ensures that the requirements and needs of the community are being considered.

Limitation and Issues of Participation

The limitations and issues of community participation in the design process has been also a subject of different research, and different authors criticized it. These limitations and issues must be considered whenever a design process project is to be implemented with a participatory approach. Obstacles that can be encountered with the participation may include: being time consuming, increasing the financial costs, no guarantee of successful result, professional superiority, uncertainty about the results of public involvement, lack of transparency, empowering specific participants, and the mistrust between the communities and experts (Cleaver 1999; Sanoff 2000; Miessen 2017b; Mubita et al. 2017).

Many researchers claimed that community participation tends to be a time-consuming process (Towers 1995; Irvin and Stansbury 2004). In the latest cited reference, the authors have recognized that in some circumstances, the community participation can be time-consuming and may be costly and ineffective (Irvin and Stansbury 2004). According to them, the community participation in the design process is more expensive than if done by a single expert. The implication of community in the design of a program became a frequent process and is now rarely questioned. However, Cleaver (Cleaver 1999) claimed that participation does not always lead to the requested solutions or result. Therefore, according to his research, there is no guarantee that the requested criteria of the community are considered in the final solution.

Urban Planning Through Different Design Processes

Urban planning is defined in theory as a space focusing on physical geometries, association, and organization of different urban functions with the purpose to shape human activity (Carmona 2010). Urban planning addresses the development, design, exploitation, and use of land spaces for human activities purposes. It is also related to social aspects with a purpose to improve human life at the social, cultural, and environmental levels (Sandercock 1997). As a terminology, the "Urban Structure" refers to the "pattern" or "arrangement" of the urban spaces considering streets, blocks, buildings, landscape, and open spaces (Llewelyn and David 2007). Mougthin defines urban planning as "the method by which man creates a built environment that fulfils his aspirations and represents his values" (Moughtin et al. 2003). He also claims that the city is an "element of people's spiritual and physical culture" and the "central to the study of urban design is man, his values and aspirations."

The process used to build and conceive urban environments has an importance in providing viable and maintainable communities (Katz et al. 1994) and ensuring that the goals of the urban structure are satisfied. Multiple studies have discussed the design processes and action stages of urban design practices. Again, Moughtin et al. (2003) agreed that following a critical evaluation of alternative solutions, the decision-making processes in urban planning are not linear; according to him, they represent a progression loop between the multiple stages; therefore, they can be called iterative processes (Moughtin et al. 2003).

Different approaches for urban planning were proposed as well (Roberts and Greed 2001; Moughtin et al. 2003; Boyko et al. 2005; Dias et al. 2014; Carmona 2015) focusing on the idea that urban planning is a sequential series of connected decisions. Each sequence has a description of actions and detailed level of decisions. This sequence is detailed as "analysis, synthesis, appraisal, and decision." The decision level is repeated for each design stage in this sequential series (Moughtin et al. 2003). Roberts and Greed (Roberts and Greed 2001) defined the urban planning process in four sequences; the process starts by defining and analyzing the problem; it continues by conducting surveys and different analysis activities. Following these analyses, the planning team develops opportunities and ideas. In the later stage, strategies and options are assessed for the decision-making process.

After this review of urban planning traditional process, we can conclude that the main components and interventions in a traditional urban planning process are related to: (1) collection of needed information; (2) general study, identification of problems, and investigation of possible solutions; (3) development of solutions; and (4) decision-making and communication (Fig. 7.1).

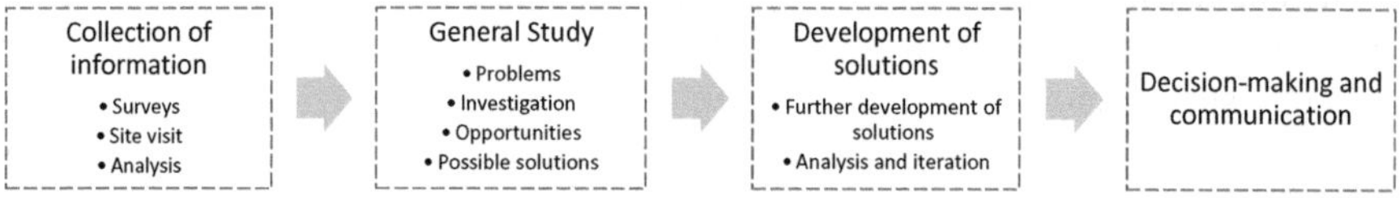

Fig. 7.1 Urban planning main design components

The Emergence of Computational Techniques in the Urban Planning Processes

Modern planning practices and theories encourage methods involving different actors in the urban planning process. The different stakeholders usually have a variety of objectives, values, and aims to accomplish (Mosadeghi et al. 2015). The applications to quantify the objectives of the stakeholders have increased with different approaches based on computational procedures and techniques that allow handling varied data and inputs. In this direction, computer-based applications applied to urban planning started to emerge in the daily practice of professionals. In particular, a high increase in computer-aided parametric design applications handling urban planning can be noticed.

Parametric Design Modelling

Parametric design modelling was used, in several studies, as both a "generative tool" and a "control function" enbaling the creation of flexible geometries based on a set of input parameters defined in algorithmic formulas (Schumacher 2008; Daniel et al. 2011).

In these applications, many research groups suggested a parametric model to establish the plots and the building shapes with consideration to parameters related to density, proportion, and alignment. As for the control function, the design was assessed in relation to environmental features such as light, noise, wind, etc. With such approaches, the designers have a possibility to modify the urban structure and pattern, preserving the relationships between the different geometrical components and controlling the assessment of desired features.

Research coupling urban design with parametric design methods increased during the past 20 years. These are including different techniques for optimizations and generations of urban spaces with different simulations and iterations. The research reported below arised from a search conducted in Google Scholar with keywords related to the association of parametric/urban planning and generative design/urban planning.

These applications deploy different technologies and methods to generate and produce urban spaces and urban structure patterns. They are based on powerful parametric and simulation capabilities or also on artificial intelligence capabilities to help designers in their urban design process.

In this context, computer-aided parametric tools provide a flexible and smart framework for the generation of urban composition. The generation of urban blocks is driven by associative and parametric approaches. These studies showed that there is a potential in applying computer-aided parametric design tools to the urban planning application. Some of the studies also showed a potential to implement a participatory approach. It should be mentioned that these applications were in most cases based on presenting parametric applications as (Dongyoun et al. 2011; Steiniger et al. 2016; Schubert et al. 2017; Chowdhury and Schnabel 2018):

1. A communication tool to inform the end-users about the design solutions
2. A controlling tool to control performance simulation of the solutions proposed
3. A generative tool to allow the generation of urban blocks based on different input
4. A tool or a methodology to capture the end-users' preferences as input parameters capable of influencing the design outcomes in few applications

The use of computer-aided parametric tools allowing a simulation and optimization of design solutions enables a performance-driven design process. Important aspects in the participation through a performance-driven design process are the following:

1. The need to have a clear and understandable interface as a common base for discussion and participation
2. The need to identify the design stages where participation of end-users will make sense and will be beneficial
3. The need to identify the methods and means of participation in the design process

Urban planning applications developed with computer-based parametric design tools, as already shown previously, are emerging in the practice of architects and urban planners (Cockey 1955; Schumacher 2008; Da Silva and Morim 2010; Saleh and Al-Hagla 2012; Muther and Halles 2015; Wang et al. 2015; Sabri et al. 2019). These applications are evolving and changing to answer different challenges related to urban planning. The functions of these applications vary from sharing information on the selected environment to discussions between end-users and other concerned stakeholders. The modalities of participation in urban planning remain complex to implement in the design process from different points of view, such as the time and cost. In the following sections, we explore these problems and identify ways to answer these issues.

Toward Digital Twins of Cities

Among the latest computational trends, digital twins are gaining a strong interest. The paradigm enables the coupling of the smart cities approach to 3D urban visualizations and is being developed to tackle urban planning, urban management, the implementation of smart systems, and several other applications ranging from technological deployment in cities to the co-creation of citizen-centric services. Among a wide number of initiatives, some of those digital twins are already in the first stages of their actual operation, such as in Helsinki (Finland) (Ruohomaki et al. 2018), Barcelona (Spain), Amsterdam (Netherlands), and others. Interestingly, such initiatives rely either on public, private, or both investment models and provide a wide range of data and applications, ranging from access to cities' open datasets to high level of consultation with citizens. These cutting-edge technologies develop rapidly and serve the purposes of cities development, while encompassing various goals including sustainability and carbon neutrality, well-being, and health as well as inclusiveness or equity.

Geodesign

The geodesign approach provides a multidisciplinary design framework and set of tools for exploring issues with a multi- and transdisciplinary view (Steinitz 2012). It consists of resolving the conflicts between the diverse points of views. This approach can be applied to different design planning processes and in particular in the urban planning application where it can be considered as a powerful tool to support the decision-making process when dealing with conflicts and issues. Carl Steinitz's geodesign framework uses a series of questions which guide the different participants of different discipline through the process of design (Fig. 7.2).

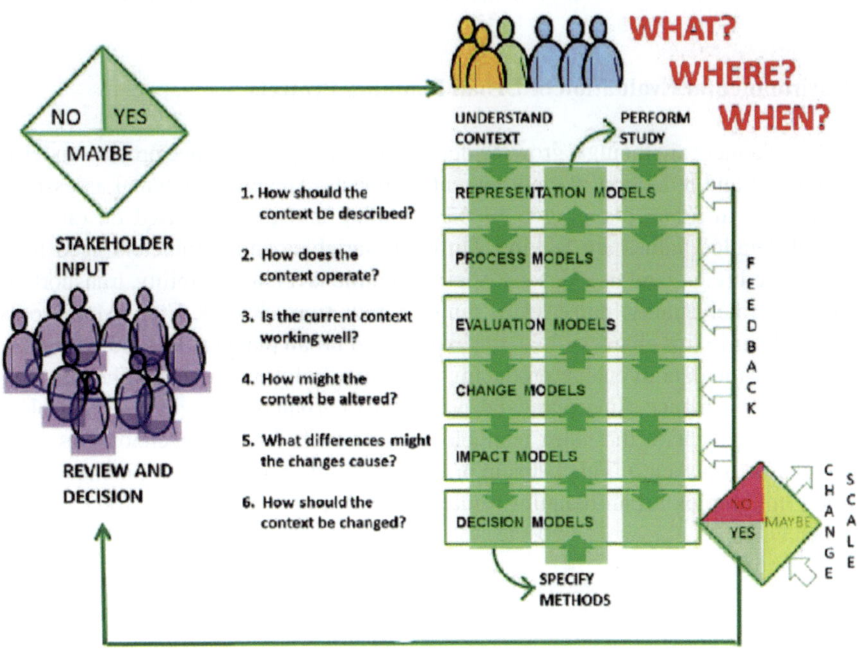

Fig. 7.2 The series of questions to guide the participants through the process of geodesign. Courtesy Carl Steinitz (Steinitz 2012)

Key Indicators of Urban Planning Performance Enabled by the Computational Techniques

Those above-mentioned goals significantly adressed through urban planning projects as well as through the management and operation of urban areas require the capability to assess and benchmark the decisions made.

A variety of initiatives address this requirement, from various perspectives:

(a) The planning of urban districts requires engineering-driven simulations to evaluate energy and water consumption, wind, solar gains, or even comfort.
(b) The development of smart cities' technology needs a careful monitoring of citizens' acceptance.
(c) Overall, the environmental impacts of territorial development and policies claim for high-level models.

This section presents some initiatives that set the scene in terms of criteria covered in the design framework. In order to insure the well-being and comfort for the inhabitants, it is important to allow them to choose and prioritize their needs in terms of performance. When the inhabitants are inputing their needs and objectives, these needs have non-physical characteristics, which mean that the inhabitants are more concerned by the performance of the urban space rather than by its physical and geometrical properties. However, these objectives are driven as well by physical parameters of the built environments that define its environmental assessments.

Environmental Evaluation of Urban Planning Projects

Urban planners are facing a growing demand for high performance projects in terms of control and reduction of environmental impacts. The environmental assessment evaluation in urban planning is time-consuming and usually based on ratios, as actual detailed values are lacking. Different variables and parameters need to be considered (e.g., land use, density, socioeconomic level, accessibility, transport, air quality, water quality, noise level, sunlight, radiation, shadow. Etc.). Many tools, shown later in this chapter, allow the experts and urban planners to handle the complexity of the data to enable producing alternative solutions with a higher environmental performance. Environmental evaluation of urban planning is now possible to conduct easily with the latest technologies and tools (sensing technologies, GIS, simulation tools, etc.). This helps in producing more accurate environmental evaluation in complex urban planning situations and in consequence to produce better urban cities.

Resiliency and Environmental Impacts Evaluation of Cities and Territories

Comprehensive frameworks appeared in the last decade to support decision-making in relation with cities' development. The EU Reference Framework for Sustainable Cities provides a set of 5 dimensions and 30 objectives "promoting a European vision of tomorrow's cities."[1] It enables policy makers, professionals, and citizens in monitoring project development while crossing several dimensions, including spatial, governance, social, economic, and environment.

Besides pure urban planning, a strong focus is put on the ability to monitor, analyze, and react to adverse event affecting urban systems. According to the Resilient Cities Network,[2] "urban resilience [is] the capacity of a city's systems, businesses, institutions, communities, and individuals to survive, adapt, and grow, no matter what chronic stresses and acute shocks they experience." Among others, the City Resilience Framework[3] is a "unique framework developed by Arup with support from the Rockefeller Foundation, based on extensive research in cities. It provides a lens to understand the complexity of cities and the drivers that contribute to their resilience."

Also, the so-called Doughnut economics principles, defined by Kate Raworth (Raworth 2010), propose a doughnut-shaped visual framework [illustrating] a safe space between "planetary boundaries" and "social boundaries" in which "humanity can thrive." Interestingly, this model enables applications beyond the scope of urban planning and even cities' managements, providing the capability to upscale the approaches at territory or even national levels, encompassing a variety of goals (Fig. 7.3).

Participation-Based Design Process Framework

To answer the gap between the actual applications used for urban design and the participation of end-users in the design process, the authors propose a conceptual framework reflecting the main steps of urban planning process and associated methods to engage the community in the process. The developed design process covers the different design stages of a traditional urban planning process and includes various levels of participation of the community/citizen/end-users and different stakeholders. The parametric capabilities integrated in the developed design process enable nonlinear interactions, loops, and modifications which help in creating more iterations and exploring more possibilities.

The framework is composed of three interactions covering different design stages of urban planning traditional process. This framework is supported by

---

[1] http://rfsc.eu/european-framework/

[2] https://resilientcitiesnetwork.org/what-is-resilience/

[3] https://www.rockefellerfoundation.org/report/city-resilience-framework/

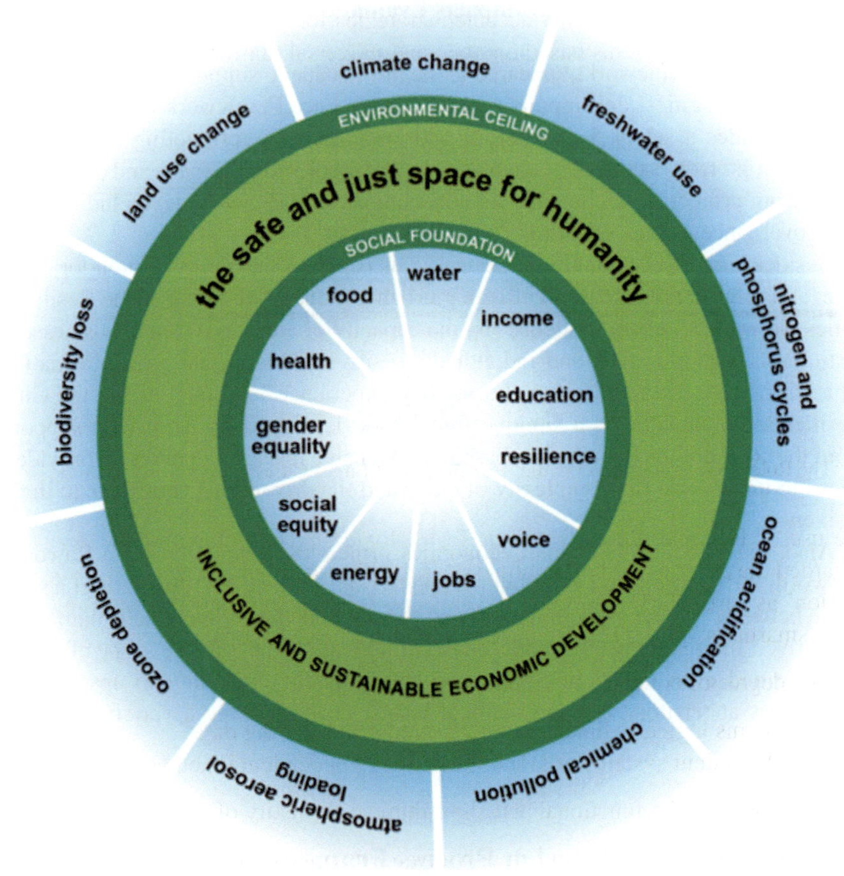

Fig. 7.3 Doughnut model. (Source: Oxfam; Sayers and Trebeck 2015)

different devices and tools which are associated with various levels of participation. This conceptual framework is characterized by recommendations of the visualization to enable a better understanding of the design project.

First Interaction: Problem Formulation and Objectives Setting

This interaction covers the stages of the urban planning where the objectives including the physical and non-physical criterions are set. In this interaction, the rank of the criteria is defined including distinctive features in terms of priorities and needs with the citizens. The criteria needed to explore the suitability of the selected land are defined to help in the site's actual conditions' identification based on previous studies and comprehension of the constraints. These studies and analysis should be

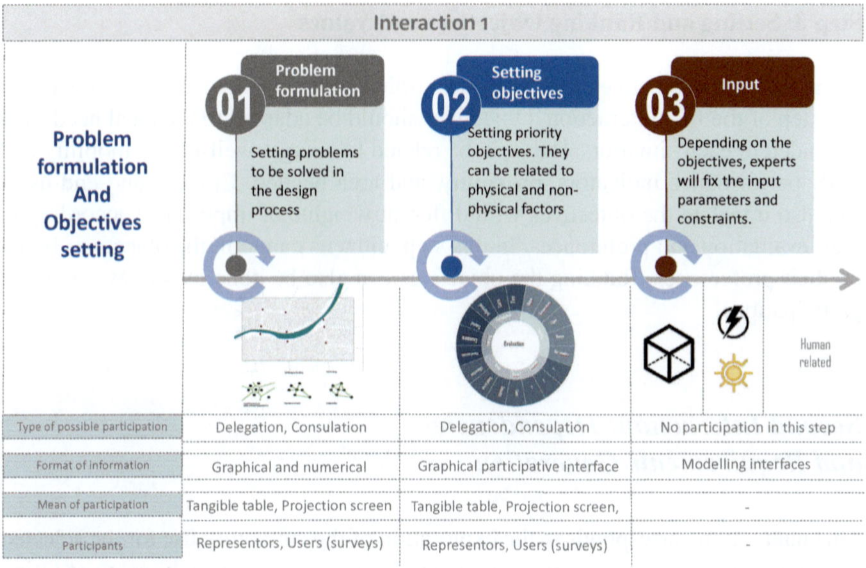

Fig. 7.4 First interaction of the framework

based on collected data related to the site or simulations based on different simulation tools. The ranking of objectives can be obtained by conducting different surveys or by a direct participation of participants in the design process (Fig. 7.4).

At the end of this interaction, a decision-making process should be conducted to pass to next steps of the framework. The steps of this interaction are the following.

Step 1: Selecting the Site, Establishing the Current, and Actual Conditions of the Site

This step consists of the selection and identification of the potential of a selected site. It considers the selection of the site and the gathering of related information which can be performed based on the actual data gathered or different simulations using computer-aided parametric design tools.

Step 2: Defining Land Constraints and Establishing Criteria for the Suitability of the Land Use

In this step, experts can perform the analysis of the site according to distinctive features and criteria (noise simulation, daylight simulation, proximity, mobility simulation, etc.). This will help in creating a diagnostic of the actual conditions of the site which will be considered in the generation of the different alternatives.

Step 3: Setting and Ranking Objectives and Values

Setting priority objectives with end-users, inhabitants, and experts is conducted in this step of the first interaction. Objectives should be adapted to the local needs and grounded to each situation. They can be related to energy, well-being, mobility and other performance indicators, proximity, and area needed. Citizens and end-users can also translate the objectives with different weights of importance according to their evaluation and preferences. In this step, citizens can rank the objectives based on their preferences. Ranking the objectives can also be done by surveys (indirect participation).

Second Interaction: Dependencies and Requirements Generation

This interaction corresponds to the exploration of the suitability use for the selected site. Indeed suitability analysis is a logic methodology to analyze the score of fitting features. In our case, the suitability analysis is performed to define the relation between the urban functions to distribute in the site and the features of the site based on fuzzy logic. The end-users can explore in this step, the site criteria and suitability use. A methodology is defined based on algorithmic formulas where the end-users will be exploring the suitability impact of the requirements. After the exploration, and since urban planning is typically concerned with assembling, organizing, and locating activities and land-use in the space, the end-users can explore the organigram of the different functions and the suitability of the organizations. The steps in this interaction are the following (Fig. 7.5).

Step 4: Site Criteria Diagnostic Communication

This step will focus on the requirements of the program function for the site. We invite in this step the end-users and delegation to discover the diagnostic of the site according to the criteria that have been set previously. The main objective of this step is to communicate the results of these criteria to make the analysis more understandable to the end-users. End-users can discover the diagnostic related to the criteria such as the daylight, the proximity of some crucial point, distance from noise, and distance from pollution sources.

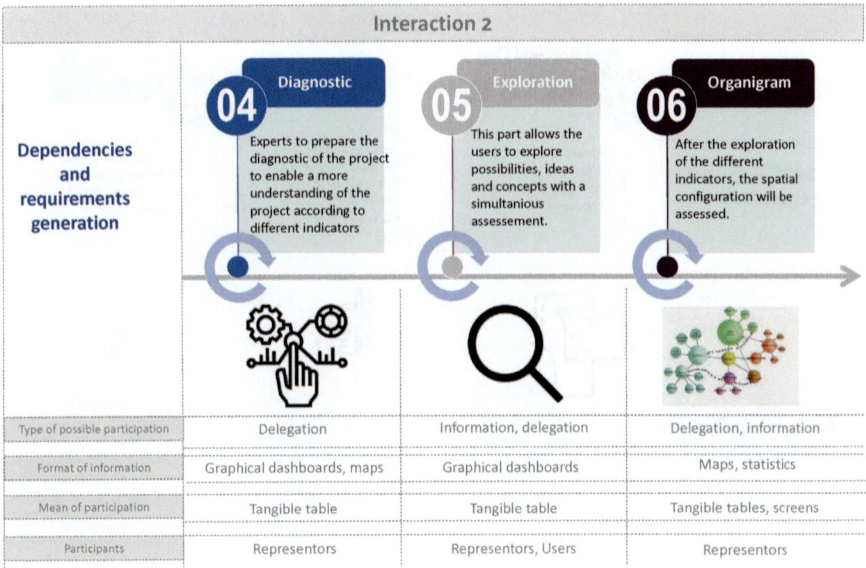

Fig. 7.5 Second interaction of the framework

Step 5: Site Exploration and Recommendation of Suitability Land Use

After the diagnosis of the site, and since we mentioned previously that we encourage a better communication and exploration process for urban planning rather than an optimized process, this part focuses on the exploration of the suitable land use according to the criteria. The main objective is to communicate and exchange between the different participants. The system will be generating recommendations and percentage of fulfilment of each land use in the selected site.

Step 6: Site Allocation and Recommendation of Fulfilment for Functions

This part will identify the organigram and distribution of the suitable land use in the selected site. Recommendations will also be made by the system to ensure that the criteria needed as required by the end-users are still respected. Participation of the citizens is particularly important in this stage and will be enabled through multiple devices, tools, and surveys.

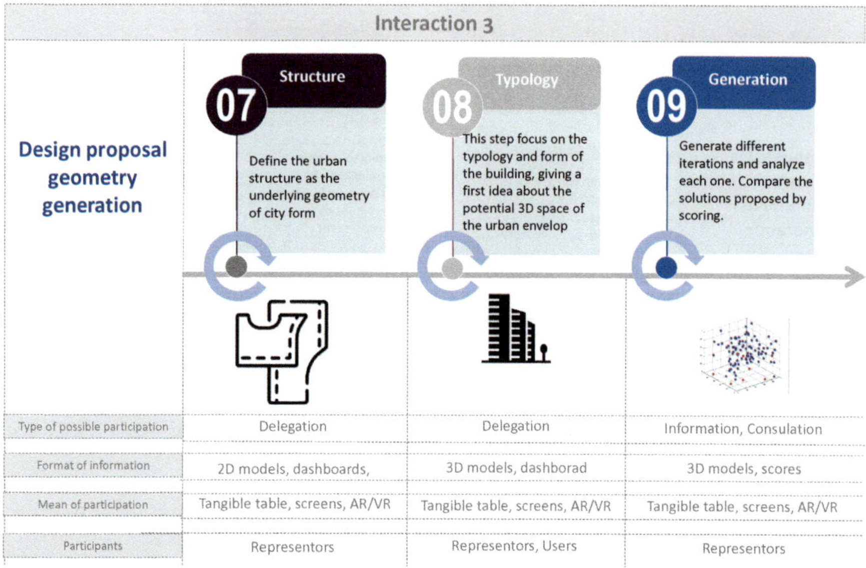

Fig. 7.6 Third interaction of the framework

Third Interaction

This interaction is related to the generation of the urban pattern and buildings' 3D geometry typology. An optimization of the building typology is addressed in this interaction in order to experiment different solutions and alternatives. This interaction enables end-users to explore different types of urban patterns and urban 3D geometry generation. The steps in this interaction are here reported in Fig. 7.6.

Step 7: Urban Structure Pattern

Urban structure is the underlying geometry of city form. This step defines the use pattern in the designing of the street networks and layouts. The professionals based on the input set in the two first interactions define this entry. Recommendations are based on cultural and other predefined parameters.

Step 8: Urban Typology

This step will focus on the typology and form of the building envelope. Different design strategies are developed to enable the creation of the 3D envelope. This step considers different criteria, objectives and constraints. The parameters responsible for the 3D generation are the maximum height and depth of the building, central court, type of the building generation and number of floors.

Step 9: Urban 3D Generation and Exploration

For each iteration of the design, end-users can analyze the performance indicators and compare between different iterations. In this interaction, the participation is enabled by giving the opportunity to understand and accept the solutions proposed. In this step, the generation and optimization of the 3D typologies is based on the requirements set by the end-users and the constraints defined by the experts (steps 1, 2, and 3 from the design framework).

Computer-Based Parametric Tools for Urban Planning

For urban planning to be effective, it is essential to approach it with a strategic lens, in order to clearly set goals, measure progress, and define and execute steps of a project. In this regard, a variety of urban planning software tools are designed to align the planning strategy with progress and support experts in the process of communicating, designing, simulating, analyzing trackable aspects of planning, and reporting.

In this chapter, a deliberate focus is put on the parametric computer-based systems, allowing for the search of solutions instead of drawing and/or simulating it.

To understand the strengths and weaknesses of these software systems, a descriptive and qualitative review of the main existing urban planning software tools is needed. This section addresses the requirements that such systems should meet; gives an overview of the four main existing groups of urban planning software tools, communication tools, designing tools, performance simulation tools, and data analysis tools; and details the devices that can be used to strenghten engagement and participation.

Requirements for Urban Planning Participatory Parametric Tools

Flexibility

The flexibility in the computer-aided parametric tools can potentially facilitate exploring different design strategies and different variations in the design solutions. A flexible design process helps in reducing the top-down design approach, and it can be enabled by giving the possibility to the end-users to set their inputs and compare different solutions through a participative process. The degree of the flexibility to modify and explore different design solution is defined by a certain range for the value of the input parameters. These values should be defined in advance depending on their types.

Easiness to Use and Visualization

From the perspective of the experience of the designed project, the visual exploration in a computer-aided parametric tool should be easy. Hence, the creation of this model can be difficult especially when dealing with complex geometries. Modelers should deliver parametric models in which users only need to manually move sliders to perform variation on the design or proceed to an automatic generation of design solutions.

Moreover, the results (geometries and associated simulations) should be presented in a way that is understandable for non-experts. The numerical values and results of the simulations should be meaningful for them so they can foresee the evaluations and impacts that are presented. To achieve it, software solutions might:

1. Present a comparative assessment of the different design solutions and their performance.
2. Translate the simulation values into more visual understandable indicators with a color range to indicate if the values are acceptable of not.
3. Present the simulation results with graphical icons and representations giving more meaningful understanding of the features of indicators being simulated.

Input and Shared Knowledge

Making the participation accessible in different design scenarios implies:

1. Collecting and understanding the needs and requirements of a group of end-users. These needs can be collected and explored with different devices; these devices enable the end-users to input their requirements and preferences. These tools should not only facilitate the experts to collect the end-users requirements but also integrate these requirements in the design solutions proposed.
2. Creating a shared information and knowledge between different actors to create a transparent design process by allowing the exchange of information and knowledge between experts and end-users in both directions.

Other Architectural Considerations

Deploying such tools on projects obviously faces more usual constraints associated with BIM and GIS tools. Indeed, besides being available, the datasets must comply with requirements associated with the purpose of their usage. In particular, urban planning requires territorial data that can either be public (e.g., cadastral GIS, often

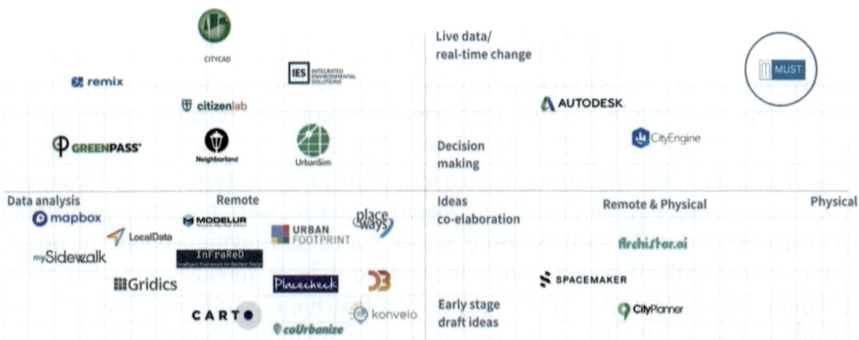

Fig. 7.7 Classification of the tools' positioning (type of consultation/stage of consultation)

providing open data), private (information on properties and land use, energy consumptions), or even data owned by public administrations but not disclosed (e.g., technical services).

It might be mentioned that such systems should rely on various sets of siloed data, thus requiring decentralized system architectures, and offer services via API interfaces connecting the underpinning systems. Moreover, relying on data standards is proven to facilitate this kind of architectures, especially when deployed ad-hoc for a given project. In relation with urban planning, CityGML (cities) and IFC (buildings) standards are applicable.

Existing Software Tools for Urban Planning

More and more systems are available on the market to support urban planning initiatives. This section attempts to describe those software tools developed and accessible in the market. The methodology underpinning this review includes a classification of the main existing software tools in the market, conduct of interviews with relevant stakeholders, and organization of an interactive online seminar to further present and assess the results achieved (Fig. 7.7).

As part of this process, identified tools were classified into four main groups, namely, (1) communication tools, (2) designing tools, (3) performance simulation tools, and (4) data analysis tools. A given tool can belong to several of these groups.

Communication Tools

Guiding an urban development within a new or existing community while considering public and environmental welfare is among the main responsibilities of experts active in planning fields. Convenience, efficiency, equitability, and sustainability are among factors that promote longevity and reduce the risk of disasters in a certain urban context. With urban planning software tools focused on communication among stakeholders, co-design, co-creation, and co-working methods are gaining momentum, resulting in creative approaches that consider broader design aspects from various perspectives. The objectives of these growing communication-based civic engagement/data platforms used in the public planning process revolve around connecting, engaging, reminding, and inspiring stakeholders via different methods while building trust by collaborating across teams and inviting public feedback. Collaborating with citizens, to better understand their preferences and aspirations, sustaining engagement by organizing events, distributing follow-up surveys, and raising awareness all address the unified goal of finding out what a place and its people can tell us.

Communication software tools within urban planning have the power to transform how planning and real estate teams connect with the community by empowering citizens to share feedback with the help of such tools. These two-way conversations are divided into four steps.

1. Idea collection: gathering suggestions per theme in a certain urban context and starting a discussion from there. Aggregating different information in an entire environment and answering the question of – "what happens if you build here? What will the impact be?"
2. Comments and votes: letting citizens react by voting for launched discussions and commenting on the ideas.
3. Surveys and Polls: collecting the feedback of citizens on a list of defined questions.
4. Scenario planning: offering citizens the opportunity to pick between several options that are aligned with the policies of the neighborhood to be further studied and implemented.

Designing Tools

Urban planning software tools operating within the physical design process steps are primarily used to build initial sketches and masses of building forms within an urban environment (e.g., Archistar and Spacemaker in Fig. 7.7). The output of these tools will eventually operate as the foundation of final products such as three-dimensional models, technical details (quantities, drawing sheets, and construction details), presentations, and walkthroughs. Along with the analytical and design skills and experiences of experts, these tools assist the planners in assigning different land uses and calculating urban control indicators such as floor area ratio or the required number of parking lots.

One of the advantages of using designing and modeling is the fast-learning curve and their wide compatibility with other software tools. This group of tools is bundled with additional software or plugins which serve different goals such as enhancing the quality of final product layout, interactive presentations, and real-time visualization and animation software. Furthermore, their compatibility with report creating and plan to graphics converting tools provide advanced design modeling potentials.

Performance Simulation Tools

Another important software tool group in urban planning is performance simulation. Performance simulation is widely implemented in automotive, aerospace, and multiple other industries. However, its adoption in urban planning and construction has been limited compared to the other fields, due to the lack of affordable tools, as well as the specialized knowledge required to successfully use simulation. It is important, however, to emphasize the emergence of cloud-based tools and their strength in reducing such barriers. Fluid flow simulation and finite element analysis hold great promise for the architecture, engineering, and construction (AEC) industry, giving architects, planners, and engineers the ability to predict, optimize, and plan the performance of buildings in the early stages of the design process both independently and within an urban neighborhood.

Efforts to make simulation affordable and accessible for everyone are turning this technology into a no longer impossible task for urban planners. Online platforms specifically designed for performance simulation and analysis of neighborhoods make simulation widely accessible for planners and users. Below are examples of types of analysis that urban planners can achieve with the assistance of performance simulations.

1. Simulate the application of urban policies: urban policies, such as densification (infill) or urban expansion (landfill), transportation systems, reforestation, risk mitigation, energy and water efficiency measures, and construction of new urban services such as schools, hospitals, parks, and other civic spaces can be simulated and measured by the assistance of performance simulation tools focused on policies and their impacts. Such tools identify which policies will help urban areas in reaching their goals, by applying different indicators, including water, energy and land consumption, infrastructure and municipal services costs, intersection and population density, greenhouse gases emissions, green areas per capita, proximity to job opportunities, and urban amenities. Results are organized in sets of presentable scenarios for further data-driven analytics.
2. Optimize thermal comfort: air velocity, temperature, and humidity can be accurately predicted and analyzed with the help of wind flow analysis streamlines and fluid flow simulations, allowing architects, planners, and engineers to visualize the airflow and evaluate temperature gradients, air distribution, or velocity plots.
3. Organizational performance simulation: Combination of design experience with digital technologies forecasts the social, economic, and environmental impacts of development within an urban context. Such simulations result in urban plan-

ning and design strategies that deliver sustainable spatial, land use, and transport networks based on land value modeling and organizational performances.
4. Decision support for urban energy simulation: Such simulations indicate the energy demand of a neighborhood, respecting the occupants' behavior and accounting for a range of commonly used heating, ventilation, and air conditioning systems. Determining the energy supplies issued from renewable sources, including the radiation exchange, is driven by a specific urban context generated by a range of commonly used energy conversion systems.

Data Analysis Tools

Mostly accessible as open platforms, data analysis tools are used by planning professionals and experts active in data science domains. Statistical analysis on geoprocessing models enables users to test and predict the behavior of an urban environment and solve complex scenario problems subjected to different conditions. Such tools use scalable numerical methods that can calculate mathematical expressions despite complex loading, geometries, and material properties.

Population movement, transport modeling based on data science, and big data within an urban context have the potential to present a flexible and transparent approach to enable stakeholders to exploit the potentials of urban networks. Below are examples of areas that data analysis tools are used to increase the resilience and adaptability of infrastructure.

1. Built environment: Assess existing conditions and land use down to the parcel level with urban, environmental, and mobility data.
2. Community resilience: Analyze and intersect vulnerabilities and policy interventions at the city or neighborhood level.
3. Climate and hazard risk: Evaluate the impacts of climate change and natural hazards on urban contexts and infrastructure.

Devices Enabling Participation

Tangible Devices

The applications based on tangible user interface (TUI) require a specific input output device called a tangible table-top device. It is accompanied by a minimization of interactive diversities reduced to the components such as the screen and interaction to what can be achieved with a mouse or keyboard (Ishii 2007). According to Ullmer and Ishii (Ishii and Ullmer 1997), tangible interfaces contribute in giving a physical and virtual form to digital information while using physical artefacts to control the "computational media" (Fig. 7.8). An early example of a TUI from the

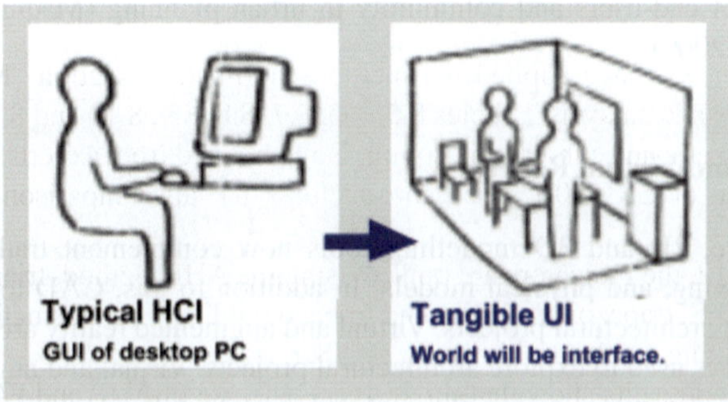

Fig. 7.8 From GUI to tangible user interfaces (Ishii and Ullmer 1997)

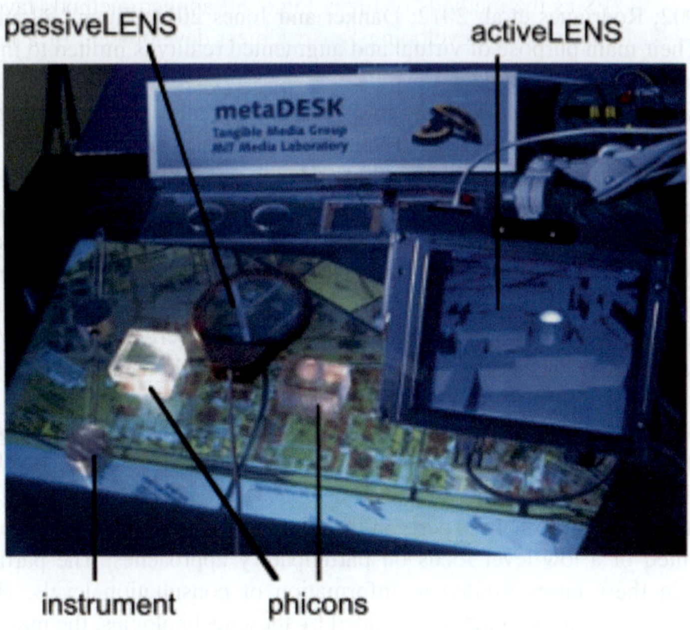

Fig. 7.9 MetaDESK (Ishii and Ullmer 1997)

1990s is the metaDESK (Fig. 7.9) in which the user could interact with a map of the MIT campus.

This technology provides a collaborative environment enabling the visualization of the data on a table-top. Adding physical shapes to the displayed information and to the controllers (sliders, inputs, etc.) enhances the human ability to identify the solutions and to control the related objects (Rodrigues et al. 2012). Digital environments such as tangible table-top combined with physical objects are important tools to improve a transparent design process (Ishii and Ullmer 1997). Tangible table-top systems can be foreseen as a participatory technology that helps to increase the engagement of end-users and community in urban planning (Maquil et al. 2008; Wagner et al. 2009).

Augmented and Virtual Reality

In architecture, 2D and 3D modelling tools now complement traditional paper sketches, drawing, and physical models. In addition to this, CAD tools allow the simulations of architectural projects. Virtual and augmented reality are considered a technique that is used to explore architectural projects. Virtual and augmented reality demonstrated a way to merge physical and digital artefacts by using virtual projection. Augmented and virtual reality started to appear in the architecture, engineering, and construction industry (Ben-Joseph et al. 2001; Morton 2001; Shen et al. 2002; Rodrigues et al. 2012; Danker and Jones 2014; Figen Gül and Halıcı 2015). Their main purpose of virtual and augmented reality is limited to improving the visualization of 3D models.

Nowadays, the urban designers and planners are also experimenting several innovative solutions for city and urban environmental visualization in virtual or augmented reality (Ben-Joseph et al. 2001; Hanzl 2007; Kaftan et al. 2011; Cirulis and Brigis 2013; Chowdhury and Schnabel 2018; Gun et al. 2020). These solutions are providing urban visualization allowing merging real cities with virtual three-dimensional (3D) buildings, with the objective to improve the immersion into urban planning solutions (Cirulis and Brigis 2013). From this perspective, a research project has demonstrated that virtual reality (VR), when properly used, allows a meaningful participation of the community or end-users (Gordon and Koo 2008). Other benefits for the application of VR and AR in urban planning were related to the time saved in the design process and the presentation of realistic solutions (Shen et al. 2002). These research studies showed the potential of computer-aided parametric tools when coupled with other tools and devices (Unity3D, ESRI CityEngine, tangible interface, etc.) to help experts in the decision-making of urban development with limited or a low-level focus on participatory approaches. The participation remains in these cases limited to information or consultation levels. However, despite the visualization function provided by these technologies, the main issue is to find how to use augmented reality to allow participation in the urban process.

Conclusion

In this chapter we have traced the evolution of participation approaches and tools in urban planning and how these aim to address key planning performance indicators at different planning stages. Based on this, we elaborated on different types of tools for civic engagement in urban planning: physical planning tools, physical/civic engagement tools, civic scenario planner tools, and data analysis methods. Major gaps have been identified among these tools, as well as the necessity to bridge them.

Building on this discussion, a framework for participative performance-driven urban planning is introduced, as an innovative method that aims to explore novel forms of citizen involvement within urban planning activities. The results of a benchmark of a wide range of tools using types and stages of consultation and the KPIs they address are then presented. This study revealed the degree to which the digital tools create an inclusive environment to exchange and implement urban planning-related ideas and the extent to which they could lead to an integrated and coherent engaging method for citizen engagement.

The framework starts first with identification of the key components to be included in terms of (1) the design process, (2) the participatory interactions to be implemented in the design process, and (3) the devices and tools to allow the participation. The framework presented in this chapter addresses the need to develop methods and tools to support the performance-driven design process with different levels of participation and end-users' feedback to be integrated in the decision-making processes for urban planning. The review of the existing tools supporting the urban planning have identified a gap between the participatory approach and the urban planning. The existing tools and platforms for urban planning are encouraging a top down design process and only allowing a lower level of participation. The modalities of participation used in architectural practice were mapped with the ladder of participation of Arnstein (1969). In Fig. 7.10, we show the possible modalities of participation used in architectural project with a map to the level of participation allowed by these modalities.

The developed framework should be considered as critical tool aiming to enhance civic engagement in urban planning processes.

This framework aims to break free from prevailing participation going off the top-down and autopilot process. It presents a theoretical and technological

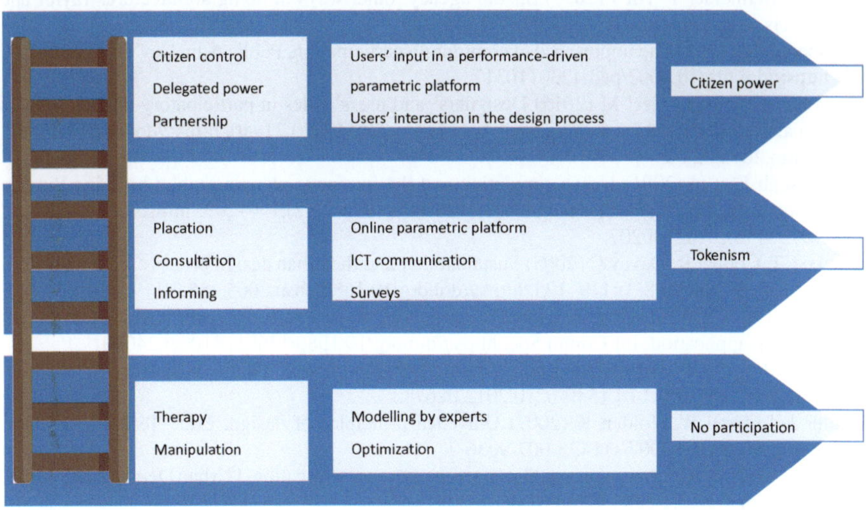

Fig. 7.10 A ladder of participation for performance-driven design. (Adapted from Arnstein (1969))

framework for including end-users' feedback and criteria in the urban planning process while also enabling the integration of social criteria, where the end-users are able to input, adjust, and define their own preferences and requirements in a participative approach.

This framework is a digital shift from a traditional design process to an experience design where the use of different participatory devices allows a better understanding of the design solutions which also keeps a transparent relation between the input and the solutions.

Another aspect enabled by the framework is the creation of a meta-design as a focused "solution space" and iterations that satisfy the needs and requirements of the end-users. The benefit of meta-design model is in the ability given for the designers to accommodate changes and modifications in the design that can be asked by the end-users with less time and effort and a guarantee that the requirements are still considered in the iterations. Another benefit of the meta-design model is the ability given to the end-users to assess, investigate, and compare between different iterations for the design solutions.

References

Abras C, Maloney-Krichmar D, Preece J (2004) DRAFT: user-centered design. Encyclop Human Comp Inter 37(4):1–14. https://doi.org/10.3233/WOR-2010-1109

Al-Kodmany K (1999) Using visualization techniques for enhancing public participation in planning and design: process, implementation, and evaluation. Landsc Urban Plan 45(1):37–45. https://doi.org/10.1016/S0169-2046(99)00024-9

Arnstein SR (1969) A ladder of citizen participation. J Am Plan Assoc 35(4):216–224. https://doi.org/10.1080/01944366908977225

Awan N, Schneider T, Till J (2013) Spatial agency: other ways of doing architecture. Taylor and Francis. https://doi.org/10.4324/9781315881249

Bamberger M (1991) The importance of community participation. Public Adm Dev 11(3):281–284. https://doi.org/10.1002/pad.4230110317

Barcellini F, Prost L, Cerf M (2015) Designers' and users' roles in participatory design: what is actually co-designed by participants? Appl Ergon 50(March):31–40. https://doi.org/10.1016/j.apergo.2015.02.005

Ben-Joseph E et al (2001) Urban simulation and the luminous planning table: bridging the gap between the digital and the tangible. J Plan Educ Res 21(2):196–203. https://doi.org/10.1177/0739456X0102100207

Boyko CT, Cooper R, Davey C (2005) Sustainability and the urban design process. Proc Insti Civil Eng Eng Sustain 158(3):119–125. https://doi.org/10.1680/ensu.2005.158.3.119

Brabham DC (2010) Moving the crowd at threadless: motivations for participation in a crowdsourcing application. Inf Comm Soc. https://doi.org/10.1080/13691181003624090

Bryson JM et al (2013) Designing public participation processes. Public Adm Rev 73(1):23–34. https://doi.org/10.1111/j.1540-6210.2012.02678.x

Butler J, Lidwell W, Holden K (2007) Universal principles of design. Educ Technol Res Dev. https://doi.org/10.1007/s11423-007-9036-7

Carmona M (2010) Contemporary public space, part two: classification. J Urban Des 15(2):157–173. https://doi.org/10.1080/13574801003638111

Carmona M (2015) Re-theorising contemporary public space: a new narrative and a new normative. J Urban 8(4):373–405. https://doi.org/10.1080/17549175.2014.909518

Chowdhury S, Schnabel MA (2018) An algorithmic methodology to predict urban form: an instrument for urban design, (May)

Cimerman B (2000) Participatory design in architecture: can computers help ? In Proceedings of the participatory design conference, (December), pp 40–48

Cirulis A, Brigis K (2013) 3D outdoor augmented reality for architecture and urban planning. Procedia Comp Sci Elsev 25:71–79. https://doi.org/10.1016/j.procs.2013.11.009

Claridge T (2004) Designing S O C I a L C a P I Ta L Sensitive Pa R T I C I Pat I O N, (June)

Cleaver F (1999) Paradoxes of participation: questioning participatory approaches to development. J Int Dev 11(4):597–612. https://doi.org/10.1002/(SICI)1099-1328(199906)11:4<597::AID-JID610>3.0.CO;2-Q

Cockey SM (1955) Software for facilitating the creation of parametric urban resource models. https://doi.org/10.1134/S1063783414100321

Da Silva RC, Morim LM (2010) Parametric urbanism: emergence, limits and perspectives of a new trend in urban design based on parametric design systems. V!Rus (3)

Daniel D, Jane B, Mark B (2011) Untangling parametric schemata: enhancing collaboration through modular programming introduction: why parametric modelling can be difficult. CAAD futures 2011: designing together, ULg, pp 55–68

Danker F, Jones O (2014) Combining augmented reality and building information. In Proceedings of the 32nd eCAADe conference, 2, pp 525–536

Davidoff P (1965) Advocacy and pluralism in planning. Journal of the American institute of planners, 1965. City Read. https://doi.org/10.1080/01944366508978187

De Smet A, Pak B, Schoonjans Y (2019) Solidary mobile housing live project. In: G Boi

Desouza KC, Bhagwatwar A (2014) Technology-enabled participatory platforms for civic engagement: the case of U.S. cities. J Urban Technol. https://doi.org/10.1080/10630732.2014.954898

Dias N, Curwell S, Bichard E (2014) The current approach of Urban Design and its implications for sustainable urban development. Proced Econ Finan Elsev 18(September):497–504. https://doi.org/10.1016/S2212-5671(14)00968-X

Dongyoun S, Muller Arisona S Schmitt G (2011) A crowdsourcing urban simulation platform using mobile devices and social sensing. CAAD futures 2011, proceeding(figure 1), pp 233–246

Evans-Cowley J, Hollander J (2010) The new generation of public participation: internet-based participation tools. Plan Pract Res. https://doi.org/10.1080/02697459.2010.503432

Figen Gül L, Halıcı SM (2015) Collaborative design with Mobile augmented reality. Complexity & simplicity – Proceedings of the 34th eCAADe conference, 1, pp 493–500

Gallo G, Pellitteri G (2018) Luigi Moretti, from history to parametric architecture. In Proceedings of the 23rd Internafional conference on computer-aided architectural design research in Asia CAADRIA, pp 209–214

Gordon E, Koo G (2008) Placeworlds: using virtual worlds to foster civic engagement. Space Cult 11(3):204–221. https://doi.org/10.1177/1206331208319743

Gün A, Demir Y, Pak B (2019) Urban design empowerment through ICT-based platforms in Europe. Int J Urban Sci. https://doi.org/10.1080/12265934.2019.1604250

Gun A, Demir Y, Pak B (2020) An ICT-integrated participatory design model for urban transformation areas under disaster risk: the case of Istanbul. Int J Urban Sust Dev 1(25). https://doi.org/10.1080/19463138.2020.1740707

Hanzl M (2007) Information technology as a tool for public participation in urban planning: a review of experiments and potentials. Des Stud 28(3):289–307. https://doi.org/10.1016/j.destud.2007.02.003

Helsper EJ (2008) Digital inclusion: an analysis of social disadvantage and the information society report

Horelli L (n.d.) New approaches to urban planning

Irvin RA, Stansbury J (2004) Citizen participation in decision making: is it worth the effort? Pub Admin Rev:55–65. https://doi.org/10.1111/j.1540-6210.2004.00346.x

Ishii H (2007) Tangible user interfaces, CHI 2006 workshop, pp 1–17
Ishii H, Ullmer B (1997) Tangible bits, (September), pp 234–241. https://doi.org/10.1145/258549.258715
Kaftan M et al (2011) Informing architecture and urban modeling with real-world data on 3D tangible interfaces and augmented displays, pp 226–233
Kanji N, Greenwood L (2001) Participatory approaches to research and development, pp 1–62
Katz P, Scully V Jr, Bressi TW (1994) The new urbanism: toward an architecture of community. McGraw-Hill, New York
Kelly D (2001) Community participation in rangeland management : a report for the Rural Industries Research and Development Corporation (RIRDC)
Kelly K, Van Vlaenderen H (1995) Evaluating participation processes in community development. Eval Program Plann
Kraft P, Bansler J (1994) The collective resource approach: the scandinavian experience. Scand J Inf Syst 6(1):4
Llewelyn, David D (2007) Urban design compendium: urban design principles. Design 125. https://doi.org/10.1080/00994480.1973.10732231
Maquil V, Psik T, Wagner I (2008) The ColorTable: a design story. Proceedings of TEI'08, pp. 97–104. https://doi.org/10.1145/1347390.1347412
Meeus B, Pak B (2018) Reflections on the counter-mapping of urban "arrival neighborhoods" through Geoweb 2.0 in Brussels and Ghent. In: Participatory design theory, Routledge research in planning and urban design: Routledge. Routledge, New York, pp 40–55. https://doi.org/10.4324/9781315110332-4
Miessen M (2017a) Crossbenching: towards a proactive mode of participation as a critical spatial practice, Doutorado
Miessen M (2017b) Crossbenching subtitle: towards a proactive mode of participation as a critical. University of London
Morton D (2001) Augmented reality in architectural studio learning : how augmented reality can be used as an exploratory tool in the design learning journey, 1, pp. 343–356
Mosadeghi R et al (2015) 'Comparison of fuzzy-AHP and AHP in a spatial multi-criteria decision making model for urban land-use planning. Comp Environ Urban Syst Elsev 49:54–65. https://doi.org/10.1016/j.compenvurbsys.2014.10.001
Moughtin C et al (2003) Urban design: method and techniques. Routledge
Mubita A, Libati M, Mulonda M (2017) The importance and limitations of participation in development projects and Programmes. Eur Sci J 13(5):238. https://doi.org/10.19044/esj.2017.v13n5p238
Muther R, Halles L (2015) Systematic layout planning – a total system of layout planning
Pak B, Verbeke J (2013) Redesigning the urban design studio: two learning experiments. J Learn Des. https://doi.org/10.5204/jld.v6i3.160
Pak B, Chua A, Vande Moere A (2017) FixMyStreet Brussels: socio-demographic inequality in crowdsourced civic participation. J Urban Technol 2(24):65–87. https://doi.org/10.1080/10630732.2016.1270047
Pateman C (1976) Pateman participoation and democratic theory
Price S, Mylius B (1991) Social analysis and community participation: guidelines and activity cycle checklist. Australian International Development Assistance Bureau. Appraisals, Evaluation and Sectoral Studies Branch
Raworth K (2010) State of the world 2003: Progress towards a sustainable society, 20th edn, pp 1–240. https://doi.org/10.4324/9781849776257
Roberts M, Greed C (2001) Approaching urban design: the design process, vol 5. Routledge
Rodrigues F et al (2012) Augmented reality and tangible user interfaces integration for enhancing the user experience. In Proceedings – VRCAI 2012: 11th ACM SIGGRAPH international conference on virtual-reality continuum and its applications in industry, (December), pp. 67–70. https://doi.org/10.1145/2407516.2407536

Ruohomaki T et al (2018) Smart city platform enabling digital twin. In 9th international conference on intelligent systems 2018: theory, research and innovation in applications, IS 2018 - proceedings, (March), pp. 155–161. https://doi.org/10.1109/IS.2018.8710517

Sabri S et al (2019) A multi-dimensional analytics platform to support planning and design for liveable and sustainable urban environment. In International archives of the photogrammetry, remote sensing and spatial information sciences - ISPRS archives, 42(4/W15), pp. 75–80. https://doi.org/10.5194/isprs-archives-XLII-4-W15-75-2019

Saleh M, Al-Hagla K (2012) Parametric urban comfort envelope an approach toward a responsive sustainable urban morphology. Waset Org 6(11):563–570

Sandercock L (1997) Towards cosmopolis: planning for multicultural cities

Sanders EB-N, Stappers PJ (2008) Co-creation and the new landscapes of design. CoDesign 4(1):5–18. https://doi.org/10.1080/15710880701875068

Sanders EB-N, Brandt E, Binder T (2010) A framework for organizing the tools and techniques of participatory design. https://doi.org/10.1145/1900441.1900476

Sanoff H (2000) Community participation methods in design and planning. Landsc Urban Plan 50(4):270–271. https://doi.org/10.1016/s0169-2046(00)00063-3

Saxton GD, Oh O, Kishore R (2013) Rules of crowdsourcing: models, issues, and systems of control. Inf Syst Manag 30(1):2–20. https://doi.org/10.1080/10580530.2013.739883

Sayers M, Trebeck K (2015) The UK Doughnut: a framework for environmental sustainability and social justice, Oxfam Research Reports, (February), p. 76. Available at http://policy-practice.oxfam.org.uk/publications/the-uk-doughnut-a-framework-for-environmental-sustainability-and-social-justice-344550 (Downloaded October 12, 2016)

Schön D (1987) A review of educating the reflective practitioner. Jossey-Bass (San Francisco, 1987) ISBN 1-55542-220-9

Schubert G, Bratoev I, Petzold F (2017) Visual programming meets tangible interfaces - generating city simulations for decision support in early design stages generating city simulations for decision support in early design stages, (September)

Schumacher P (2008) Parametricism as style-parametricist manifesto

Senbel M, Church SP (2011) Design empowerment. J Plan Educ Res. https://doi.org/10.1177/0739456x11417830

Shen BJ, Wu Y, Liu H (2002) Urban planning using augmented reality. Manager 127(3):118–125

Steiniger S, Poorazizi ME, Hunter AJS (2016) Planning with citizens: implementation of an e-planning platform and analysis of research needs. Urban Plann 1(2):49–64. https://doi.org/10.17645/up.v1i2.607

Steinitz C (2012) A framework for geodesign: changing geography by design. Illustrate. Edited by Esri Press

Toker Z (2007) Recent trends in community design: the eminence of participation. Des Stud 28(3):309–323. https://doi.org/10.1016/j.destud.2007.02.008

Towers G (1995) Building democracy. Edited by Universtiy Press

Turner JFC (1963) Architectural design

Vicente MR, Novo A (2014) An empirical analysis of e-participation. The role of social networks and e-government over citizens' online engagement. Gov Inf Q. https://doi.org/10.1016/j.giq.2013.12.006

Wagner, I. et al. (2009) Supporting community engagement in the city: urban planning in the MR tent. In Proceedings of the fourth international conference on communities and technologies, pp. 185–194

Wang W et al (2015) Parametric studies of urban morphologies of high density cities and their air ventilation performance under neutral and unstable atmospheric conditions using advanced large-eddy simulations. In 9th international conference on urban climate jointly with 12th symposium on the urban environment, (July), pp. 1–18

White A (1981) Community participation in water and sanitation: concepts, strategies and methods. IRC International Water and Sanitation Centre, The Hague

Chapter 8
Making Opportunities for Developing Smart Cities Using Artificial Intelligence

Mimica Milošević, Dušan Milošević, and Ana Stanojević

Abstract Smart Cities are one of the non-exclusive layers of modern social policy, circular economy, and urban development and the subject of contemporary society and scientific researchers. The significant lifestyle changes have invoked us to think to build more sustainable and Smart Cities, the ones able to withstand the rapid evolution of our surroundings. This chapter aims to explore key indicators as the opportunities for the development of Smart Cities in a pandemic COVID-19 using artificial intelligence. It provides a theoretical overview of scientific insights into the development and application of the Smart City concept, listing possible obstacles. In the identification of key indicators that are important prerequisites for the development of a Smart City, multi-criteria decision-making (MCDM) were applied: analytic hierarchy process (AHP), triangular fuzzy analytic hierarchy process (FAHP), triangular interval type-2 fuzzy sets (IT2FS), trapezoidal FAHP, and trapezoidal IT2FS as methods of fuzzy logic. Based on six groups of the criterion, and a large number of sub-criteria, the dominant indicators were singled out as the development of the legislative and strategic framework of the Smart City platform and its application in the COVID-19 pandemic circumstances, as well as the standardization of ICT and ICT management. The presented model could help in the policy-making process as the starting point of discussion between stakeholders, as well as citizens in the final decision of adoption measures and best-evaluated options.

Keywords Smart Cities · Artificial intelligence · Multi-criteria decision-making

M. Milošević (✉)
Faculty of Business Economics and Entrepreneurship, Belgrade, Serbia

D. Milošević
Faculty of Electronic Engineering, University of Niš, Niš, Serbia
e-mail: dusan.milosevic@elfak.ni.ac.rs

A. Stanojević
Faculty of Civil Engineering and Architecture, University of Niš, Niš, Serbia
e-mail: ana.stanojevic@gaf.ni.ac.rs

G. C. Lazaroiu et al. (eds.), *Holistic Approach for Decision Making Towards Designing Smart Cities*, Future City 18,
https://doi.org/10.1007/978-3-030-85566-6_8

Introduction

Although the centers of culture, science, and education, cities are increasingly mentioned in a negative context compared to rural areas. They are in the public focus, primarily due to the problems that accompany urban environments – from being the biggest environment polluters to being the biggest energy consumers (Moser et al. 2014). The increased needs of contemporary man in the context of comfort, high standard of living, and financial security have significantly contributed to the concentration of population in cities, shifting from rural to urban areas, but also attracting migration to economically developed areas, even outside the borders of the country where people used to live. The consequences caused by the continuous urbanization process, neglect of care for the surroundings, and irrational consumption of resources, some of which endanger the quality of the environment, are just some of the problems that urban areas face (Wang et al. 2020). Rigorous lifestyle changes have prompted us to consider innovative ways in creating a more sustainable society, able to withstand the rapid development of our environment. Qualitative management based on sustainable strategies, accountability, transparency, citizen participation, and reducing emissions, energy efficiency, waste management, and mobility are significant factors in ensuring sustainable progress (Salvioni and Almici 2020).

The Smart City concept is one step toward it. Smart Cities are a subject of contemporary society and scientific researchers since they have become one of the non-exclusive layers in modern social policy conducting, circular economy, and urban development. The Fourth Industrial Revolution and digitalization are a prerequisite for all planned activities in the managing framework of urban environments, their sectors, and infrastructure. In the search for solutions that will accelerate the monitoring of urban processes, improve infrastructure systems, and simplify daily activities, information-communicative technologies (ICT) are becoming an important tool in creating future technical patents and smart grids (Monzon 2015).

The planet has become a global market in which countries compete in innovation and sustainability of created technological products, personnel expertise, scientific achievements, and strong companies. The Smart City model varies from country to country, depending on the different political, social, and economic factors. While developed countries are proud of inventions that use sensors and artificial intelligence, require almost no manual human control, and make extensive use of available resources, many developing countries are still nowhere near strategies that accurately implement the Smart City concept in future urban development.

The whole world is currently actively committed to the fight against the pandemic caused by the COVID-19 virus, and this disorder temporarily reduces development plans in creating smart environments. Simultaneously, the pandemic is increasingly fostering reflections on developing existing digital infrastructure more innovatively and the search for smart solutions that will help, both in the long-term way to eliminate the virus and in the short-term when it comes to treating patients. The pandemic has shifted boundaries in the way we think about sustainable future development and has significantly changed priorities in creating sustainable living environments.

This chapter examines the opportunities for developing Smart Cities using artificial intelligence. It provides a theoretical overview of current scientific insights on the development and implementation of the Smart City concept, providing a discussion of possible barriers. The research tries to identify crucial indicators as significant prerequisites for Smart City development using multi-criteria analysis. Given the complexity of the problem we are investigating and its multidimensionality, we opted for the application of the AHP method, triangular FAHP, triangular IT2FS, trapezoidal FAHP, and trapezoidal IT2FS as the methods of fuzzy logic.

The chapter is structured as follows. In the next section, the Smart City concept from idea to reality, defining indicators for the formation of Smart City, is presented. The third section gives the relation between Smart City and artificial intelligence. The fourth section assesses new COVID challenges for urban areas and COVID as a catalyst for Smart Cities. The fifth chapter deals with the methodology of fuzzy numbers and analysis, a tool of artificial intelligence in the selection of key indicators. The sixth chapter gives numerical results, and the seventh contains discussion and recommendations. The conclusion is in the last section.

Smart City Concept: From Idea to Reality

In response to the growing challenges of urbanization, the Smart City movement connects urban theories with new scientific and technological achievements, offering solutions to contemporary urban problems. Developed in the 1990s in the USA, the concept was first mentioned in the context of "growing cities in a smart way" and promoting the "compact city" model to prevent agglomeration growth of urban spaces and develop an awareness of environmental issues (Wey and Hsu 2014). In that sense, experts dealing with urban planning have recognized important indicators in creating new and remodeling existing urban areas, which are still considered the main postulates of urban development (Kotharkar et al. 2014; Al-Shouk and Al-Khfaji 2018):

- The formation of mixed-use urban centers
- Differentiating motor traffic from a pedestrian in a narrow city center and other public areas promoting pedestrian movement and the use of bicycles as a means of transport, promoting healthy living and reducing emissions
- Socioeconomic diversity of the population with planning that offers all social classes the same urban conditions and facilities, without the possibility of gentrification
- Efficient use of all resources.

Linking the epithet "smart" to the city, as the core from which sustainability must be drawn, happened in 1994 (Dameri and Cocchia 2013). Since the EU started using the "smart" label in 2010 to qualify sustainable urban development projects, the concept has moved to a wider application (Susantia et al. 2016). "Smart" devices, applications, roads, phones, lights, and buildings have become part of our everyday life. Internet search has completely suppressed the importance of libraries as institutions (Koca-Baltić and Momčilović-Petronijević 2020), sensors built into phones inform us about weather-meteorological conditions, applications for free parking space, the arrival time of public transport, and traffic route that is less loaded (Park et al. 2018).

Technological development has pushed the boundaries when thinking about smart sustainable cities. Innovative solutions have become a tool to gather the information that would help detect current urban problems. Besides, they can be a successful tool to improve urban sectors and infrastructure. With the notion of a Smart City, artificial intelligence has developed to justify the innovative solutions, reached with the creativity and knowledge of man, but also with the application of devices and robotics (Voda and Radu 2019). Smart City has materialized in the film industry and seen through the automatization, the application of aircraft, and the replacement of humans with robotic power.

The interpretation of a Smart City in the literature is different – from representing entities completely dependent on modern technologies to a form of sustainable city that has a clearly defined planning strategy, quality management, good communication between government and citizens, and the "intelligent" way for self-improvement and resilience (Stratigea 2012; Milošević et al. 2021a). There is no precise definition of a Smart City. According to Hall (Mosannenzade and Vettorato 2014), "a city that monitors and integrates conditions of all of its critical infrastructure, including roads, bridges, tunnels, rail, subways, airports, seaports, communications, water, power, even major buildings, can better optimize its resources, plan its preventive maintenance activities, and monitor security aspects while maximizing services to its citizens" can be considered as the smart one. Generally accepted (Li et al. 2019), "the city can be considered as smart when investments in human and social capital, and traditional (transport) and modern (ICT) communication infrastructure fuel sustainable economic growth and high quality of life, with a wise management of natural resources through participatory governance."

According to different authors and studies, a Smart City represents:

- A global movement developed parallel with the Fourth Industrial Revolution and reflected largely in the scientific field of urbanism and spatial planning (Bibri 2019).
- A city, usually of medium size (100–500 k inhabitants) (Lazaroiu and Roscia 2012), created by various actors: government, public-private partnership, investors, IT and other types of companies, and scientists but also the citizens themselves.
- A sustainable system that improves the quality of life in the city, urban water, energy, transport, construction, and infrastructure management sectors (Talari et al. 2017).

- A product of the Internet of Things (IoT) platform, which uses big data, artificial intelligence platforms, sensors, and applications to improve existing urban environments (Yadav et al. 2019).
- Livable urban area, with a high degree of resistance to current changes (Milošević et al. 2019). Technological innovations within the development of a Smart City in different segments of everyday life can be systematized as follows (Milošević et al. 2020a; Milošević et al. 2017; Milošević et al. 2021b):
- *Technological innovations in education* – the creation of online educational platforms for distance learning, digital book and lecture formats, online knowledge testing.
- *Technological innovations in healthcare* – the creation of an integrated information system, development of telemedicine, providing medical care online, monitoring the patient's condition using GPS bracelets.
- *Technological innovations in housing* – construction of "smart" homes, application of "smart" energy-saving devices, kinetic furniture, control of heating, cooling, air conditioning systems using sensors.
- *Technological innovations in mobility* – GPS applications for tracking traffic jams, available parking spaces, and public transport, automated starting of motor vehicles, "smart" transport logistics, and "smart" services.
- *Technological innovations in architecture* – digitization of cultural heritage, application of augmented reality software to investigate the characteristics of monuments that were previously demolished, application of modern software for detection of endangering the stability of existing structures, application of GIS and BIM systems for preserving and restoring architectural heritage in urban areas.
- *Technological innovations for environmental monitoring* – monitoring of water and energy consumption, innovative systems for the conversion of one form of energy into another.
- *Technological innovations in culture and tourism* – virtual sightseeing, digital presentation of the tourist offer.
- *Technological innovations in governing* – e-government and e-portals for communication with citizens, digital public administration, and services. The development strategy is a precondition for transforming the existing city into a "smart" one. In creating a platform for Smart City development and finding new opportunities to overcome barriers, there are different levels at which "smart" solutions can be implemented. Degree from general to more specific is proposed, starting from the development of the real estate, through the development of basic infrastructure, to improving life services and lifestyle.

Levels represent an attempt to introduce a hierarchical approach to the development of a Smart City concept as follows (Milošević et al. 2019):

- Development of real estate – includes implementation of spatial and urban planning, considers the needs of citizens, geolocation plans, and urban development of the city in terms of basic needs for accommodation and life.

- Development of basic infrastructure – the task is to develop the infrastructure for contemporary life and business, electricity, water, CT infrastructure, gas, sewerage, and road network.
- Development of "smart" infrastructure – the task is to add sensor systems and ICT, so the original infrastructure is transformed into a "smart" infrastructure that is capable to collect and analyze all the data for higher levels of management.
- Improving life service – development of classic and "smart" infrastructure and opportunities that can provide advanced services for citizens and organizations in the urban area.
- Improving lifestyle – a high standard of living provides low levels of pollution, acceptable economic costs, and higher living needs, as well as meeting individual needs with lifestyle, art, and culture

Defining Indicators for the Formation of Smart City

According to Griffing (Giffinger et al. 2007), a Smart City defines six dimensions: "smart" governing (A), "smart" economy (B), "smart" citizens (C), "smart" livability (D), "smart" mobility (E), and "smart" environment (F). Each dimension of a Smart City refers to several indicators that must be taken into account when creating Smart City development strategies.

Smart governing (A) refers to indicators related to planning and the legal framework for the implementation of management procedures. This group of indicators includes various forms of management and coordination of processes and actors in the city, with continuous monitoring of problems, and work on attempts to combat them and overcome them by new plans and strategies. Smart governing implies collaboration between different actors at different levels of the hierarchy, of which the most important is the relationship between government and citizens and the development of governance models that combine centrally defined regulation with actions and citizen participation (Glasmeier and Christopherson 2015; Mellouli et al. 2014).

It also refers to management in the ICT sector and the development of design and implementation of innovative solutions, as well as enabling transparent access to all data and services by citizens. Participatory decision-making within the framework of urban planning is extremely important for the management of smart environments. These factors are introduced as follows:

- Development of a legislative and strategic framework of Smart City platform and its implementation in COVID-19 pandemic circumstances – A_1
- Standardization of ICT and ICT management – A_2
- Public-private stakeholder partnership with active citizen's participation – A_3
- Higher data openness and transparency – A_4

Advances in technology follow the economic progress of the area. Smart economy (B) as one of the drivers of Smart City development in developing countries indicates economic factors related to financial analysis of investments and returns. Market growth, its productivity, and transformation, entrepreneurship, primarily created in the field of innovation and development of the IT sector, as well as various forms of financing from state and private budget funds aimed at designing and implementing new solutions and placing them on the national and international market, are the subject of numerous studies (Anand and Navio-Marco 2018; Sujata et al. 2016). Smart economy obtains:

- Entrepreneurship and innovation – B_1
- ICT sector development and job opportunities within it – B_2
- Higher funding for design and implementation of local and national "smart" solutions and initiatives – B_3
- Higher commercialization of innovative technologies assessment – B_4
- Higher technology competition on the national and international market – B_5
- Higher external funding for the Smart City platforms – B_6
- Development of e-commerce and e-business platforms – B_7
- Flexibility and market transformations – B_8

For the development of *smart governing* and *smart economy*, human capital is recognized through a group of indicators of *smart citizens* (C). Smart Cities need smart citizens who live in harmony with technology. The studies most often emphasize the importance of citizen participation in the activities of the urban area to create some support for current strategies for the development of a Smart City (Batty et al. 2012; Monfaredzadeh and Krueger 2015). The following indicators related to smart citizens were selected:

- Higher awareness level in the community – C_1
- High level of qualification and education – C_2
- Expressed flexibility, creativity, and openness to innovation – C_3
- Public confidence in modern solutions – C_4
- Higher citizen engagement – C_5
- Social and ethnic diversity – C_6

Smart livability in Smart Cities includes improved standards in many segments of everyday life (Lin et al. 2019; Sofeska 2017) and relates to:

- Personal safety – D_1
- Affordable housing – D_2
- Utilities, resource availability, and infrastructure equipment – D_3
- Job opportunities for all – D_4
- Improvement of the health, education, tourism, and culture sectors – D_5
- Regulation of privacy data – D_6
- Social integration – D_7

The smart mobility should include indicators that indicate cooperation and interconnection of all available modes of transport and infrastructure, fast exchange of information and data, and complete user orientation. They include (Cassandras 2017; Stephanedes et al. 2019):

- ICT infrastructure integration – E_1
- An innovative transportation system that favors non-motorized vehicles – E_2
- Local and international accessibility – E_3

Concern for the environment indicates various indicators necessary for sustainable development, as well as environmental imperative in the development of Smart Cities (Chen and Han 2018; Dostal and Ladanyi 2018). Smart environment includes different ways for energy-saving and natural resources protection:

- Sustainability consideration – F_1
- Environment protection and quality monitoring – F_2
- Urban recycling – F_3
- Use of renewable energy sources – F_4
- Construction of energy-efficient and smart facilities – F_5
- The decrease in energy consumption associated with new technologies – F_6
- Management and protection of the natural resource – F_7

Defining the Success for the Development of a Smart City

The successful transformation of cities into smart ones, and the further development of cities that already have the epithet smart, will result if the significant success factors of digital transformation are used, namely, people, processes, and technologies. The concept of a Smart City exists in recent decades, but so far, only a few ideas reach the stage of realization in terms of the public service that the city provides to its citizens. There are many reasons, but the omissions have mainly been attributed to infrastructure mismatches and legal barriers. It is necessary to hold wide-ranging postulates for any initiative, in the concept of a Smart City, to overcome this practice (Milošević et al. 2019):

- It is necessary to continuously engage citizens in creating and establishing a vision, using digital technologies, social networks, media.
- Wishes should not replace the defined project framework, and it is necessary to set measurable goals and choose appropriate indicators.
- The lean methodology should be applied wherever possible, especially if the development of new expansive infrastructure is needed.

Guided by Smart City postulates, it is necessary to harmonize the ambitions of local authorities. The process can be successful if the following conditions are obligatory tasks:

- It is necessary to choose the relevant indicators of Smart City development. This choice determines the quality of the results of the transformation process achieving benefits for citizens.

- In the Smart City establishing process, the responsibility and the ability to make quick decisions are capital to succeed. The initiative must have a powerful political consensus within local authorities.
- It is necessary to define the goals and enable continuous evaluation and quantification.
- The public-private cooperation model, especially of society, universities, and local government, could be a success guarantee when identifying the mutual benefits, and the economic effects are significant.
- In practice, the use of opportunities for Smart City development and long-term sustainability of all planning segments have to be achieved.
- Citizens as service users can provide valuable data, ideas, and comments such they are significant participants in a Smart City.

There are three steps to implementing a Smart City strategy moving forward:

1. They need to have a vision of what they want to deploy.
2. They have to make sure they have a monetization plan as to which areas they want to digitize: schools, universities, hospital.
3. They have to understand how to monetize this.

There is no need to constantly define Smart Cities and continue with the topic if it includes resilience or learning. It is necessary to focus on supporting government business and meet all kinds of economic, environmental, and social goals, using technology in the best way.

Smart City and Artificial Intelligence

Artificial intelligence (AI) is increasingly present in everyday life, which facilitates an accelerated lifestyle. In some routine tasks, AI makes life simpler and can make daily life in the city relieve, while consistent implementation and acceptance by the citizens are necessary. In the age of uncertainty and complexity that lies ahead, the continued adoption of AI is expected to continue and thus its impact on the sustainability of our cities (Yigitcanlar and Cugurullo 2020). The benefits of artificial intelligence are becoming more and more apparent thanks to the development of Smart Cities. The synergy of artificial intelligence and Smart City leads to the improvement of city life and the facilitation of human lives because that is their common goal. People are not even aware of how much artificial intelligence is present in their lives. A large amount of data generated in a Smart City would be useless without the use of AI. AI processes and analyzes data generated from machine-to-machine interaction in the context of Smart Cities, intelligent stores, and city infrastructure. There are countless smart apps in the city where AI can play a significant role: from improving traffic through smart parking management to the secure integration of autonomous vehicles and shuttles. Artificial intelligence is one of the main segments of Smart Cities and raises the city to a whole new level. AI collects a large amount of data to make recommendations, predict future events, and help make decisions. The future of Smart Cities is expected to be closely linked to AI use and will be present in almost all areas of the urban community, from traffic situations to citizen safety. Some research shows that respondents perceive artistic intelligence as a significant aspect influencing Smart City development (Voda and Radu 2018). The technology used has intended for people who use the technology, and the user experience is not universal, but everyone experiences it differently. The goal is for citizens to use the devices regardless of their education and technological knowledge. Significant opportunities are the intersection of AI development with Smart City development in theory and practice (Khan et al. 2018).

Big data as a part of artificial intelligence contributes to making the next decisions, determining the necessary resources, and identifying critical places thanks to the large amount of data it collects. Thanks to using AI in Smart Cities, it makes life easier for people; saves electricity; reduces exhaust fumes, the number of traffic accidents, and the number of thefts; and contributes to the quality of life in various areas. The application of artificial intelligence must be reliable. To safely implement AI respecting ethical principles and law, it is necessary to meet some requirements. The first requirement is human action and supervision, which provides that the system does not compromise human autonomy. It is imperative to provide an opportunity for people to intervene, although it is not always possible and desirable. The systems must be safe, accurate, and reliable, and technical stability and security should ensure safety checks at all times. Privacy and data management have guaranteed at all stages of the AI life cycle, and access to data involves a certain level of control. Transparency allows the user to explain the work of artificial intelligence. The application of artificial intelligence must affect social and environmental well-being in such a way as to promote the sustainability and ecological responsibility of the system and must take into account social performance as a whole.

COVID as a Catalyst for Developing Smart Cities

In 2020 the planet faced a new challenge that affects cities and completely changed our lives. The global pandemic caused by the COVID-19 virus forced us to make changes. Social distance has caused a shift from face-to-face to online communication and affected various aspects of urban life. A huge number of people all over the planet work from the space of their homes, while students continue their education mostly on the principle of distance learning. Living spaces are becoming working. It changes people's daily habits and creates the need for better organization of space and activities. The lack of socialization, as well as the impossibility of intimacy-closeness, significantly disturbed the mental health of people, affecting the personal dissatisfaction of citizens, but also insecurity and increased fear of health and life.

Most countries are in lock-down, closing their borders to visitors and introducing a limited movement of citizens. Restaurants, hotels, malls, spas, gyms, theaters, and museums have been out of order for months. This significantly influenced the economic downturn, causing job losses and the bankruptcy of smaller family businesses. Many states have also limited the number of people who gather outdoors, so city parks, squares, and other public areas are almost empty. The public transport reduces, while a healthy lifestyle, walking, and cycling are increasingly being promoted, raising the question of the future creation of urban solutions for public spaces. The need to create virtual activities that would occupy people's free time is recognized.

Developing Smart Cities are facing economic recovery, considering the huge budgets currently focused on helping citizens, small businesses, and the healthcare system, investing in medical equipment, drugs, and vaccines to immunize the population (Allam and Jones 2020). The reduction of the finance aimed at the technology innovation projects must be overcome by a more precise urban strategy. The question of the real possibilities for implementation of the sustainable "smart" solution in nearly future arise. Table 8.1 gives an overview of COVID challenges for Smart City development.

Firstly, it seemed that the concept of a Smart City threatens to stagnate in a COVID pandemic. Many governments have used the global situation as an advantage. Moving jobs to living spaces, and shifting business meetings, sales, shopping, education, entertainment, information to the online mode, there is a need to improve existing digital technology systems and develop new smart solutions. Paradoxically, social isolation has necessitated digital connectivity. Work and studying from home have caused the need for more online data, fewer system failures, faster networking, secure systems, easily accessible web portals, and high-quality available data (Abusaada and Elshater 2020).

COVID pandemic demonstrates how cities are resilient in crises when the existing database, information, and digital infrastructure enable. An increasing number of ICT companies offer platforms for learning and communication between

Table 8.1 COVID challenges for Smart City development

Governing	**Economy**
Limited access to certain services	Financial crisis
The need for different city management	Job loss/bankruptcy
The need for stronger ICT management	Redirecting funds to healthcare
Mobility	**People**
Reduced public transport	Increased need for ICT professionals
Reduced international transport	The need for health workers
The need for urban solutions that promote pedestrian and bicycle traffic	The need for new scientific advances
Livability	**Environment**
A sense of fear and insecurity	The need to preserve natural environments
The need to reorganize the healthcare and education sectors	Development of smart sustainable solutions
The need to renew culture and tourism	
Lack of social integration	

teachers and students, for business and other meetings, overcoming the barrier set by the lack of social gathering and travel. "Smart" applications are used around the world as a form of sending information to citizens when they may go outside their homes, as a way of communication between citizens and police authorities. Due to the impossibilities or limited possibilities of travel, but also visits to museums, theaters, and other cultural institutions, digitalization has enabled online transmission of numerous performances, film premiers, and virtual tours of museums and famous buildings around the world from the comfort of home.

COVID has become a powerful catalyst for the development of a Smart City. The application of "smart" solutions in the direct suppression of the pandemic is of importance. Artificial intelligence has enabled the production of "smart helmets" equipped with thermal cameras to detect potentially infected in the United Arab Emirates (Smart City Transformation in a post-COVID world 2020). The Australian government has launched the COVIDSafe App to track the contacts of the infected (Smart City Transformation in a post-COVID world 2020). Similar applications have been developed in the United Kingdom (England, NHS Test and Trace; Scotland, Test and Protect; Wales, Test, Trace, Protect; Northern Ireland, contact tracing service) (https://www.adalovelaceinstitute.org/blog/uk-contact-tracing-apps-the-view-from-northern-ireland-and-scotland/ n.d.). These applications use the principle of informing the citizens on a positive test, by asking them to log in to the application, leaving data and names of all persons with whom they had close contact. Many other platforms created by the urban health system management sector enable interactive daily monitoring of cases of infections, making online reports with data streamed on case numbers, age, gender, place of residence, and vaccination system, as well as the number of recovered, dead, and tested.

The pandemic has become a polygon for testing technological possibilities and moving the existing boundaries of digitalization. Due to the citizen's distrust of governance and new technologies, there is a need to protect personal data. At the same time, we must not neglect the transparency of events and the engagement of citizens in the management of cities. On the other hand, the impossibility of meeting and talking live and discussing current issues may still be detrimental to the development of Smart City strategies, so there is a need for these projects to be fully digitized. Many online urban and architectural competitions encourage citizens to give their vision of post-COVID urban environments and thus offer new urban, architectural, technological, and other types of solutions that nurture partial social distancing, without reducing the quality of life in cities.

The pandemic has provided a new perspective on the way we plan our infrastructure. It depends on our abilities to re-invent and refine Smart City models to adapt to the precarious situation we are facing today. One option is an upgraded Smart City model – Smart^p City (p = pandemic ready). Smart City concepts must consider the following ways to resolve an uncertain situation:

- Touchless delivery of goods and food items
- Smart supply chain set up
- Touch-free inspection of patients

- Wireless inspection of civic violations
- Tackling cybercriminals and fake news
- Online education and examination infrastructure
- Touch-free sanitation and waste management
- A blockchain-enabled citizen tracking system
- E-voting infrastructure
- Smart intensive care units and isolation wards

The flexibility of the Smart City model is imperative. If the upgraded models are not synchronized and fail to adapt to new situations, the overall effort will fail further.

Methodology: Fuzzy Numbers and Analysis

In this paper, in addition to the crisp numbers used in the Analytic Hierarchy Process (AHP) method (Saaty 1980; Selimi et al. 2018; Milošević et al. 2016), triangular and trapezoidal fuzzy numbers are used when applying the Fuzzy Analytic Hierarchy Process (FAHP) (Srdjević and Medeiros 2008; Kahraman et al. 2014). The fuzzy number is a special fuzzy set $F = \{(x, \mu_F(x)), x \in R\}$ where $\mu_F(x)$: $R \rightarrow [0, 1]$ is a continuous function. The triangular fuzzy number (TFN) is denoted with $\tilde{T} = (l, m, u)$ (Milošević et al. 2020b). The membership function for TFN is:

$$\mu_F(x) = \begin{cases} \dfrac{x-l}{m-l}, & x \in \{l, m\} \\ \dfrac{u-x}{u-m}, & x \in \{m, u\} \\ 0, & otherwise. \end{cases} \tag{8.1}$$

In the trapezoidal FAHP method, trapezoidal fuzzy numbers (TrFN) denoted with $\tilde{M} = (l, m^l, m^h, u)$ are used (Ozkok 2019). The corresponding membership function is now

$$\mu_F(x) = \begin{cases} \dfrac{x-l}{m^l-l}, & x \in (l, m^l) \\ 1, & x \in (m^l, m^h) \\ \dfrac{u-x}{u-m^h}, & x \in (m^h, u) \\ 0, & otherwise. \end{cases} \tag{8.2}$$

If $m^l = m^h$, the trapezoidal fuzzy number is reduced to the triangular fuzzy number.

The laws for operations for an arbitrary two trapezoidal fuzzy numbers $\tilde{M}_1 = (l_1, m_1^l, m_1^h, u_1)$ and $\tilde{M}_2 = (l_2, m_2^l, m_2^h, u_2)$ are like in (Oztaysi 2015):

$$\tilde{M}_1 \oplus \tilde{M}_2 = (l_1 + l_2, m_1^l + m_2^l, m_1^h + m_2^h, u_1 + u_2) \quad (8.3)$$

$$\tilde{M} \Theta \tilde{M}_2 = (l_1 - u_2, m_1^l - m_2^h, m_1^h - m_2^l, u_1 - l_2) \quad (8.4)$$

$$\tilde{M}_1 \odot \tilde{M}_2 = (l_1 \cdot l_2, m_1^l \cdot m_2^l, m_1^h \cdot m_2^h, u_1 \cdot u_2) \quad (8.5)$$

$$\tilde{M}_1 \oslash \tilde{M}_2 = (l_1/u_2,\ m_1^l/m_2^h,\ m_1^h/m_2^l,\ u_1/l_2) \quad (8.6)$$

$$k\tilde{M}_1 = (kl_1, km_1^l, km_1^h, ku_1) \quad (8.7)$$

$$\sqrt[n]{\tilde{M}_1} = \left(\sqrt[n]{l_1}, \sqrt[n]{m_1^l}, \sqrt[n]{m_1^h}, \sqrt[n]{u_1}\right) \quad (8.8)$$

The operations for triangular fuzzy numbers are similarly defined.

Interval type-2 fuzzy sets (IT2FS) have been introduced in which the required number of computational operations is significantly reduced compared to T2FS while maintaining the good features possessed by T2FS. This chapter presents a comparative study of the FAHP with both independent approaches.

Type-2 fuzzy number (T2FN) is defined as a fuzzy set $G = \{((x, u), \mu_G(x, u))\ \forall\, x \in X, \forall u \in I_x \in [0, 1], 0 \le \mu_G(x, u) \le 1\}$ where I_x denotes an interval in [0,1] (Chen and Lee 2010). Interval type-2 fuzzy number (IT2FN) is a special case of T2FN when the membership function is $\mu_G(x, u) = 1$.

The trapezoidal IT2FN is represented with

$$\tilde{\tilde{M}} = \left(\left(\tilde{M}^U : H_1\left(\tilde{M}^U\right), H_2\left(\tilde{M}^U\right)\right), \left(\tilde{M}^L;, H_1\left(\tilde{M}^L\right);, H_2\left(\tilde{M}^L\right)\right)\right), \quad (8.9)$$

where $\tilde{M}^U = \left(l^U, \left(m^l\right)^U, \left(m^h\right)^U, u^U\right)$ and $\tilde{M}^L = \left(l^L, \left(m^l\right)^L, \left(m^h\right)^L, u^L\right)$ are TrFNs, while $H_1\left(\tilde{M}^U\right)$, $H_2\left(\tilde{M}^U\right)$, $H_1\left(\tilde{M}^L\right)$ and $H_2\left(\tilde{M}^L\right)$ represent the middle left and middle right vertex heights of the upper and the lower trapeze, respectively (see Fig. 8.1). Heights $H_1\left(\tilde{M}^U\right)$, $H_2\left(\tilde{M}^U\right)$, $H_1\left(\tilde{M}^L\right)$ and $H_2\left(\tilde{M}^L\right)$ belong to the interval [0,1]. For two trapezoidal IT2FNs, $\tilde{\tilde{M}}_1 = \left(\left(\tilde{M}_1^U; H_1\left(\tilde{M}_1^U\right); H_2\left(\tilde{M}_1^U\right)\right), \left(\tilde{M}_1^L; H_1\left(\tilde{M}_1^L\right); H_2\left(\tilde{M}_1^L\right)\right)\right)$ and $\tilde{\tilde{M}}_2 = \left(\left(\tilde{M}_2^U; H_1\left(\tilde{M}_2^U\right); H_2\left(\tilde{M}_2^U\right)\right), \left(\tilde{M}_2^L; H_1\left(\tilde{M}_2^L\right); H_2\left(\tilde{M}_2^L\right)\right)\right)$ arithmetic operations are as follows:

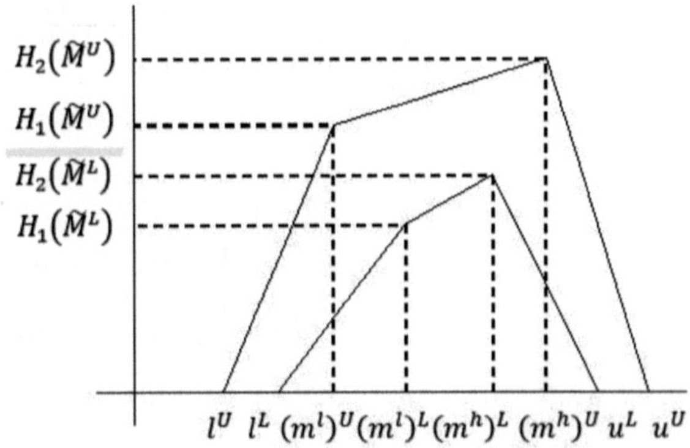

Fig. 8.1 Graphical representation of trapezoidal IT2FN

$$\widetilde{\widetilde{M}}_1 \oplus \widetilde{\widetilde{M}}_2 = \begin{pmatrix} \left(\tilde{M}_1^U \oplus \tilde{M}_2^U;, \min\left(H_1\left(\tilde{M}_1^U\right), H_1\left(\tilde{M}_2^U\right)\right);, \min\left(H_2\left(\tilde{M}_1^U\right), H_2\left(\tilde{M}_2^U\right)\right)\right), \\ \left(\tilde{M}_1^L \oplus \tilde{M}_2^L;, \min\left(H_1\left(\tilde{M}_1^L\right), H_1\left(\tilde{M}_2^L\right)\right);, \min\left(H_2\left(\tilde{M}_1^L\right), H_2\left(\tilde{M}_2^L\right)\right)\right) \end{pmatrix} \tag{8.10}$$

$$\widetilde{\widetilde{M}}_1 \Theta \widetilde{\widetilde{M}}_2 = \begin{pmatrix} \left(\tilde{M}_1^U \Theta \tilde{M}_2^U;, \min\left(H_1\left(\tilde{M}_1^U\right), H_1\left(\tilde{M}_2^U\right)\right);, \min\left(H_2\left(\tilde{M}_1^U\right), H_2\left(\tilde{M}_2^U\right)\right)\right), \\ \left(\tilde{M}_1^L \Theta \tilde{M}_2^L;, \min\left(H_1\left(\tilde{M}_1^L\right), H_1\left(\tilde{M}_2^L\right)\right);, \min\left(H_2\left(\tilde{M}_1^L\right), H_2\left(\tilde{M}_2^L\right)\right)\right) \end{pmatrix} \tag{8.11}$$

$$\widetilde{\widetilde{M}}_1 \odot \widetilde{\widetilde{M}}_2 = \begin{pmatrix} \left(\tilde{M}_1^U \odot \tilde{M}_2^U;, \min\left(H_1\left(\tilde{M}_1^U\right), H_1\left(\tilde{M}_2^U\right)\right);, \min\left(H_2\left(\tilde{M}_1^U\right), H_2\left(\tilde{M}_2^U\right)\right)\right), \\ \left(\tilde{M}_1^L \odot \tilde{M}_2^L;, \min\left(H_1\left(\tilde{M}_1^L\right), H_1\left(\tilde{M}_2^L\right)\right);, \min\left(H_2\left(\tilde{M}_1^L\right), H_2\left(\tilde{M}_2^L\right)\right)\right) \end{pmatrix} \tag{8.12}$$

$$\widetilde{\widetilde{M}}_1 \oslash \widetilde{\widetilde{M}}_2 = \begin{pmatrix} \left(\tilde{M}_1^U \oslash \tilde{M}_2^U;, \min\left(H_1\left(\tilde{M}_1^U\right), H_1\left(\tilde{M}_2^U\right)\right);, \min\left(H_2\left(\tilde{M}_1^U\right), H_2\left(\tilde{M}_2^U\right)\right)\right), \\ \left(\tilde{M}_1^L \oslash \bar{M}_2^L;, \min\left(H_1\left(\tilde{M}_1^L\right), H_1\left(\tilde{M}_2^L\right)\right);, \min\left(H_2\left(\tilde{M}_1^L\right), H_2\left(\tilde{M}_2^L\right)\right)\right) \end{pmatrix} \tag{8.13}$$

$$\widetilde{\widetilde{M}}_1 = \left(\left(k\tilde{M}_1^U;, H_1\left(\tilde{M}_1^U\right);, H_2\left(\tilde{M}_1^U\right)\right), \left(k\tilde{M}_1^L;, H_1\left(\tilde{M}_1^L\right);, H_2\left(\tilde{M}_1^L\right)\right)\right) \tag{8.14}$$

$$\sqrt[n]{\widetilde{\widetilde{M}}_1} = \left(\left(\sqrt[n]{\tilde{M}_1^U}; H_1\left(\tilde{M}_1^U\right); H_2\left(\tilde{M}_1^U\right)\right), \left(\sqrt[n]{\tilde{M}_1^L}; H_1\left(\tilde{M}_1^L\right); H_2\left(\tilde{M}_1^L\right)\right)\right) \tag{8.15}$$

Similarly, the triangular IT2FN can be represented as follows:

$$\tilde{\tilde{T}} = \left(\left(\tilde{T}^U; H\left(\tilde{T}^U\right)\right), \left(\tilde{T}^L; H\left(\tilde{T}^L\right)\right)\right). \tag{8.16}$$

In formula (8.16), $\tilde{T}^U = \left(l^U, m^U, u^U\right)$ and $\tilde{T}^L = \left(l^L, m^L, u^L\right)$ are TFNs, while $H\left(\tilde{T}^U\right)$ and $H\left(\tilde{T}^L\right)$ are triangle heights of the upper and the lower triangle, respectively. Like in the trapezoidal case, heights $H\left(\tilde{T}^U\right)$ and $H\left(\tilde{T}^L\right)$ belong to the interval [0, 1]. Arithmetic operations for the triangular IT2FNs are similar to arithmetic operations for trapezoidal IT2FNs.

Tables 8.2, 8.3 and 8.4 give the linguistic meanings of triangular and trapezoidal fuzzy numbers and triangular and trapezoidal IT2FNs.

Comparison matrices

$$A = \begin{bmatrix} 1 & a_{12} & \ldots & a_{1n} \\ \frac{1}{a_{12}} & 1 & \ldots & a_{2n} \\ \vdots & \vdots & \ddots & \vdots \\ \frac{1}{a_{1n}} & \frac{1}{a_{2n}} & \ldots & 1 \end{bmatrix} \tag{8.17}$$

are constructed first for all considered preference criteria. In formula (8.17), a_{ij}, $i, j = 1, 2, \ldots, n$ is a crisp number in the AHP method or corresponding triangular

Table 8.2 Crisp value, TFN and TrFN with linguistic variables

Crisp Value	TFN	TrFN	Linguistic variables
1	(1,1,3)	(1,1,1,3)	Equally important (EI)
2	(1,2,3)	(1,1.5,2.5,3)	Intermediate value (I_1)
3	(1,3,5)	(1,2,4,5)	Weakly important (WI)
4	(3,4,5)	(3,3.5,4.5,5)	Intermediate value (I_2)
5	(3,5,7)	(3,4,6,7)	Fairly important (FI)
6	(5,6,7)	(5,5.5,6.5,7)	Intermediate value (I_3)
7	(5,7,9)	(5,6,8,9)	Strongly important (SI)
8	(7,8,9)	(7,7.5,8.5,9)	Intermediate value (I_4)
9	(7,9,9)	(7,9,9,9)	Absolutely important (AI)

Table 8.3 Definition of interval type-2 fuzzy scale of the linguistic variables for the triangular IT2FN

Triangular interval type-2 fuzzy scale		Linguistic variables
(1,1,3;1)	(1,1,2;0.9)	Equally important (EI)
(1,2,3;1)	(1.5,2,2.5;0.9)	Intermediate value (I_1)
(1,3,5;1)	(2,3,4;0.9)	Weakly important (WI)
(3,4,5;1)	(3.5,4.4,5;0.9)	Intermediate value (I_2)
(3,5,7;1)	(4,5,6;0.9)	Fairly important (FI)
(5,6,7;1)	(5.5,6,6.5;0.9)	Intermediate value (I_3)
(5,7,9;1)	(6,7,8;0.9)	Strongly important (SI)
(7,8,9;1)	(7.5,8,8.5;0.9)	Intermediate value (I_4)
(7,9,9;1)	(8,9,9;0.9)	Absolutely important (AI)

Table 8.4 Definition of interval type-2 fuzzy scale of the linguistic variables for the trapezoidal IT2FN

Trapezoidal interval type-2 fuzzy scale		Linguistic variables
(1,1,1,3;1,1)	(1,1,1,2;0.9,0.9)	Equally important (EI)
(1,1.5,2.5,3;1,1)	(1.5,1.75,2.25,2.5;0.9,0.9)	Intermediate value (I_1)
(1,2,4,5;1,1)	(2,2.5,3.5,4;0.9,0.9)	Weakly important (WI)
(3,3.5,4.5,5;1,1)	(3.5,3.75,4.25,4.5;0.9,0.9)	Intermediate value (I_2)
(3,4,6,7;1,1)	(4,4.5,5.5,6;0.9,0.9)	Fairly important (FI)
(5,5.5,6.5,7;1,1)	(5.5,5.75,6.25,6.5;0.9,0.9)	Intermediate value (I_3)
(5,6,8,9;1,1)	(6,6.5,7.5,8;0.9,0.9)	Strongly important (SI)
(7,7.5,8.5,9;1,1)	(7.5,7.75,8.25,8.5;0.9,0.9)	Intermediate value (I_4)
(7,9,9,9;1,1)	(8,9,9,9;0.9,0.9)	Absolutely important (AI)

fuzzy number $\tilde{T}$, trapezoidal fuzzy number $\tilde{M}$ triangular IT2FN $\tilde{\tilde{T}}$, and trapezoidal IT2FN $\tilde{\tilde{M}}$ in FAHP.

The aggregation of different experts' opinions is calculated by the averaging method. Based on the linguistic assessments of k experts, for triangular numbers, their fuzzy presentation $e_i = (l_i, m_i, u_i)$, $i = 1, \ldots, k$ are obtained. The corresponding aggregated crisp value has then calculated by the formula

$$c_v = \frac{1}{k}\sum_{i=1}^{k} m_i$$

rounded to the nearest integer. Similarly, in the case of trapezoidal numbers, based on the opinion of experts $e_i = \left(l_i, m_i^l, m_i^h, u_i\right), i = 1, \ldots, k$ crisp value has calculated by the formula

$$c_v = \frac{1}{2k}\sum_{i=1}^{k}\left(m_i^l + m_i^h\right)$$

rounded to the nearest integer. The corresponding linguistic value of the aggregate opinion, in both cases, is obtained from Table 8.2.

After that, the consistency index $CI = \frac{\lambda_{\max} - n}{n-1}$ and consistency ratio $CR = \frac{CI}{RI}$ are calculated. The value λ_{max} represents the maximal eigenvalue of the comparison crisp matrix A, RI is known random index, while n is the dimension of the matrix. If $CR < 0.1$, the comparison matrix is consistent and the estimates of the relative importance of the criteria are counted as acceptable. Otherwise, when $CR \geq 0.1$, experts must correct their assessments.

For assigned a_{ij} the geometric mean of each row is calculated as follows:

$$r = \left[a_{11} \odot a_{12} \odot \ldots \odot a_{1n}\right]^{\frac{1}{n}}, i = \overline{1, n}, \tag{8.18}$$

According to that, fuzzy weights of each criterion are obtained

$$w_j = r \odot \left[r_1 \oplus r_2 \oplus \ldots \oplus r_n \right]^{-1}, j = \overline{1, n}. \tag{8.19}$$

The defuzzified values in the FAHP methods are obtained using the center area method (Do et al. 2015). In the case of triangular fuzzy number $\tilde{T} = (l, m, u)$, the defuzzified value is $\frac{1}{4}(l + 2m + u)$. When $\tilde{M} = (l, m^l, m^h, u)$ is a trapezoidal fuzzy number, the defuzzified value is $\frac{1}{4}(l + m^l + m^h + u)$. For the triangular IT2FN, denoted by $\tilde{\tilde{T}}$ and given by (16), the defuzzified value is $\frac{1}{8}\left(l^U + u^U + l^L + u^L + 2H\left(\tilde{T}^U\right)m^U + 2H\left(\tilde{T}^L\right)m^L\right)$, and for the trapezoidal IT2FN, denoted by $\tilde{\tilde{M}}$ and given by (9), the defuzzified value is $\frac{1}{8}\left(l^U + u^U + l^L + u^L + H\left(\tilde{M}^U\right)\left(\left(m^l\right)^U + \left(m^r\right)^U\right) + H\left(\tilde{M}^L\right)\left(\left(m^l\right)^L + \left(m^r\right)^L\right)\right)$.

Results

In this section, we have applied the methods outlined in section "Methodology: fuzzy numbers and analysis". Tables 8.5, 8.6, 8.7, 8.8, 8.9, 8.10 and 8.11 give fuzzy matrices of criteria and sub-criteria comparison obtained from experts. Based on the obtained value of CR < 0.1, one can conclude that all comparison matrices are consistent.

In Table 8.12, the indicators are ranked using the AHP method, triangular FAHP, triangular IT2FS, trapezoidal FAHP, and trapezoidal IT2FS as the methods of fuzzy logic. The ranking results are shown in Fig. 8.2.

The obtained results show that, in this case, the use of AHP and FAHP methods does not prioritize the same indicators. Comparing the finally ranked indicators using the AHP method, triangular FAHP, triangular IT2FS, trapezoidal FAHP, and trapezoidal IT2FS as the methods of fuzzy logic, all applied methods favor as the most crucial indicators for developing Smart Cities: development of a legislative and strategic framework of Smart City platform and its implementation in COVID-19 pandemic circumstances (A_1), standardization of ICT and ICT management (A_2), and entrepreneurship and innovation (B_1). The results of applying the AHP method slightly over the FAHP methods favor the ICT sector development and job

Table 8.5 Pairwise evaluation matrix of criteria, CI = 0.00827004, CR = 0.00666939

	A	B	C	D	E	F
A	EI	I_1	WI	WI	I_2	I_2
B	$1/I_1$	EI	I_1	I_1	WI	WI
C	1/WI	$1/I_1$	EI	I_1	I_1	I_1
D	1/WI	$1/I_1$	1/EI	EI	I_1	I_1
E	$1/I_2$	1/WI	$1/I_1$	$1/I_1$	EI	EI
F	$1/I_2$	1/WI	$1/I_1$	$1/I_1$	1/EI	EI

Table 8.6 Pairwise evaluation matrix of sub-criteria A, CI = 0.0034543, CR = 0.00383811

	A_1	A_2	A_3	A_4
A_1	EI	I_1	WI	WI
A_2	$1/I_1$	EI	I_1	I_1
A_3	1/WI	$1/I_1$	EI	I_1
A_4	1/ WI	$1/I_1$	1/EI	EI

Table 8.7 Pairwise evaluation matrix of sub-criteria B, CI = 0.0140206, CR = 0.0099437

	B_1	B_2	B_3	B_4	B_5	B_6	B_7	B_8
B_1	EI	I_1	I_1	WI	I_2	I_2	FI	FI
B_2	$1/I_1$	EI	EI	I_1	WI	WI	I_2	I_2
B_3	$1/I_1$	1/EI	EI	I_1	I_1	WI	I_2	I_2
B_4	1/WI	$1/I_1$	$1/ I_1$	EI	I_1	I_1	WI	WI
B_5	$1/I_2$	1/WI	1/WI	$1/I_1$	EI	EI	I_1	I_1
B_6	$1/I_2$	1/WI	1/WI	$1/I_1$	1/EI	EI	I_1	I_1
B_7	1/FI	$1/I_2$	$1/I_2$	1/WI	$1/I_1$	$1/I_1$	EI	EI
B_8	1/FI	$1/I_2$	$1/I_2$	1/WI	$1/I_1$	$1/I_1$	1/EI	EI

Table 8.8 Pairwise evaluation matrix of sub-criteria C, CI = 0.0289418, CR = 0.0233402

	C_1	C_2	C_3	C_4	C_5	C_6
C_1	EI	I_1	WI	WI	FI	SI
C_2	$1/ I_1$	EI	I_1	I_1	I_2	I_3
C_3	1/WI	$1/I_1$	EI	EI	WI	FI
C_4	1/WI	$1/I_1$	1/EI	EI	WI	FI
C_5	1/FI	$1/I_2$	1/WI	1/WI	EI	WI
C_6	1/SI	$1/I_3$	1/ FI	1/FI	1/WI	EI

Table 8.9 Pairwise evaluation matrix of sub-criteria D, CI = 0.014781, CR = 0.0111977

	D_1	D_2	D_3	D_4	D_5	D_6	D_7
D_1	EI	I_1	I_1	WI	I_2	FI	FI
D_2	$1/I_1$	EI	EI	I_1	WI	I_2	I_2
D_3	$1/I_1$	1/EI	EI	I_1	I_1	I_2	I_2
D_4	1/WI	$1/I_1$	$1/ I_1$	EI	I_1	WI	WI
D_5	$1/I_2$	1/WI	1/WI	$1/I_1$	EI	I_1	I_1
D_6	1/FI	$1/I_2$	$1/I_2$	1/WI	$1/I_1$	EI	EI
D_7	1/FI	$1/I_2$	$1/I_2$	1/WI	$1/I_1$	1/EI	EI

opportunities within it (B_2) compared to public-private stakeholder partnership with active citizen's participation (A_3). According to importance, the higher data openness and transparency (A_4) has the same rank for all applied methods. Triangular IT_2FS slightly favors indicator personal safety (D_1) compared to indicator higher

Table 8.10 Pairwise evaluation matrix of sub-criteria E, CI = 0.00914735, CR = 0.0157713

	E_1	E_2	E_3
E_1	EI	I_1	I_2
E_2	$1/I_1$	EI	WI
E_3	$1/I_2$	1/WI	EI

Table 8.11 Pairwise evaluation matrix of sub-criteria F, CI = 0.0280297, CR = 0.0212346

	F_1	F_2	F_3	F_4	F_5	F_6	F_7
F_1	EI	I_1	WI	WI	I_2	FI	SI
F_2	$1/I_1$	EI	I_1	I_1	WI	I_2	I_3
F_3	1/WI	$1/I_1$	EI	EI	I_1	I_2	FI
F_4	1/WI	$1/I_1$	1/EI	1/EI	I_1	WI	FI
F_5	$1/I_2$	1/WI	$1/I_1$	$1/I_1$	EI	I_1	I_2
F_6	1/FI	$1/I_2$	1/WI	1/WI	$1/I_1$	EI	WI
F_7	1/SI	$1/I_3$	1/FI	1/FI	$1/I_2$	1/EI	EI

awareness level in the community (C_1). Further, the AHP method favors personal safety over ICT infrastructure integration (E_1).

Discussion

Multi-criteria decision analysis (MCDA) is necessary to identify key indicators developed to create optimal conditions for the cities' evolution into smart ones. The obtained results may differ from each other regarding the research aspect, the existence of an appropriate strategy for creating a Smart City, or the degree of its application in practice.

Previous research has mainly focused on urban areas that, by their characteristics, have implemented urban policies and implemented technological advances and strategies that are already classified as a positive example of Smart City development.

Our research as a case study takes cities that are in the process of developing to become smart. These cities are often part of developing countries, and therefore the preconditions for the development of Smart Cities are often not oriented toward a particular urban sector, which is recognized as underdeveloped and needs to be improved but is a problem of a broader nature and requires through the parallel development of different sectors.

The proposed methods rank the criteria, using all available information to take advantage of opportunities and overcome barriers to the development Smart City concept. By precise implementation of the applied methods, the following significant indicators have been obtained.

The eight most significant Smart City indicators are given in Fig. 8.3.

Table 8.12 The final ranking of sub-criteria

AHP		Triangular FAHP, IT2FS				Trapezoidal FAHP, IT2FS			
A_1	0.16535	A_1	0.14588	A_1	0.15216	A_1	0.14465	A_1	0.15117
A_2	0.09543	A_2	0.08682	A_2	0.08956	A_2	0.08644	A_2	0.08924
B_1	0.08452	B_1	0.07871	B_1	0.08058	B_1	0.07855	B_1	0.08044
B_2	0.05436	A_3	0.05741	A_3	0.05597	A_3	0.05714	A_3	0.05583
A_3	0.05115	B_2	0.05178	B_2	0.05279	B_2	0.05174	B_2	0.05275
A_4	0.05115	A_4	0.04820	A_4	0.04857	A_4	0.04827	A_4	0.04861
D_1	0.04137	E_1	0.04554	E_1	0.04446	E_1	0.04541	E_1	0.04441
E_1	0.04113	C_1	0.03992	C_1	0.03962	D_1	0.04151	C_1	0.03952
C_1	0.03811	D_1	0.03909	D_1	0.03951	C_1	0.03975	D_1	0.03951
B_3	0.03320	B_3	0.03701	B_3	0.03609	B_3	0.03711	B_3	0.03618
B_4	0.03320	B_4	0.03306	B_4	0.03294	B_4	0.03328	B_4	0.03308
D_2	0.02580	C_2	0.02788	C_2	0.02712	D_2	0.02817	C_2	0.02705
D_3	0.02580	E_2	0.02783	E_2	0.02653	E_2	0.02806	E_2	0.02671
F_1	0.02483	D_2	0.02671	D_2	0.02649	C_2	0.02775	D_2	0.02647
C_2	0.02429	C_3	0.02572	C_3	0.02540	C_3	0.02565	C_3	0.02536
C_3	0.02429	D_3	0.02438	D_3	0.02459	F_1	0.02395	D_3	0.02460
E_2	0.02354	F_1	0.02390	F_1	0.02398	D_3	0.01995	F_1	0.02402
F_2	0.01629	C_4	0.01778	C_4	0.01695	D_4	0.01818	C_4	0.01709
D_4	0.01582	D_4	0.01696	D_4	0.01643	C_4	0.01796	D_4	0.01661
C_4	0.01503	F_2	0.01618	F_2	0.01612	F_2	0.01626	F_2	0.01618
B_5	0.01469	B_5	0.01568	B_5	0.01525	B_5	0.01590	B_5	0.01541
F_3	0.01003	C_5	0.01159	F_3	0.01097	D_5	0.01177	F_3	0.01104
F_4	0.01003	F_3	0.01143	C_5	0.01086	C_5	0.01166	C_5	0.01092
D_5	0.00988	E_3	0.01123	E_3	0.01048	F_3	0.01151	E_3	0.01060
C_5	0.00909	D_5	0.01079	D_5	0.01038	E_3	0.01139	D_5	0.01050
C_6	0.00909	C_6	0.01064	F_4	0.01015	C_6	0.01073	F_4	0.01023
E_3	0.00898	F_4	0.01038	C_6	0.01013	F_4	0.01049	C_6	0.01020
B_6	0.00764	B_6	0.00763	B_6	0.00758	D_6	0.00815	B_6	0.00762
F_5	0.00623	C_7	0.00692	D_6	0.00662	B_6	0.00769	D_6	0.00666
D_6	0.00617	D_6	0.00686	C_7	0.00653	C_7	0.00695	C_7	0.00656
D_7	0.00617	F_5	0.00669	F_5	0.00647	A_1	0.14465	F_5	0.00654
C_7	0.00555	C_8	0.00635	D_7	0.00613	A_2	0.08644	D_7	0.00617
C_8	0.00555	D_7	0.00623	C_8	0.00609	B_1	0.07855	C_8	0.00613
F_6	0.00410	F_6	0.00461	F_6	0.00437	A_3	0.05714	F_6	0.00444
F_7	0.00214	F_7	0.00219	F_7	0.00214	B_2	0.05174	F_7	0.00215

Our research shows that, without the satisfactory legislative and strategic framework of the Smart City platform and its implementation and challenges, Smart City development will not be possible.

A clear vision and course of development of urban areas have to be accompanied by appropriate strategic and planning documents. Simultaneously, it implies the implementation and standardization of IC technologies, i.e., the introduction of

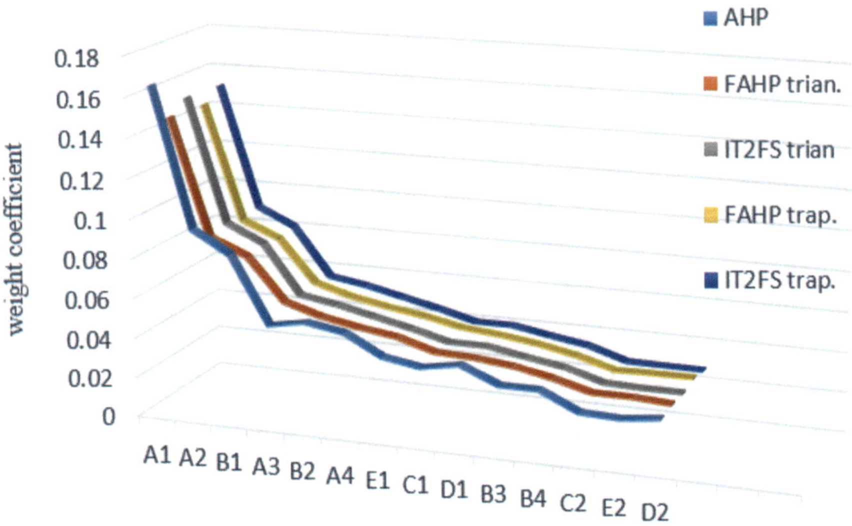

Fig. 8.2 Weights of key indicators for Smart City

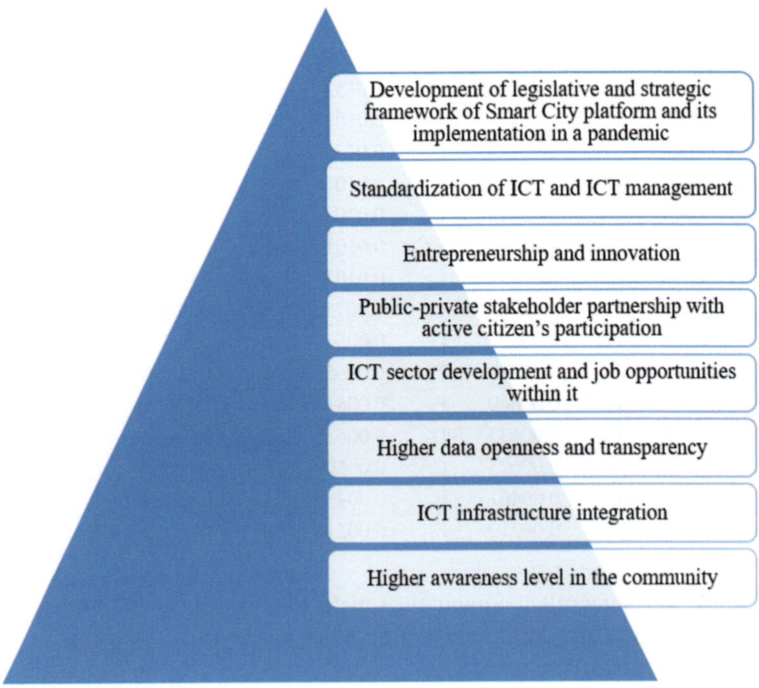

Fig. 8.3 Key indicators of opportunities for the development of Smart City using AI

"layers" of technological infrastructure through all levels of urban space as a system, in a way that has been precisely defined by the institutional framework, urban and planning regulations. The management of innovative technologies is related to the development of the ICT sector in the local and regional economy and the development of innovations within entrepreneurial activities (opening companies, production of smart devices and sensors, etc.). The expansion of urban development strategies for Smart Cities and the ICT sector requires the support of appropriate actors, as indicated by the dominant ranking of indicators obtained in this research. A public-private partnership, achieved through collaboration between urban government, local government, representatives of educational and scientific institutions, ICT experts, relevant planners, urban planners, architects and other engineers, private investors, and non-profit organizations, with active citizen participation, is a significant planning segment.

Given the role and importance of AI in a Smart City, to maximize the benefits of AI, it is necessary to use the following recommendations:

- To achieve close cooperation between the public and private sectors, between government organizations and citizens, and between universities and businesses to ensure the development of the necessary artificial intelligence technologies and their acceptance naturally.
- To create programs to inform citizens about the benefits and risks of using information and communication technologies, including artificial intelligence, by the level of education, age, and interests.
- Delineate through regulations the responsibilities of developers and users of artificial intelligence technologies.
- Conduct rigorous research about the side effects of ubiquitous sensors on human health and disseminate the results.

The role and participation of citizens in creating and improving the live environments are of particular significance because the planning process enables problem identifications that citizens face every day and the implementation of solutions that provides them with higher quality and standard of living.

Legislative and strategic framework for smart city development, with appropriate private-public partnership, citizen participation, development, and management of the ICT sector and infrastructure, is particularly important to align with the challenges of the COVID-19 pandemic. During the pandemic, cities turned to network tools and collaboration tools, digital communications platforms, and rapid development to move employees into virtual environments. The following stages in the continuum are budget reduction, economic slowdown, and return to growth. The final stage should involve cloud and infrastructure optimization, service deployment, and data analysis to help make decisions relevant to the future. The efforts of Smart Cities must rely on the digitization of data and content management. Without it, cities cannot get real-time insight into the data.

The ranking of indicators is carried out by taking into account the circumstances caused by the pandemic spread. This research is one of the contributions that can make decision-making easier for decision-makers.

Conclusion

Rigorous changes in the way of life in cities, which are happening every day, have made us think of ways to build a more sustainable society, one that will be able to resist the rapid development of our environment. The question is whether the existing model of the smart city is agile enough to cope with a situation similar to a pandemic. It also depends on our abilities to re-invent and refine Smart City models to adapt to the precarious situation we are facing today.

In this chapter, we explore the possibilities for the development of Smart Cities relying on AI. The theoretical background is primarily an attempt to explain the most important aspects and challenges of a Smart City. The task was to provide a theoretical overview of current scientific insights into the development and application of the Smart City concept, developing a discussion of a possible obstacle. In terms of applied methodology, the contribution lies in the comparative application of AHP methods, triangular and trapezoidal FAHP methods, and the corresponding hybrid IT2FS methods in the field of Smart City development. The research identifies crucial indicators as significant prerequisites for the development of the Smart City. By defining indicators of the Smart City formation, the concept of a Smart City, from idea to reality, is presented by emphasizing the symbiosis between the Smart City and AI and AI's role in Smart City development.

COVID's new challenges for urban areas indicate that an analysis of the current situation and quality of life indicators, a long-term vision of the city, and measurable goals are necessary. Human society must think and live more smartly, not only because of the pandemic but also because of globalization, environmental catastrophes, and overcrowding in general. This research is an attempt to model the concept of a Smart City in a pandemic using AI. Based on six criterion groups and thirty-five sub-criteria, the main dominant indicators were the development of a legislative and strategic framework of the Smart City platform and its implementation in COVID-19 pandemic circumstances, standardization of ICT and ICT management, entrepreneurship and innovation, public-private stakeholder partnership with active citizen's participation, and higher data openness and transparency. Making opportunities for developing Smart Cities can be achieved with well-defined goals with transparent regulatory frameworks. This research could be the basis for further research efforts relying on results related to the main indicators, proposed levels, and key tasks.

References

Abusaada H, Elshater A (2020) COVID-19 challenge, information technologies, and smart cities: considerations for well-being. Int J Com WB:1–8. https://doi.org/10.1007/s42413-020-00068-5

Allam Z, Jones DS (2020) On the Coronavirus (COVID-19) outbreak and the smart city network: universal data sharing standards coupled with Artificial Intelligence (AI) to benefit urban health monitoring and management. Healthcare 8:46. https://doi.org/10.3390/healthcare8010046

Al-Shouk ND, Al-Khfaji AS (2018) Toward sustainable compact city: (Study in Convert Traditional Najaf City to Sustainable Compact City). KnE Engineering / Postmodern Urban and Regional Planning in Iraq:167–192. https://doi.org/10.18502/keg.v3i4.2168

Anand PB, Navio-Marco J (2018) Governance and economics of smart cities: opportunities and challenge. Telecommun Policy 42:795–799. https://doi.org/10.1016/j.telpol.2018.10.001

Batty M, Axhausen KW, Giannotti F, Pozdnoukhov A, Bazzani A, Wachowicz M, Ouzounis G, Portugali Y (2012) Smart cities of the future. Eur Phys J Spec Top 214:481–518. https://doi.org/10.1140/epjst/e2012-01703-3

Bibri SE (2019) On the sustainability of smart and smarter cities in the era of big data: an interdisciplinary and transdisciplinary literature review. J Big Data 6:25. https://doi.org/10.1186/s40537-019-0182-7

Cassandras CG (2017) Automating mobility in smart cities. Annu Rev Control 44:1–8. https://doi.org/10.1016/j.arcontrol.2017.10.001

Chen Y, Han D (2018) Water quality monitoring in the smart city: a pilot project. Autom Constr 89:307–316. https://research-information.bris.ac.uk/en/publications/water-quality-monitoring-in-a-smart-city-a-pilot-project

Chen SM, Lee LW (2010) Fuzzy multiple attributes group decision-making based on the interval type-2 TOPSIS method. Expert Syst Appl 37:2790–2798

Dameri RP, Cocchia A (2013) Smart city and digital city: twenty years of terminology evolution. In: Proceedings of the 10th Conference of the Italian Chapter of AIS (ITAIS), Universita Commercial Luigi Bocooni, Milan, Italy, 14 December 2013, pp 1–8

Do QH, Chen JF, Hsieh HN (2015) Trapezoidal Fuzzy AHP and fuzzy comprehensive evaluation approaches for evaluating academic library service. WSEAS Trans Comput 14:2224–2872

Dostal Z, Ladanyi L (2018) Demands on energy storage for renewable power sources. J Energy Storage 18:250–255. https://doi.org/10.3390/en13133421

Giffinger R, Fertner C, Kramar H, Kalasek R, Pichler-Milanovic N, Meijers E (2007) Smart cities – ranking of European Medium-Sized. Centre of Regional Science, Vienna

Glasmeier A, Christopherson S (2015) Thinking about smart cities. Camb J Reg Econ Soc 8:3–12. https://doi.org/10.1093/cjres/rsu034

https://www.adalovelaceinstitute.org/blog/uk-contact-tracing-apps-the-view-from-northern-ireland-and-scotland/

Kahraman C, Süder A, Kaya I (2014) Fuzzy multicriteria evaluation of health research investments. Technol Econ Dev Econ 20:210–226. https://doi.org/10.3846/20294913.2013.876560

Khan S, Paul D, Momtahan P, Aloqaily M (2018) Artificial intelligence framework for smart city microgrids: state of the art challenges and opportunities. In: 2018 Third International Conference on Fog and Mobile Edge Computing (FMEC), pp 283–288

Koca-Baltić I, Momčilović-Petronijević A (2020) Revitalization as protection instrument of the cultural heritage of libraries. Facta Univ Ser Architect Civil Eng 18(1):1–14. https://doi.org/10.2298/FUACE190911001K

Kotharkar R, Bahadur P, Sarda N (2014) Measuring Compact Urban Form: A Case of Nagpur City, India. Sustainability 6:4246–4272. https://doi.org/10.3390/su6074246

Lazaroiu GC, Roscia M (2012) Definition methodology for the smart cities model. Energy 47(1):326–332. https://doi.org/10.1016/j.energy.2012.09.028

Li CX, Fong PSW, Dai S, Li Y (2019) Towards sustainable smart cities: An empirical comparative assessment and development pattern optimization in China. J Clean Prod 215:730–743. https://doi.org/10.1016/j.jclepro.2019.01.046

Lin C, Zhao G, Yu C, Wu YJ (2019) Smart city development and residents' well-being. Sustainability 11:676. https://doi.org/10.3390/su11030676

Mellouli S, Luna-Reyes LF, Zhang J (2014) Smart government, citizen participation and open data. Inf Pol 19:1–4. https://doi.org/10.3233/IP-140334

Milošević A, Milošević M, Milošević D, Selimi A (2016) Ahp multi-criteria method for sustainable development in construction. In: Proceedings of the 4th International Conference, Contemporary Achievements in Civil Engineering, Subotica, Serbia, 22 April 2016, pp 929–938

Milošević D, Stanojević A, Milošević M (2017) AHP method in the function of logistic in development of smart cities model. In: Proceeding 6th International Conference Transport and logistic TiL, Niš, Serbia, 25–26. May 2017, ISBN 978-86-6055-088-2, pp 287–294

Milošević M, Milošević D, Stević D, Stanojević A (2019) Smart city: modeling key indicators in Serbia Using IT2FS. Sustainability 11(13):3536. https://doi.org/10.3390/su11133536

Milošević M, Milošević D, Stanojević A, Simjanović D (2020a) IAHP kao podrška primeni tehnoloških inovacija u razvoju pametnih gradova, ECOLOGICA, Naučno-stručno društvo za zaštitu životne sredine Srbije, no. 99. ISSN 0354-3285, pp 407–413

Milošević DM, Milošević MR, Simjanović DJ (2020b) Implementation of Adjusted Fuzzy AHP Method in the Assessment for Reuse of Industrial Buildings. Mathematics 8:1697. https://doi.org/10.3390/math8101697

Milošević MR, Milošević DM, Stanojević AD, Stević DM, Simjanović DJ (2021a) Fuzzy and interval AHP approaches in sustainable management for the architectural heritage in smart cities. Mathematics 9(4):304. https://doi.org/10.3390/math9040304

Milošević M, Milošević D, Stanojević A (2021b) Managing cultural built heritage in smart cities using fuzzy and interval multi-criteria decision making. In: Kahraman C, Onar SC, Oztaysi B, Sari IU, Cebi S, Tolga AC (eds) Advances in Intelligent Systems and Computing – Proceedings of the INFUS 2020 Conference – Intelligent and Fuzzy Techniques: Smart and Innovative Solutions, Turkey, Istanbul 21–23 July 2021, vol 1197, pp 599–607. ISBN 978-3-030-51156-2. https://doi.org/10.1007/978-3-030-51156-2

Monfaredzadeh T, Krueger K (2015) Investigating social factor of sustainability in a smart city. Proc Eng 118:1112–1118. https://doi.org/10.1016/j.proeng.2015.08.452

Monzon A (2015) Smart cities concept and challenges: bases for the assessment of smart city projects. Commun Comput Inf Sci 579:17–31. https://doi.org/10.1007/978-3-319-27753-0_2

Mosannenzade F, Vettorato D (2014) Defining smart city: a conceptual framework based on keyword analysis. TEMA J Land Use Mobil Environ 12:683–694. https://doi.org/10.6092/1970-9870/2523

Moser C, Wendel T, Carabias-Hütter V (2014) Scientific and practical understandings of smart cities. In: Proceedings of the Conference REAL CORP, Vienna, Austria, 21–23 May 2014, pp 507–514

Ozkok BA (2019) Finding fuzzy optimal and approximate fuzzy optimal solution of fully fuzzy linear programming problems with trapezoidal fuzzy numbers. J Intell Fuzzy Syst 36:1389–1400. https://doi.org/10.3233/jifs-18016

Oztaysi BA (2015) Group decision making approach using interval Type-2 fuzzy AHP for enterprise information systems project selection. J Mult Valued Log Soft Comput 24:475–500. https://doi.org/10.3390/su8020118

Park E, Pobil AP, Kwon SJ (2018) The role of Internet of Things (IoT) in smart cities: technology roadmap-oriented approaches. Sustainability 10:1388. https://doi.org/10.3390/su10051388

Saaty T (1980) The Analytic Hierarchy Process. Mcgraw Hill, New York, NY, USA

Salvioni DM, Almici A (2020) Transitioning toward a circular economy: the impact of stakeholder engagement on sustainability culture. Sustainability 12:8641. https://doi.org/10.3390/su12208641

Selimi A, Milošević M, Saračević M (2018) AHP–TOPSIS model as a mathematical support in the selection of project from aspect of mobility–case study. J Appl Math Comput 2:257–265. https://doi.org/10.26855/jamc.2018.06.004

Smart City Transformation in a post-COVID world – Reimagining the digital transformation of our cities on the road to recovery from COVID-19. KPMG, 2020

Sofeska E (2017) Understanding the livability in a city through smart solutions and urban planning toward developing sustainable livable future of the city of Skopje. Procedia Environ Sci 37:442–453. https://doi.org/10.1016/j.proenv.2017.03.014

Srdjević B, Medeiros Y (2008) Fuzzy AHP assessment of water management plans. Water Resour Manag 22:877–894

Stephanedes JY, Golias MPE, Dedes G, Douligeris C, Mishra S (2019) Challenges, risks and opportunities for connected vehicle services in smart cities and communities. IFAC Pap 51:139–144. https://doi.org/10.3390/su12072789

Stratigea A (2012) The concept of 'smart cities'. Towards community development? Netcom. Réseaux, communication et territoires:375–388. https://doi.org/10.4000/netcom.1105

Sujata J, Saksham S, Tanvi G (2016) Shreya: developing smart cities: an integrated framework. Proc Comput Sci 93:902–909. https://doi.org/10.1016/j.procs.2016.07.258

Susantia R, Soetomo S, Buchori I, Brotosunaryo PM (2016) Smart growth, smart cities and density: in search of the appropriate indicator for residential density in Indonesia. Proc Soc Behav Sci 227:194–201. https://doi.org/10.1016/j.sbspro.2016.06.062

Talari S, Shafie-khah M, Siano P, Loia V, Tommasetti A, Catalão JPS (2017) A review of smart cities based on the Internet of Things concept. Energies 10:421. https://doi.org/10.3390/en10040421

Voda AI, Radu LD (2018) Artificial intelligence and the future of smart cities. Broad Res Artif Intell Neurosci 9:110–127

Voda AI, Radu LD (2019) Chapter 12: How can artificial intelligence respond to smart cities challenges? In: Visvizi A, Lytras MD (eds) Smart cities: issues and challenges – mapping political, social and economic risks and threats, pp 199–216., Elsevier. https://doi.org/10.1016/B978-0-12-816639-0.00012-0

Wang W, Wu T, Li Y, Xie S, Han B, Zheng H, Ouyang Z (2020) Urbanization impacts on natural habitat and ecosystem services in the Guangdong-Hong Kong-Macao "Megacity". Sustainability 12:6675. https://doi.org/10.3390/su12166675

Wey WM, Hsu J (2014) New urbanism and smart growth: toward achieving a smart National Taipei University District. Habitat Int 42:164–174. https://doi.org/10.1016/j.habitatint.2013.12.001

Yadav G, Mangla SK, Luthra S, Rai DP (2019) Developing a sustainable smart city frame-work for developing economies: an Indian context. Sustain Cities Soc 47:101462. https://doi.org/10.1016/j.scs.2019.101462

Yigitcanlar T, Cugurullo F (2020) The sustainability of artificial intelligence: an urbanistic viewpoint from the lens of smart and sustainable cities. Sustainability 12:8548. https://doi.org/10.3390/su12208548

Chapter 9
Blockchain and Smart Charging Infrastructure of Electric Vehicles

George Cristian Lazaroiu, Mariacristina Roscia, Virgil Dumbrava, and Korhan Kayisli

Abstract The rapid spread of electric vehicles will be one of the key elements for the transition to sustainable energy and at the same time will lead to the reduction of air pollution in cities. Obviously, the forecast estimates of the demand for electricity will have to be reconsidered and above all the functioning of the electricity grid and the two-way energy exchange methods, also to ensure that everyone is aware of this opportunity. The remodeling of electric mobility must therefore be reconsidered to reduce renewable energy costs, increase the flexibility of the smart grid, and make EVs available to the greatest number of prosumers, even in shared mode. Another point to consider is the implementation of the charging points, considering the risk that if not carefully controlled and increasing the peak load in the smart grid, and proper planning of the charging stations is necessary to avoid network congestion; for example, if a large number of electric vehicles were charged at the same time, it could lead to an excessive increase in costs, due to an increase in unnecessary available power. In this, smart grids have a fundamental task, to obtain new models for managing energy flows, recharging EVs, and related cost profiles.

Another crucial factor, supporting this evolution toward electric mobility, will be the ability to generate business models, which, in addition to supporting the needs of food and a profit for the supply of ancillary services and power systems, will also have to support the needs of EVs fleets, considering vehicle and battery consumption. Then, it will be necessary to analyze factors such as the charging time, the duration, and the state of the battery, since, in the case of shared fleets of electric

G. C. Lazaroiu (✉) · V. Dumbrava
University POLITEHNICA of Bucharest, Bucharest, Romania
e-mail: cristian.lazaroiu@upb.ro; virgilius.dumbrava@upb.ro

M. Roscia
University of Bergamo, Bergamo, Italy
e-mail: cristina.roscia@unibg.it

K. Kayisli
Gazi University, Ankara, Turkey
e-mail: korhankayisli@gazi.edu.tr

G. C. Lazaroiu et al. (eds.), *Holistic Approach for Decision Making Towards Designing Smart Cities*, Future City 18,
https://doi.org/10.1007/978-3-030-85566-6_9

vehicles, an operating model is established "on demand," with an energy market that involves fluctuations cost and therefore appropriate charge-discharge, which reduces the life of the batteries.

Finally, since a battery of an electric vehicle alone would not be able to bring advantages to the network, it is necessary that they are aggravated, as in shared systems, or that there are many electric vehicles in circulation, whose owners want to provide the transfer of service energy from the battery to the grid, in exchange for a fiscal or economic advantage. This chapter aims to analyze scenarios for a new electric mobility, an integral part of Smart Cities.

Keywords Charging infrastructure · Blockchain · Electric vehicle · Smart grid · New business models

Introduction

A pillar of Smart Cities is certainly sustainable mobility, a crucial factor for reducing pollution and improving the quality of life. This aspect becomes even more relevant, if we consider that the cities are already densely populated, and it is expected that in the next 2 years 70% of the inhabitants will live in urban contexts.

The synergy between the use of electric vehicles and Smart Cities is two-way and represents an opportunity for the management of smart grids in an efficient way, considering all the elements of randomness, such as renewable sources and smart appliances in smart homes, which interact with the electricity grid.

The German Center for Solar Energy and Hydrogen Research (ZSW) estimated that around 5.6 million of EVs were circulating globally in 2019. Mobility related to electric vehicles is evolving very dynamically; China and the United States represent the largest markets, with about 2.5 million and 1.2 million of EVs, respectively. Around 300,000 electric vehicles were sold in Europe in 2019, approximately double compared to 2018. If from 2040 only electric-powered vehicles were to be produced and sold, it can be estimated that more than 1 billion EVs can circulate by 2050.

Electric vehicles can thus constitute an electricity storage capacity, equal to large power plants. However, to allow this diffusion, intelligent charging models will have to be used that can manage energy flows, to balance the demand in peak load moments, which would entail an investment advantage for distributed generation from renewable sources, which is well suited combined with the fact that vehicles normally for 90% of their life are parked, these periods of inactivity, combined with storage systems, will represent a flexibility of the network, which are the basis of business models and which must fall into frameworks regulatory.

Electric vehicles integrated into smart grids, through Internet of Things (IoT), information and communication technologies, and blockchain open economic scenarios that should be carefully supported and regulated, also from the point of view of privacy.

Electric Vehicle Dynamic Market

The size of the global electric vehicle market is expected to grow from 4,093,000 units in 2021 to 34,756,000 units by 2030; Japan and South Korea have also strengthened the market for their EVs, thanks to the intervention of governments, which, through the widespread diffusion of electric charging systems, deadlines and taxes for ICE-type vehicles, have induced consumers to purchase electric vehicles.

The growing demand, linked to the need for zero emissions, has led politicians and decision makers to new support policies through tax reductions, incentives, and investments for charging stations, which always remains the crucial issue and which could leads to a halt in sales, after a first general enthusiasm. Obviously, almost all activities, in the COVID 19 period, even that linked to electric mobility, had to wait for the revocation of the lockdown periods, to resume its distribution process; however, it was one of the sectors that lost the least, since the demand for EVs also in 2020 has grown all over the world and at the same time many countries have equipped themselves to introduce more charging stations.

Many factors affect the distribution of EVs, for example:

- The spread of recharging stations, to reduce the risk of anti-exit range
- Battery performance and costs
- Access to the smart grid
- The business models implemented to provide the consumer
- An economic return in the face of battery sharing
- User acceptance of new vehicle types
- Political will

Electric vehicles have very low emissions, are very efficient, can be recharged from home, and are eco-sustainable, making their diffusion easier. The batteries, which is about 70% of the weight of an EV, are the most important and most expensive part, as they supply power to the electric vehicle engine; the battery management system (BMS) manages the output and the charging-discharge of the battery; then it will be possible to establish how much electricity should flow through the battery at what speed; the on-board chargers are used to convert alternating current into direct current, into low voltage, also to avoid overheating.

The on-board instruments guarantee vital information to the driver, as well as manage the battery, also in terms of autonomy in relation to energy consumption and the type of driving of the driver.

Governments will have to pay close attention to the standard of charging infrastructure; for example, Europe, the United States, and Korea use CCS; Japan uses CHAdeMO; and China uses GB/T.

Charging stations can only be compatible with a certain type of voltage: AC stations provide a voltage of 120 V AC and 240 V AC, while DC charging stations provide fast charging via 480 V AC.

It is known that one of the key limits to the diffusion of electric vehicles is access to recharging points; therefore, also the charging infrastructures in the workplace is essential to accelerate the adoption of electric mobility; in fact the charging stations in companies, in the workplace, can represent a valid solution both for those who cannot have a column at home yet and because parking times are long. Obviously it will be necessary to provide a valid tariff, with affordable prices, which make the driver choose to recharge his car at work.

To reduce polluting emissions, at least locally, i.e., in cities, due to transport systems, the sale of electric vehicles is being encouraged, especially those of the PEV type, or plug-in, that allow the driver to still use the ICE-type engine, combined with the electric one, to overcome the anxiety range, i.e., the fear of not being able to recharge the car in time. This is a fundamental step, as long as there is no widespread distribution of the charging infrastructures on the roads, which give consumers confidence. The various governments around the world are therefore implementing various support policies to push the consumer to purchase (Alahmad et al. 2016, 1).

Smart Charging and Impact of EV on Electricity Systems in Smart Cities

Smart charging must be able to receive, process, and respond to the signals of the energy system managers or the actors of the energy system, as they must highlight when it is the right time to charge or discharge the batteries of the EVs, in relation to the specificity of each system, power supply, and generation profile. For example, where it is possible to have charging systems with a solar base, to make the most of the production, which will be in the hours between 11 am and 3 pm, it will be appropriate that they be located in the workplace or in the presence of commercial premises.

If, on the other hand, they are linked to the wind-type source, it will be necessary to make an assessment linked to the territory, rather than to the time slot. In this way it is possible to have fleets of vehicles, which with their batteries can create a considerable capacity for storing renewable electricity.

With smart charging, it will be possible to have a control of the charging and discharging process, through different prices for the charging operations, encouraging the consumers to shift the charge from peak to off-peak periods. Complex approaches will be needed for the provision of ancillary services and real-time grid balancing (Lazaroiu and Roscia 2019a, b, 371).

By 2030 the electric vehicles that will be able to draw energy from the electricity grid could be about 160 million; this will lead to congestion of the network at the local level, if the recharge were uncontrolled. For example, for the United Kingdom, in 2035 about ten million EVs are expected, and this would lead to an increase of about 3 GW, with uncontrolled recharging, for the evening demand (IRENA 2019). If it is possible to control it, the increase would only be of 0.5 GW, (see Fig. 9.1) drastically reducing investments to upgrade the network. Furthermore, intelligent recharging would lead to an increase in demand by 7 GW in the hours with the lowest price ranges, flattening the load curve, to the advantage of installed power.

The charging process can be carried on with slow battery chargers (< 22 kW) in single-phase (3–7 kW) or three-phase configuration (11–22 kW), with fast chargers (50 kW–100 kW) and ultra-fast chargers (>150 kW) in direct current supply systems. The charging time varies from around 10 min to more than 10 h in case of single-phase charger

A key element will be the assessment of the impact of slow or rapid recharging on the electricity grid, in order to appropriately choose the places to install the recharging systems, i.e., in areas with low demand impact and low levels of congestion.

Another factor that could affect it is in the areas where it is installed in a high diffusion of electric heating, even if, precisely because of this presence, they can already be robust networks with a high dimension, which could already represent a natural installation point for recharging systems, with already adequate cables and power (Tayarani et al. 2019, 1; Khaki et al. 2019, 15).

The charging technology is mature to respond to market demands, but they have a cost:

- The fast charging infrastructures require strengthening the electrical supply equipment.
- The electronic equipment are more expensive in a charging station, as well as cooling equipment and protection devices.
- If small section charging cables are chosen, their active cooling is required.

The congestions in the mains supply can occur, and a factor should be considered for the electric vehicles to charge simultaneously; the value of this factor depends on the size of the distribution network. In networks with high share of renewable energy sources, the charging processes can be prioritized to overlap the production

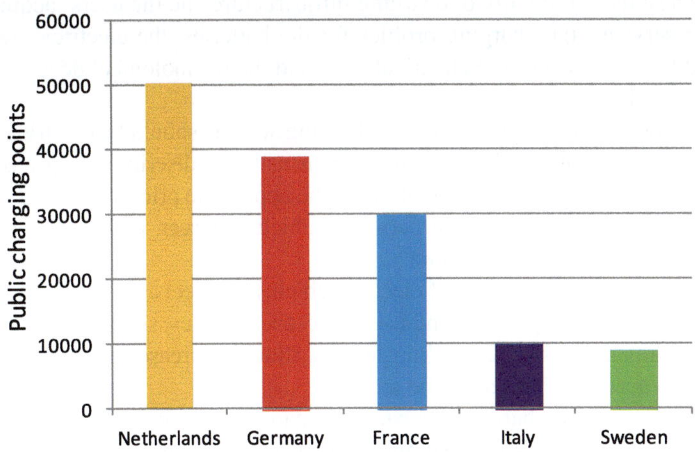

Fig. 9.1 Public charging points for electric vehicles in Europe

from renewable energy sources. Various national codes are in place all over the world, and the distribution operators must comply with the admissible bandwidth for variation of frequency and voltage. Thus, in distribution networks with high share of electric vehicles, excursions of voltage and frequency can occur, and the operator is required to make further investments to accommodate these vehicles.

Public Policies Impact and Incentives

The penetration of electric vehicles, until now, has been possible thanks to the support of governments, through incentives to purchase electric vehicles. In terms of economic incentives, Germany, Norway, and the Netherlands have increased internal combustion engine (ICE) vehicle taxes and at the same time granted tax breaks or exemptions for electric vehicles. The United Kingdom, Italy, and France have introduced one-off incentives for the purchase of an electric car. The United States and Norway preferred to use non-monetary incentives, such as the ability to use carpool lanes or preferential lanes, to avoid traffic. However, these policies can only have a temporary value, as it would not be possible to give the possibility of accessing the preferential lanes to all electric vehicles, when the numbers are at the opening of the ICE vehicles.

Therefore, government incentives and tax cuts on electric vehicles are a starting point to launch the market, even if those with an EV pay about 38% less tax than an ICE-type vehicle. However, these measures are not sufficient. The public charging stations of electric vehicles will be equipped with a two-way smart metering and communication system, in order to provide information on prices, both to system operators and to prosumers. Another crucial factor in these systems is that they should be harmonized across countries.

Many factors are influencing the electrical demand in the smart grids with electric vehicles: the availability of charging infrastructure and the users' acquisition of charging services, the charging profiles for the batteries, the electrical generation time and the influence of the electrical grid, and the technological developments in electric vehicles.

In distribution networks, the public charging sockets should be electrically available all day long, all week days on an open and non-discriminatory base for all customers, and should give information on the applicable prices; standard protocols for charging should be available between vehicle, charger, and management systems, for example, with blockchain.

For fostering the integration of electric vehicles, the decision makers should consider incentives for acquisition and use of electric vehicles, should plan sustainable urban mobility and dedicated parking spaces, should decrease registration fees, and should give access to limited traffic areas.

The tariffs must be harmonious and transparent to avoid disparities, and therefore not based only on kWh consumption but also on recharge duration to determine the vehicle owner to move and to allow another driver to travel toward the electric charging (Lucas et al. 2018, 1).

Various incentives for electric vehicles are applied in Europe, and the different cases are reported in Table 9.1.

The strategies to incentivize the electric vehicle market must in general consider that they are first and foremost systems for mobility and not to decrease load peaks and network regulation; therefore, it remains the enabling factor for the economic link of electric vehicles with different types of columns: charging, data management, and measurement. The recharging of electric vehicles will impact the distribution grid, which will determine various network investments (including cables and transformers) and will be linked to the types of the onsite network, to the distribution system operators (DSOs), and to the possibility of using the production from renewable energy sources (RESs).

Targets vary by country, and the Paris agreement for climate preservation plays an important role in guiding countries' commitments on reducing the emissions generated by the transportation sector; in fact the governments of Norway, Germany, France, Spain, the Netherlands, and the United Kingdom have deadlines set for ICE car sale bans.

Furthermore, as it is clear that charging stations need to be widespread, there is a lot of pressure on installation at residential level, in workplaces, and in public places such as shopping centers.

New Business Models for EVs Recharge

The market for sustainable mobility, in particular electric, is very fragmented and includes:

- Sale of electric vehicles, of the hybrid or pure electric type
- Electricity sales, both from private and public partners
- Sale of services related to mobility, such as fleet management, shared cars, data collection and analysis, support for recharging, and management of recharging systems
- Installation and maintenance of columns and stations for electric recharging
- E-price: interoperability between different operators, for the purchase of energy
- Services related to the smart grid, intelligent charging, discharging, peak load management, etc.

An emerging business model is that which involves the integration of both the sales service of the EV and electricity.

Currently, top-up prices are not used to prevent grid load peaks or to determine recharging overlap with the production from RESs. The rates are defined by the aggregators because individual prosumers do not have yet the possibility of making single transitions on the local network energy market.

Table 9.1 Incentives for electric vehicles in European countries

Country	Buying incentives	Advantages on road taxes	Advantages for property taxes	Tax advantages for company	VAT benefits	Other economic benefits	Local benefits	Incentives for infrastructure
Austria	•	•	•	•	•		•	•
Belgium		•	•	•		•		
Denmark		•	•	•		•		•
Finland	•	•	•			•	•	
France	•	•	•	•		•	•	
Germany	•		•	•		•	•	•
Greece		•	•	•		•	•	
Hungary	•	•	•	•		•	•	•
Ireland	•	•	•	•			•	•
Italy	•		•			•		•
Luxembourg	•		•	•				
Netherlands		•	•	•			•	•
Norway			•	•		•		
Poland	•				•	•	•	
Portugal	•	•	•	•	•	•	•	
Romania	•	•	•					•
Slovakia	•	•				•	•	•
Spain	•	•	•			•	•	•
Sweden	•		•	•		•	•	•
Switzerland		•	•			•	•	
UK	•	•	•	•		•	•	•

To facilitate the electric mobility, it must be considered as a service both for the fleets and for the shared vehicles. Car sharing and carpooling have already changed the habits of drivers, who are starting to consider their cars to be no longer true, especially in big cities, and mobility as a service (MaaS) will increase with digitization. This leads to the acquisition of data for forecasting analysis on the use of electric vehicles, optimizing the recharging schedules and avoiding them in the moments of peak load, and at the same time delivering services to prosumers like recognition of bonus tokens or cooling/heating of the free car function to the various habits of the drivers.

Since the services related to electric mobility are not centralized, with on demand services, the mechanisms linked to blockchains (BC) can be used for market operators, which can represent the solution for the optimized management and scheduling of charging, linked to the needs of the smart grid. Prosumers will have real-time access to advice on prices and economic transactions related to the energy market, to choose if and when to charge/discharge car batteries. The charging methods can be divided by category by time, one-off, flat rate or by kWh, you can also think of creating rates for family groups or companies, exactly like the rates of telephone operators.

Prosumers will have the possibility to recharge from different suppliers, integrating with additional services like parking, car sharing, etc. Thus, it will be possible to generate dynamic contracts where prices follow the real-time trend of energy, generated with a platform that calculates cash flows, designed to use the blockchain register controlled through a mobile app. All members of this network can monitor and control all transactions due.

Car batteries will be able to give a quick response useful for some ancillary services, even if the power capacity is limited. Thus, a single electric vehicle cannot deliver these services for a longer period required to supply the system, but electric vehicles such as fleets transform into a VPP (virtual power plant) with a fast answer and ability to deliver services for the time needed.

Boundaries are due:

- Safety, so as not to compromise electric vehicles.
- Interoperability standards for electric vehicles to recharge from different operators.
- Privacy, blockchains are public records, and information on the location and movements of users must be confidential.

Since flat tariffs do not guarantee any type of incentive for intelligent charging, dynamic tariff plans must be designed, in such a way that they favor intelligent transitions, i.e., prices fluctuate according to the demand and supply of electricity. The setting of prices must be so dynamic, as to also consider the driving behavior of the drivers, e.g., some are willing to pay extra for the availability of the charging station.

There are charging station operators, electric vehicle drivers, RESs producers, smart grid managers, parking operators, etc.; the dynamic definition of tariffs should account all the interests and also advantageous for drivers, but this could be a major limitation to the penetration of electric vehicles (Lazaroiu et al. 2020, 363).

Blockchain for EV Literature Review

The possibility of introducing storage systems to reduce electricity has always been seen as a possible scenario to compensate for the variability of electricity production, especially with the massive introduction of renewable sources (Manjunatha et al. 2013, 1; Doukas et al. 2013, 1). First, a typical summer-winter load profile for UK domestic customers, in which 100 customers were referenced, was assessed (Putrus et al. 2009, 827).

With this research activity, different scenarios are evaluated:

(a) Uncontrolled home charging
(b) Non-peak domestic top-up
(c) Charging planning

In case a), it was imagined that uncontrolled domestic recharging has no possibility of controlling the scheduling of loads. This will induce users to power their loads, in this case electric vehicles, as soon as they return from work to their homes, approximately around 6 pm. An example of charging optimization for electric delivery vehicles based on dynamic programming is formulated in Skugor and Deur (2014, 1). An example of an analysis of the impact that charging can have on the electrical grid was highlighted in a regional design application for the city of Los Angeles, conducted in Jiang et al. (2016, 1). The goal is to present the monitoring of the actual charging behavior of electric vehicles of owners of 64 electric vehicles (divided into 5 brands and with 8 different models) and charging stations at the Los Angeles Department of Water and Energy for just over a year. What previously reported in the literature concerns in the specific context of electric vehicles and its different charging methodologies; the scope of the blockchain is subsequently reviewed. Hertz-Shargel and Livingston (2019, 32) treats blockchain as new and perhaps even adaptive financial instruments, as they can take on a precise value in cases where multiple actors are involved; it remains to be verified whether the various technical, privacy, and security considerations necessary to manage it effectively bring about such a situation. For the management of demand response programs in smart grids, a decentralized solution was proposed in Pop et al. (2018, 1).

The elements of the smart grid were integrated with a blockchain architecture and obtained smart contracts to guarantee the expected levels of energy flexibility, the validation of DR (demand response) agreements, and the balance between demand and production. A methodology for processing power consumption and processing overheads using Bitcoin BC has been proposed by Dorri et al. (2017, 618).

Integration with a smart home was analyzed as a representative case study. Andoni et al. (2019, 143) analyzed various academic and industrial sources, so as to be able to present a vision of the fundamentals of BC technologies, considering system architectures and distributed consensus algorithms, which represent critical components for the performance of all BC matrix ecosystems. They also illustrated several energy use cases, with a careful analysis of the advantages and disadvantages that BC technology solutions are handling for each application. The use of CBs was also the basis of a decentralized control mechanism, for regulate flexibility on smart electricity grids, proposed by D'Oriano et al. (2018, 39).

To improve power system data security against cyberattacks, a blockchain-based distributed data protection framework was proposed in Liang et al. (2019, 3162).

Using the distributed security features of BC technology, used in the Bitcoin cryptocurrency exchange, it is possible to improve the self-defense capabilities of power systems against cyberattacks. A blockchain-based system that has the characteristics of guaranteeing origin, transparency, and immutability is thus applied to smart grid systems. This also helps in the numerous inconsistencies between consumers and electricity companies, on the use of electricity and the corresponding bills. Thanks to smart metering, consumers/prosumers can finally know the real amount of energy consumed and also know which appliances consume them the most and in which time slots (Gao et al. 2018, 9917; Lazaroiu and Roscia 2019a, b, 337).

An innovative approach to smart grids' electric vehicles management was presented in Samuel et al. 2019, 1) A system in which the drivers themselves are transformed into prosumers, participants of a Smart Grid Community (SGC) is presented. Thanks to electric vehicles, they can exchange energy in case of network need or buckling (Samuel et al. 2019, 1).

A demonstration of a clever and privacy-protecting data aggregation system was evaluated in Zhitao Guan et al. (2018, 82). Alladi et al. (2019, 1) analyzed the commercial implementations of BC applied to the smart grid and addressing the integration of these two technologies.

Blockchain

Blockchain technology allows the creation, coordination, and synchronization of a distributed database, consisting of blocks containing the transactions that took place between the nodes of a network. Its decentralized structure of databases defined according to blockchain technology is particularly suitable for the control and implementation of economic processes in local power grids such as Microgrids. Through smart contracts and so-called DApps (decentralized applications), it is possible to implement automated peer-peer trading mechanisms (P2P) whose object of exchange is represented by energy.

The growth of distributed generation based on renewable sources is exponential, leading to innovative grid configurations such as nanogrids and microgrids.

To fully utilize the potential, local market mechanisms must be introduced in the individual smart districts.

This entails the need for specialized solutions for energy transactions and delivery of services between all the players in the blockchain energy system, no longer based on centralized logic, but with distributed and horizontal information. The backbone of the BC is the reliability of the transactions as once a certain transaction has been entered; it cannot be modified or canceled.

The use of blockchain technology within electricity grids, along with other factors, could contribute to the evolution of the latter toward the smart grid model. This term defines a network that incorporates information and communication technologies (ICT) into every aspect of the generation of electricity, minimizing the environmental impact, reducing costs, and improving reliability, service and efficiency.

Blockchain technology provides a level of trust by integrating business interests between charging service providers, utilities, and electric vehicle drivers. Furthermore, the BC allows the creation of a shared register of data, a single source of information of primary importance to facilitate interactions, a safe and reliable market in which transactions can be facilitated without intermediaries through smart contracts.

The advantages of applying BC are especially evident in the peer-to-peer sector:

- **Optimization**: to monitor the consumption and to control the quantities and prices
- **Automatic execution**: smart contract to reduce time, costs, and control
- **Transaction costs**: possible cost reduction thanks to the presence of third parties (brokers)
- **Exchange energy**: prosumers in the same district
- **Security:** information is spread across multiple nodes, making a cyberattack more difficult.

Further analysis is necessary in relation to the security of the IoT device, which will have to collect data from the system.

Blockchain is a distributed ledger technology (DLT) allowing to record and share data across multiple archives (ledgers). Each of them has the same data records and are managed and controlled by a distributed network of computer servers, defined nodes.

BC technology is a system that uses an encryption method and mathematical algorithms to build and control an ever-growing data structure – data that can be added and from which existing data cannot be removed – and takes the form of a chain of "transaction blocks," which works like a distributed ledger (Houben and Alexander 2018, 1).

The BC data can only be updated if there is consent between the various actors in the system and, when new data have been entered, they cannot then be removed. It is a technology that can be written only once and more times; in this way a registration is verifiable at every single transaction. There are risks of serious bugs in the software used on top of the blockchain and as some companies have already been able to verify, it is not the definitive resolution to many technological issues (Mearian 2020).

Members of a distributed ledger compete to complete and add new blocks. The blocks affect what follows through a unique hash code that serves to identify both the block and its content, so if there are changes in the content of a block, the hash code of the previous block will then be modified.

To manage the content of each individual block, the hash codes of all previous blocks must be changed, an operation that requires time and energy.

This organization is known as "cryptographic hashing" and serves to contain the risk of cyberattacks. Together with cryptographic hashing, BCs use different algorithms to create blocks that ensure, in real time, that the communication network is secure.

The following are the algorithms most used by the BC currently.

1. **Proof of Work (PoW)**

To defend themselves against attacks, the BCs use the "proof of work," the consensus algorithm underlying the blockchain network. The reason for using PoW is to slow down the generation of new blocks.

Therefore, within the distributed register, each member will have to solve computational problems, which however require time for a new block to be added to the chain.

However, the problems should not be overly complex, as in that case generating new blocks would take too long, transactions would not be processed, and the network flow would be blocked. If the problem does not have a well-defined resolution time, generating new blocks would be virtually impossible. PoW imposes several limits on the actions that can be taken on the network, and an efficient attack would require a lot of time and incredible computing power. Therefore, although attacks on a blockchain are in theory possible, in practice the results would be disappointing and the costs extremely high, because to manipulate the content of a block, it is necessary to modify the content of the subsequent blocks in the chain.

For example, proof-of-work calculations to add a new block to the chain for a bitcoin take 10 min. In case of malicious activity, making changes to the block is only possible if the hacker performed proof-of-work and would modify the contents of a block and entire precedents.

A hacker must be able to modify more than 51% of the consensus network to take care of adding blocks, but it would take incredible computing power to control the entire BC. If this information reaches the rest of the users, the network is considered compromised and promptly abandoned.

The complexity of this problem is the function of the number of users, the available computing capacity, and the network demand. The hash of each block contains the hash of the previous block, increasing security and preventing all sorts of computer breaches (see Fig. 9.2).

2. **Proof of Stake (PoS)**

However, the mining process requires highly specialized machines, capable of solving extremely complex algorithms in a short time. These devices are very expensive and consume large quantities of electrical network, further driving up prices. This is a dangerous threat to the decentralization of the system, as a reduced portion of users can afford this investment.

In PoS, the physical process through which supercomputers compete with each other to solve complex mathematical problems, i.e., mining, is replaced by a system

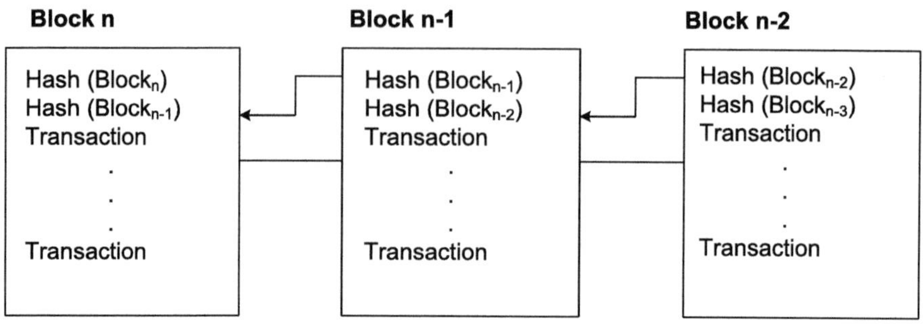

Fig. 9.2 Implementation of blockchain algorithm

in which the so-called validators guarantee the validity of the operations carried out by pledging a portion of their cryptocurrencies (so-called stake). The blocks in PoS systems, unlike PoW blocks, are not extracted (mining) but coined (forge). In order to become validators, nodes must deposit a share of their cryptocurrencies within the network, engaging it as a sort of guarantee or security deposit. The deposited fee cannot be used or spent (Pilkington 2015, 1).

This can potentially lead to faster BCs with lower electricity consumption and a lower chance of an attack (Nguyen et al. 2019, 85,727).

PoS systems also seem to be able to favor the nodes with greater economic resources, making the richest of the network rich. However, the greater the stake deposited and the greater the chances of being selected as a validator, this parameter is not the only one to be taken into consideration by the algorithm which, in any case, consists of a random component, such as a wheel needle luck, and it is not possible to know with certainty in which area of the wheel it will stop, but obviously the greater the area (stake) possessed by a node, the greater the possibility that the needle will stop in that area, but this eventual inequality is counterbalanced by the presence of preferential criteria, a sort of weights affixed to the needle, which guide its direction.

In addition, PoS is a more environmentally friendly technology, as it does not need to power powerful mining machines, greatly reducing energy consumption. Even if the proof of stake to date is not yet particularly widespread and is not used by the main existing blockchains, thanks to its potential and the advantages it entails, it is establishing itself as a preferential consent method, highlighting the abuses and shortcomings of it now. The PoW method can be overcome (Nair and Dorai 2021, 279).

3. **Delegated Proof of Stake (DPoS)**

The Delegated Proof of Stake (DPoS) consensus algorithm was proposed by Daniel Larimer in 2014. Bitshares, Steem, Ark, and Lisk are some of the cryptocurrencies that use the DPoS algorithm.

A DPoS uses a "voting system," in which interested parties delegate their work to a third party: they can vote for delegates, who protect the network for them.

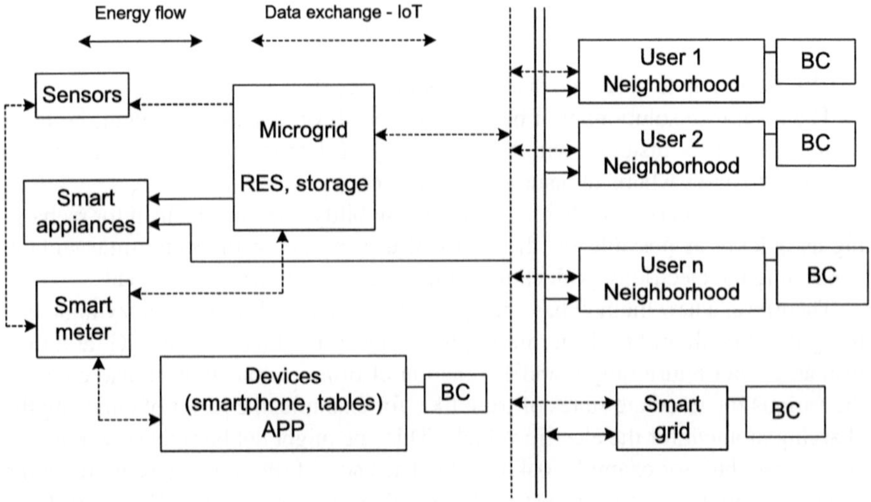

Fig. 9.3 Blockchain platform distributed ledgers

Delegates are also called witnesses and are tasked with reaching consensus during the generation and validation of new blocks. The voting capacity is proportional with the number of coins of each user.

The voting system may vary function of the project but, generally, each delegate formulates an individual proposal that network users can evaluate before choosing who to vote for. The rewards obtained by the delegates are distributed proportionally to the respective voters.

Hence, the DPoS algorithm realizes a voting system that directly depends on the reputation of the delegates. If an elected node does something wrong or doesn't work efficiently, they are quickly kicked out and replaced by a new delegate. As for performance, DPoS blockchains offer greater scalability, managing to process many more transactions per second (TPS) than PoW and PoS (Ferdous et al. 2021, 14).

The DPoS consensus algorithm is considered to be a more democratic and efficient version of its predecessor, the PoS, both of which require fewer resources and are more sustainable and environmentally friendly than PoW. DPoS is a technology with higher speed, large transaction rates, and reduced power consumption (Castor 2017). Unfortunately, DPoS algorithms are vulnerable to centralization given the small number of witnesses (see Fig. 9.3).

Electric Mobility Blockchain

The electric vehicle market is increasing exponentially, this has generated the need for innovative tools to mitigate network congestion, manage the priorities of EVs charging, etc. The BC allows to record in a transparent and immutable way the data of the sensors of a vehicle in a decentralized network; the smart contracts will allow to process and implement the data in an insurance plan, which together with a black box BC, will help to solve the disputes after an accident.

The insurance solutions will be peer-to-peer, flexible, and convenient, both for customers and for insurers, with the possibility of determining the payment to the customer in cases where an insurance event occurs.

One of the main issues inherent in electric mobility is the mechanism for recharging as quickly as possible for the needs of drivers, but at the same time without congesting the electricity grid, causing further critical issues in peak loads.

Therefore, a new method has been proposed to charge EVs with fuzzy logic and using the PoS algorithm. Determining the connection of a car to the SG, at a time instant, cannot return only 0 and 1, function of off/on connection. It also contains the various intermediate states of truth like, for example, the result of choosing the charging moment for the electric vehicle. This one might not be right or wrong in a strict sense, but, for example, 0.476 right. The use of fuzzy logic appears to be the most appropriate, as it is closest to the brain's way of reasoning. The fuzzy logic tool deals with uncertainty. The fuzzy theory allows a mechanism for transforming linguistic constructs such as "many," "low," "medium," "often," "few." In general, the fuzzy logic gives an inference structure enabling appropriate human reasoning capabilities. The utility of fuzzy sets is in their capacity to model uncertainties or ambiguous data facing in real life (Sivanandam et al. 2007, 2).

By aggregating data, it is possible to provide a series of partial truths that further aggregate into higher truths that can cause certain further results.

Using BC-PoS technology, members with a higher stake will be picked in each turn to forge their block and earn the prize. This results in the members with the most stakes or cryptocurrencies ruling the hole network.

Fuzzy logic in Matlab can be handled very easily thanks to the fuzzy logic toolbox; it provides a complete set of functions for designing and implementing various fuzzy logic processes. The main fuzzy logic operation includes fuzzification, defuzzification, and fuzzy inference. They run through various functions and can also be realized using the graphical user interface. Many applications can be realized using the Simulink "fuzzy logic controller" block available in Matlab-Simulink toolbox (Sivanandam et al. 2007, 2).

In the case study, concerning electric mobility, the merit nodes are chosen on the basis of their status.

Three thresholds to take advantage of the fuzzy logic were evaluated as follows:

1. SoC: charge status of the electric vehicle
2. Load profile (kWh) of the charging station where the electric vehicle is connected
3. Energy production (kWh) from RES

EVs with lower SoCs and load profiles, with higher RESs in their charging locations, are more likely to be chosen as the preferred node for charging in the following round due to their higher weights. The membership functions are assumed to have a Gaussian distribution of the inputs (thresholds), since the thresholds are not uniformly dispersed as illustrated in Figs. 9.5, 9.6 and 9.7.

The individual charging infrastructures have a different load profile, and the priority of each EV in the single socket to which they are connected is evaluated based on the load profile, SoC, and RES.

Later, when each member has agreed on the preferred nodes and the transaction has been executed, a copy of the generated block is transmitted to all nodes due to the inheritance of BC technology operation.

In the decision-making process for charging-discharging electric vehicles, the charging profile of the charging locations and the energy from RESs present in that charging area play a decisive role. If there are high demands for electricity, fleets will avoid charging the EVs, to reduce the risk of grid congestion. Furthermore, since with the enormous amount of energy produced from renewable sources, at the same time (as with solar energy), there would be a potentially harmful situation for the equipment installed on the electricity grid. Thus, to manage this problem, the unused energy can be accumulated in electric vehicles and storage installations present at homes or in the grid.

Thus, the method presented above was applied in two different scenarios.

1. **First scenario**

Exploiting the fuzzy logic, at the start of each round (every 5 min), fuzzy weights between 0 and 1 will be produced for each participant belonging to the network, based on their SoC, load profile, and RESs.

Higher weighted members will appear as preferable nodes in the distributed ledger and will be given permission to link their blocks to the chain. Afterwards, the modified block will be sent to all participants and only after verification will it be added thus to the chain.

The blocks hold the hash code of the previous block, the identifying information, and the power demand of the corresponding EV. The charging of the EVs, in this scenario, is charged in a fair way, based on their status, and is not causing problems to the smart grid.

2. **Second Scenario**

A slightly modified algorithm is used compared to the previous one.

The choice consent mechanism remains the same, but instead to vote the nodes with the highest weights, at each turn those with fuzzy weights above the average of all the weights of the net will be chosen. Hence, electric vehicles, which resulted in higher than average fuzzy weights, enable block forging, resulting in the charge permit. The possibility of having a fuzzy weight that can exceed the fixed threshold, to make the process democratic and transparent, as per BC technology, is distributed evenly over all the network nodes. Thus, when a node is loaded, the fuzzy weight associated with it in the next turn would be smaller determined by the higher SoC of that ones that did not have a chance for charging. At this point, the possibility of grid congestion during high demand periods and decreases of voltage levels would be smaller if electric vehicles in the grid were in the grid. The knot selection mechanism in each round is shown in Fig. 9.8.

Dynamic Pricing Through Blockchain Technology

Blockchain technology allows you to have safe and economic transactions, acting as distributed databases capable of having a growing list of data records (blocks). Transactions are ensured by nodes, which are basically computers, which can also be managed by future prosumers on the network, exactly as is happening for BitCoins, without the need for third parties to guarantee the correct transaction, which in this case would take the name of RESCoin, digital currency linked to RES production. Given the possibility of contributing to the balancing of the network, with the batteries of electric vehicles, instead of contributing to the imbalance, there would be a compensation for the prosumer, with the RESCoins managed by the blockchain in an individual wallet for each driver. Figure 9.9 shows a schematic of the business model (Lazaroiu and Roscia 2018, 1196).

The open public network (unpermissioned ledgers) does not have a "property" or a key actor. The objective is to permit each member (nodes) of the network to give a contribution to update the data on the ledger and to have all the copies of operations or to have the identical copies of all approvals through consent.

With this BC application, no member will be able to block a transaction, and it will be added to the ledger once the consensus has been obtained among all the nodes of the blockchain. This is thanks to the fact that the network confers immutability of all documents or transitions, which require high security in terms of consent, such as smart energy contracts (RESCoin) (see Fig. 9.10).

Thus, the blockchain protocol should contain in the ledger the period between an offer is formulated and the request acceptance (**real-time clock** (RTC)). The RTC should always remain appropriate with the average confirmation time for reducing the computing requirements. The ledger should have a **data storage** (by 8 to 32 bytes per transaction), the **transaction value** (that varies with the characteristics of the transaction), and the **transaction fees** (not to interrupt the network from unnecessary transitions, and a penalty of the transition if this one does not succeed). The **total hash rate** is the processing capacity of the RESCoin network, and greater the hashing power, the larger is the security and resistance to attacks.

The framework for the blockchain dedicated to EVs is shown in Fig. 9.4, where the various "participants" can be highlighted, who can both sell and buy energy, in this specific case only from renewable sources, in relation to the fact that the energy offered is greater or less on the sales market; at the moment in for the following quarter of an hour (clock), a specific price is defined, (RESCoin for kWh), which can be accumulated in a virtual money box/wallet or are spent, again through BC (Lima 2018a, 1; Lima 2018b, 106).

The meeting point is linked to the fact that the incentives are of a public nature, while the charging stations are of private individuals, who want to earn, to obtain an affordable price for all the players in the intelligent electric mobility sector.

The protocol used must be designed to "be lightweight," i.e., to minimize the quantity of data to be stored in the BC. For this reason the "offers" are kept for a defined period of time, which gives the possibility to the driver/prosumer of an EV

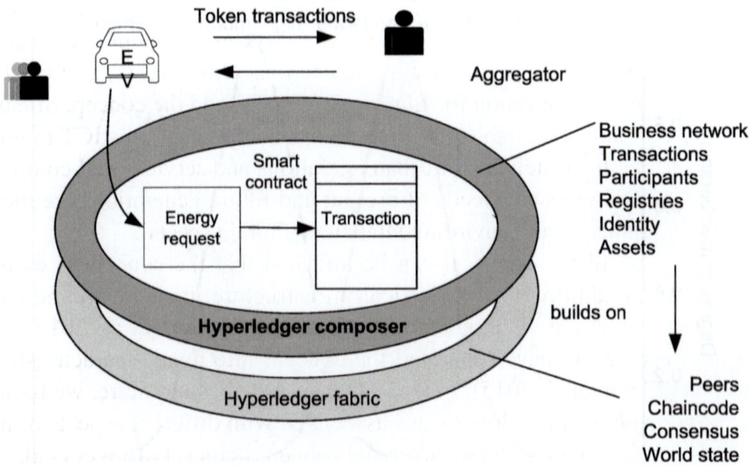

Fig. 9.4 Layout of the proposed blockchain scheme

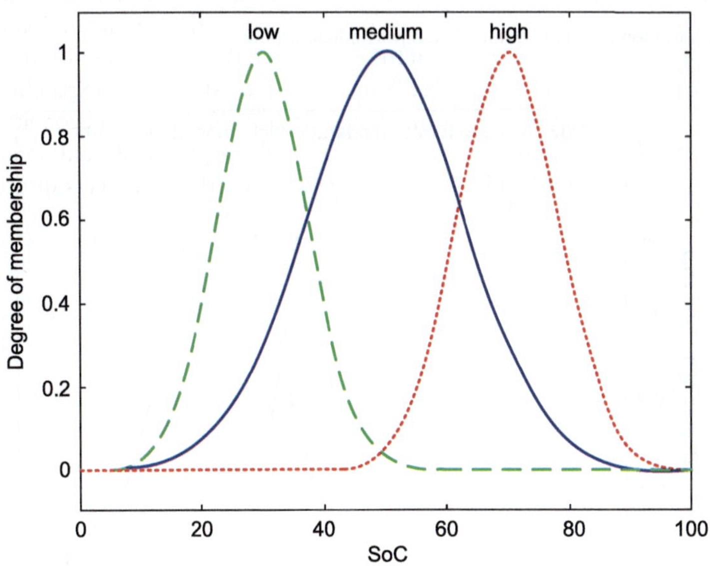

Fig. 9.5 Variation of degree of membership with SoC

to accept, but, after the time deemed acceptable, the process will be interrupted and the offers must be reshaped, such as an auction (Knirsch et al. 2018, 71).

The participants can be both the direct users or owners of the electric vehicles, and the managers of the charging stations themselves, as each of them is part of the blockchain structure, using parameters, as for an app on a smartphone, for example, "ID Name," "email," "wallet," "smartphone number," etc.

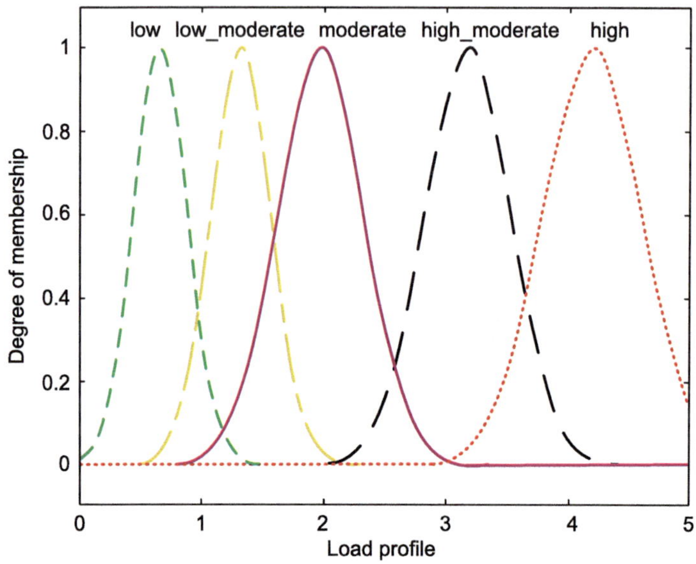

Fig. 9.6 Variation of degree of membership with load profile

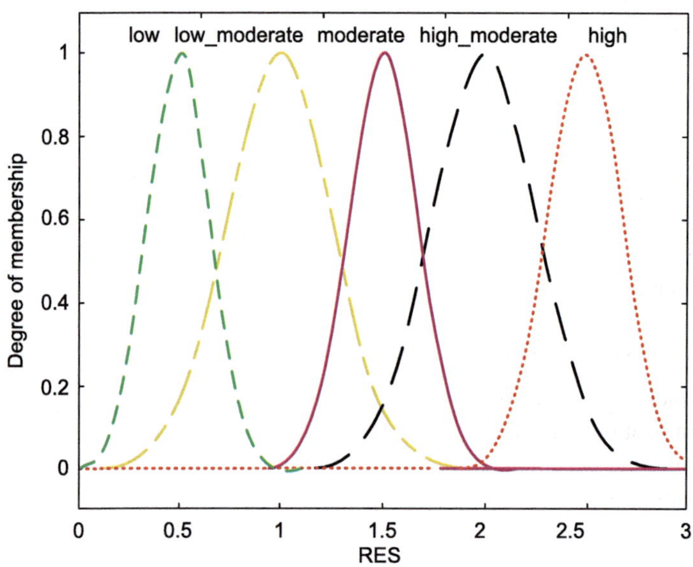

Fig. 9.7 Variation of degree of membership with RES

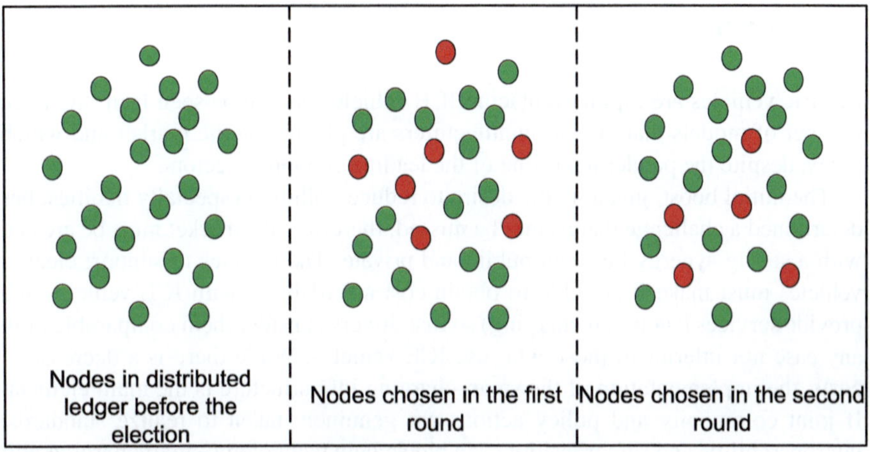

Fig. 9.8 Choosing mechanism in each round

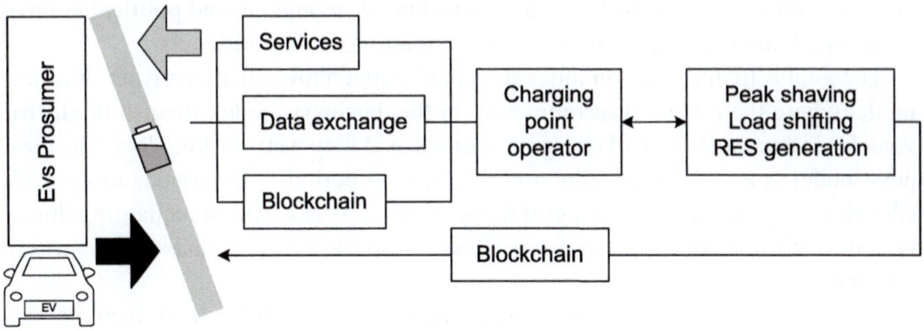

Fig. 9.9 Scheme of the business model

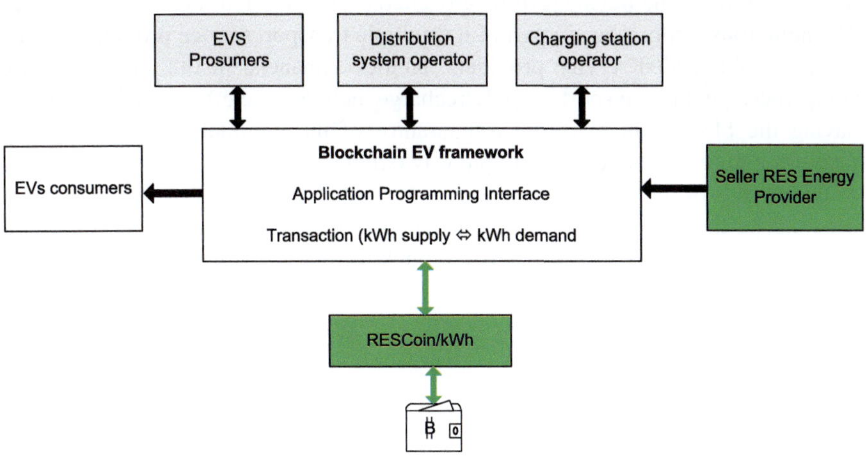

Fig. 9.10 Blockchain EVs framework

Conclusions

Electric vehicles are rapidly replacing ICE vehicles, as can be seen from the large number of models that all car manufacturers are placing on the market and which today, despite the pandemic, is one of the leading economic sectors.

The initial boost, given by the desire to reduce pollution especially in cities, has determined a challenge that cannot be missed; therefore, the market must be created with a strong synergy between public and private. The policies to support electric vehicles must make it possible to obtain cost affordability with ICE vehicles and provide services (such as recharging) so that drivers consider them comparable or in any case not inferior to those who use ICE vehicles. While there is a decrease in costs, the implementation of the smart charging infrastructure is the main element. If joint community and policy actions are genuinely taken to realize standards, interoperability, and safe transitions via blockchain technology, smart charging will shift from promise to practice by 2030.

The aspects of the future mobility system linked to the massive introduction of electric vehicles were considered, from a technical, economic, and political point of view, which are the basis of the decision-making skills of politicians.

The goal is to introduce an innovative public (but above all also private) business model where there is no discrimination. In this business model, the use of electric vehicles is encouraged in all charging stations as if they were parking lots. The business model is encouraging reduction of charging periods and monetization of the EVs charging transferred to the grid in the event of a peak load or recharging during periods with high productions from RESs based on a safe, transparent, and reliable model.

Blockchain creates scenarios of great interest in the mobility field, from its simplest forms (a vehicle – multiple owners with ownership management in charge of the platform) to the more complex forms of integrated mobility (management of a mobility coin for the payment of heterogeneous vehicles and means of transport). The numerous actors who are part of it (multiple transport service providers, users/citizens, value-added service providers, builders, financial institutions, insurance companies, public transport, supply/recharge networks, etc.) could interoperate having the blockchain platform as guarantor of the immutability of ownership records and use, to operate in a simple way (single interface), secure (immutability of the records of the trips/services used), integrated (the record of the trip/service used allows the relative cost to be calculated in real time by activating the reimbursement to the service provider), and quick to follow changes to the ecosystem (e.g., new services/actors).

References

Alahmad A, Rayyan A'a, Al Juheshi F, Sharif H, Alahmad M, Shuaib K, Aljuhaishi N (2016) Overview of ICT in the Advancement of Electric Vehicle Penetration. 2016 12th Int. Conf. Innov. Inf. Technol:1–6

Alladi T, Chamola V, Rodrigues JJPC, Kozlov SA (2019) Blockchain in smart grids: a review on different use cases. Sensors 19(22):1–25

Andoni M, Robu V, Flynn D, Abram S, Geach D, Jenkins D, McCallum P, Peacock A (2019) Blockchain technology in the energy sector: a systematic review of challenges and opportunities. Renew Sust Energ Rev 100:143–174

Castor A (2017) A (Short) guide to Blockchain consensus protocols. https://www.coindesk.com/short-guide-blockchain-consensus-protocols. Accessed 10 Apr 2021

D'Oriano L, Mastandrea G, Rana G, Raveduto G, Croce V, Verber M, Bertoncini M (2018) Decentralized blockchain flexibility system for Smart Grids: Requirements engineering and use cases. In: 2018 International IEEE Conference and Workshop in Óbuda on Electrical and Power Engineering (CANDO-EPE), pp 39–44

Dorri A, Kanhere SS, Jurdak R, Gauravaram P (2017) Blockchain for IoT security and privacy: the case study of a smart home. In: 2017 IEEE International Conference on Pervasive Computing and Communications Workshops (PerCom Workshops), pp 618–623

Doukas DI, Bakas P, Marinopoulos A, Kim H (2013) Energy storage integration into grid connected utility-scale photovoltaic (PV) systems. In: 2013 IEEE Grenoble Conference, pp 1–6

Ferdous MS, Chowdhury MJM, Hoque MA (2021) A survey of consensus algorithms in public blockchain systems for crypto-currencies. J Netw Comput Appl 182:1–28

Gao J, Asamoah KO, Sifah EB, Smahi A, Xia Q, Hu X, Zhang X, Guishan D (2018) GridMonitoring: secured sovereign blockchain based monitoring on smart grid. IEEE Access 6:9917–9925

Guan Z, Si G, Zhang X, Longfei W, Guizani N, Xiaojiang D, Yinglong M (2018) Privacy-preserving and efficient aggregation based on Blockchain for power grid communications in smart communities. IEEE Commun Mag 56(7):82–88

Hertz-Shargel B, Livingston D (2019) Assessing blockchain's future in transactive energy. Atlantic Council:1–35

Houben R, Alexander S (2018) Cryptocurrencies and blockchain. In: Paper prepared for the European Parliament's Special Committee on Financial Crimes, Tax Evasion and Tax Avoidance. European Parliament, pp 1–86. https://www.europarl.europa.eu/cmsdata/150761/TAX3%20Study%20on%20cryptocurrencies%20and%20blockchain.pdf. Accessed 10 Apr 2021

IRENA (2019) Innovation outlook: smart charging for electric vehicles. International Renewable Energy Agency Abu Dhabi. https://www.irena.org/-/media/Files/IRENA/Agency/Publication/2019/May/IRENA_Innovation_Outlook_EV_smart_charging_2019_Old.pdf?la=en&hash=080C746B6183DF9C528FA4C4C4FAE801F07D93C2. Accessed 10 Apr 2021

Jiang Z, Tian H, Beshir MJ, Vohra S, Mazloomzadeh A (2016) Analysis of electric vehicle charging impact on the electric power grid: Based on smart grid regional demonstration project — Los Angeles. In: 2016 IEEE PES Transmission & Distribution Conference and Exposition-Latin America (PES T&D-LA), pp 1–5

Khaki B, Chung Y-W, Chu C, Gadh R (2019) Probabilistic electric vehicle load management in distribution grids. In: ITEC 2019–2019 IEEE Transp. Electrif Conf Expo, pp 15–20

Knirsch F, Unterweger A, Engel D (2018) Privacy-preserving blockchain-based electric vehicle charging with dynamic tariff decisions. Comput Sci Res Dev 33:71–79

Lazaroiu GC, Roscia M (2018) RESCoin to improve Prosumer Side Management into Smart City. In: 2018 7th International Conference on Renewable Energy Research and Applications (ICRERA), pp 1196–1201

Lazaroiu GC, Roscia M (2019a) New approach for Smart Community Grid through Blockchain and smart charging infrastructure of EVs. In: 2019 8th International Conference on Renewable Energy Research and Applications (ICRERA), pp 337–341

Lazaroiu GC, Roscia M (2019b) New approach for smart community grid through Blockchain and smart charging infrastrcuture of EVs. In: 8th International Conference on Renewable Energy Research and Applications, pp 337–341

Lazaroiu GC, Roscia M, Saatmandi S (2020) Blockchain strategies and policies for sustainable electric mobility into Smart City. In: 2020 International Symposium on Power Electronics, Electrical Drives, Automation and Motion (SPEEDAM), pp 363–368

Liang G, Weller SR, Luo F, Zhao J, Dong ZY (2019) Distributed Blockchain-based data protection framework for modern power systems against cyber attacks. IEEE Trans Smart Grid 10(3):3162–3173

Lima C (2018a) DLT/Blockchain architectures and framework – system-of-system approach. IEEE NIST Blockchain Global Summit:1–30. https://blockchain.ieee.org/images/files/pdf/20180917-blockchain-architecture-and-reference-frameworks_-_c-lima.pdf. Accessed 10 Apr 2021

Lima C (2018b) Developing open and interoperable DLT\/Blockchain Standards [Standards]. Computer 51(11):106–111

Lucas A, Prettico G, Flammini MG, Kotsakis E, Fulli G, Masera M (2018) Indicator-based methodology for assessing EV charging infrastructure using exploratory data analysis. Energies 11:1–18

Manjunatha AP, Korba P, Stauch V (2013) Integration of large battery storage system into distribution grid with renewable generation. In: 2013 IEEE Grenoble Conference, pp 1–6

Mearian L (2020) What is blockchain? The complete guide. Computerworld. https://www.computerworld.com/article/3191077/what-is-blockchain-the-complete-guide.html. Accessed 10 Apr 2021

Nair PR, Dorai DR (2021) Evaluation of performance and security of proof of work and proof of stake using blockchain. In: 2021 Third International Conference on Intelligent Communication Technologies and Virtual Mobile Networks (ICICV), pp 279–283

Nguyen CT, Hoang DT, Nguyen DN, Niyato D, Nguyen HT, Dutkiewicz E (2019) Proof-of-stake consensus mechanisms for future Blockchain networks: fundamentals, applications and opportunities. IEEE Access 7:85727–85745

Pilkington M (2015) Blockchain technology: principles and applications. In: Olleros FX, Zhegu M (eds) Research handbook on digital transformations. Edward Elgar, 2016

Pop C, Cioara T, Antal M, Anghel I, Salomie I, Bertoncini M (2018) Blockchain based decentralized management of demand response programs in smart energy grids. Sensors 18(1):1–21

Putrus GA, Suwanapingkarl P, Johnston D, Bentley EC, Narayana M (2009) Impact of electric vehicles on power distribution networks. In: 2009 IEEE Vehicle Power and Propulsion Conference, pp 827–831

Samuel O, Javaid N, Awais M, Ahmed Z, Imran M, Mohsen G (2019) A Blockchain model for fair data sharing in deregulated smart grids. In: 2019 IEEE Global Communications Conference (GLOBECOM), pp 1–7

Sivanandam SN, Sumathi S, Deepa SN (2007) Introducction to Fuzzy Logic using Matlab. Springer, Berlin

Skugor B, Deur J (2014) Dynamic programming-based optimization of electric vehicle fleet charging. In: 2014 IEEE International Electric Vehicle Conference, pp 1–8

Tayarani H, Jahangir H, Nadafianshahamabadi R, Golkar MA, Ahmadian A, Elkamel A (2019) Optimal charging of plug-in electric vehicle: considering travel behavior uncertainties and battery degradation. Appl Sci 9:1–15

Chapter 10
Smart Mobility in Africa

Alexandra Catalina Sima, Mariacristina Roscia, Claudia Laurenta Popescu, Mihai Octavian Popescu, and Jacques Jansen

Abstract The urban population of Africa is increasing exponentially compared to Europe, and in fact it is forecasted that the urban inhabitants will grow by 62% until 2050. Nowadays, more than 30% of the population does not have access to electrical and thermal energy. The Global Sustainable Development Goals have set established worldwide access to sustainable energy by 2030. At this particular historical moment, in Africa, equitable, sustainable energy and responsible energy industries should be promoted to ensure access to clean energy for the population but also to services such as transport systems and the water. The World Bank reports "that a country's growth in measured gross domestic product is directly curtailed by poor public transport infrastructure." Access to these key elements is essential to reduce deficit and drive economic development, and a sustainable approach is also needed to plan energy services and smart mobility. Africa will become "the next big market" determined by fast rise of emerging economies. The smart cities, thanks also to sustainable mobility, will represent the key element in Africa to ensure improved quality of life for Africans, decreased energy use, and environmental influence, bringing benefits globally. Fundamental in the ongoing development countries and beyond will become the implementation of smart grids and Internet of Things (IoT), electric vehicles, and charging systems, the spread of the 5G network, a new holistic vision of the territory where the user is transformed into a hub, and being a prosumer and a user of new technologies, thanks to the renewable energy widespread throughout the territory (solar, hydroelectric, wind). The number of electric vehicles in Africa has the potential to significantly increase by 2030, and the challenge will

A. C. Sima (✉) · C. L. Popescu · M. O. Popescu
University POLITEHNICA of Bucharest, Bucharest, Romania
e-mail: claudia.popescu@upb.ro; mihaioctavian.popescu@upb.ro

M. Roscia
University of Bergamo, Bergamo, Italy
e-mail: cristina.roscia@unibg.it

J. Jansen
University of Pretoria, Pretoria, South Africa

G. C. Lazaroiu et al. (eds.), *Holistic Approach for Decision Making Towards Designing Smart Cities*, Future City 18,
https://doi.org/10.1007/978-3-030-85566-6_10

also be in charging infrastructure, powered by grid-connected or off-grid renewable solutions, in step with market interests. This chapter wants to give a holistic view of the transportation system in African cities, so that decision makers and future research can build smart cities that can give real opportunities for improvement for developing countries such as Africa. A key consideration is the challenge of knowing what needs to be developed first; there are so many priorities to be addressed, so it is not clear to the decision makers themselves, to politicians, on what needs to be focused, even in terms of investments, and yet, this effort could be the most effective way for Africa to seize the opportunities that technology can offer, guaranteeing their position as the energy supplier of tomorrow, all over Europe.

Keywords Africa · Electric mobility · Smart grids · Sustainable development

Introduction

The world economic forum describes the future of energy as one that would consider the long-term carbon goals to relieve the current negative environmental impact as in Africa the air pollution has high impact on human health (2013 Bank World). This would mean the installation of renewable energy sources at homes and businesses. In essence, causing the decentralization of electricity generation and subsequently decentralization of filling stations. Electric vehicles would be recharged at home from their own generated systems. Decentralized energy would require that the home become "smarter" to effectively manage energy to all the demands. Smart homes would not only manage their resources but would also manage occupants' lives, entertainment, and comforts. Optimally managed smart homes would from a bottom-up approach help in achieving the goals of smarter cities, goals of maximizing energy and the optimum usage of the available resources and infrastructure such as the management of traffic from electric vehicles on the roads. The smart mobility which is diffused worldwide will be integrated in the transport infrastructure.

The World Bank emphasizes that "a country's growth in measured gross domestic product is directly curtailed by poor public transport infrastructure," and the emerging markets have the opportunity to change the future for the next generations by rethinking and redesigning their transportation.

The concept of a sustainable city became global in 1990 (Sorensen et al. 2014, 10), although there is still no clear definition of what defines urban areas as really sustainable (Herbert 2004, 45). Cities have various characteristics: number of populations; urban services like water, electrical energy, education, and health; so as to transform cities by integrated systems available to all for increasing life's quality ((Satterthwaite 1997, 1667; Timothy et al. 1991, 15). This must be guaranteed by economic and social policies for the sustainable development (Anastasiadis and Metaxas 2010, 64).

About 70% of the world inhabitants will move to urban area by 2050 which will lead to the development of cities (UN 2019). This is expected majorly in countries under development, which will face an exponential growth over the following decades. Between 1990 and 2005 the urban population increased with 1.2 mil/week, which is 165 k per day, and this value is expected to double (WHO 2010, 4). The African cities had the highest increase determined by economic necessities.

Africa continent will reach 2.4 billion people in the following decades, and this will bring more and more needs and requirements in cities and village areas. The fully connected smart cities will be planned to improve the quality of life, distributing energy, water, and health, creating jobs, redeveloping the territory, managing demographic challenges, reducing emissions, promoting green sources, and redistributing the richness.

The centric element of smart cities will be the citizen and thanks to the IoT will ensure information from various objects to obtain data, for the management of the SC, this without any proprietary software favoring a hardware supplier. The fast-increasing urban population puts a strain on the transport and energy infrastructure. In 2015, 1.5 million electric and plug-in hybrid vehicles were adopted worldwide, and Africa is the continent that is urbanizing very fast, by 4% each year compared to the 2% world average.

Electric vehicles offer the potential to reduce pollution, improve the environment, and also create new job opportunities; obviously, this needs to be supported by a network of electric charging stations large enough to meet the growing number of electric vehicles. Rooftop solar panels can be installed all over Africa, enabling the location of electric vehicle charging stations, thanks to the abundant supply of sunlight. One challenge Africa will face is electric load and capacity, with electric vehicles putting additional pressure on the electricity grid. Therefore, intelligent vehicle-to-grid charging systems are fundamental to maximize energy optimization.

Africa Energy Sector

Africa requires to install more than 7GW of new generation annually to meet demand (Foster and Briceno-Garmendia 2010, 2) or up to a tenfold increase in generation to meet the requirements by 2065. Some data indicate that demand will grow by 30% by 2050.

The largest electricity producer, South Africa, is facing a supply crisis. Many of the existing generation plants are outdated or lack sufficient maintenance and excessive failure rates. Of the installed capacity of 58,195 MW, power generation has decreased by 8.8% since 2010. Availability factor went from 71.9% in 2018 to 65% in 2021 and around 20% in total over the past 20 years.

In the rest of the SSA, the installed capacity remains relatively low. After South Africa, the countries with the highest installed capacity are Nigeria with 12,522 MW (although only around 3000 MW of these are almost every day), Ghana with

4399 MW, Ethiopia with 4244 MW, Kenya with 2819 MW, Democratic Republic of Congo with 2677 MW, Zimbabwe with 2240 MW, Botswana with 920 MW, and Senegal with 864 MW; Chad has 125 MW of installed capacity, Rwanda has 218 MW, and South Sudan generates 130 MW.

There is therefore a huge disparity between the generation capacity and the rural nature of many communities, meaning that about 600 million inhabitants have energy access. The main among these are South Sudan (5.1% access), Chad (6.4%), and Burundi (6.5%). Currently, one-third of the African population has no access to electricity, as shown in Fig. 10.1.

In some countries like Egypt and South Africa, the electrical power system is supplying the designated customers, but in other countries, the population access to an electrical network is about 20%, with some countries having only a 5% value. Thus, classic fueled generators and remote grids are in place.

The Africa energy sources are gas and oil in North and West parts of Africa, hydropower in Central and Eastern Africa, and coal in South Africa. The integration of renewable energy sources is increasing. The hydropower has high potential in sub-Saharan Africa, but only 45% is used which represents a share of this resources (Juld et al. 2005, 2990). Other resources could be solar and biomass which are used in residential facilities and not for large-scale energy production (Deichmann et al. 2011, 215).

In Africa, the element of distributed generation is consolidated; in fact, domestic and solar systems have been used for many years for rural electrification, and this makes it easier to implement distributed energy, especially solar photovoltaics, in urban and rural centers. This leads to more efficient and low cost power generation, effectively creating prosumers naturally and reducing carbon emissions.

Electricity costs remain high, the costs for generation with diesel ranges from 0.35 $/kWh to more than 1 $/kWh on islands in the Pacific and some remote location on the continent. Thus, the generation capacity is not enough to cover the growth demand and low security of supply. Unfortunately, despite the considerable

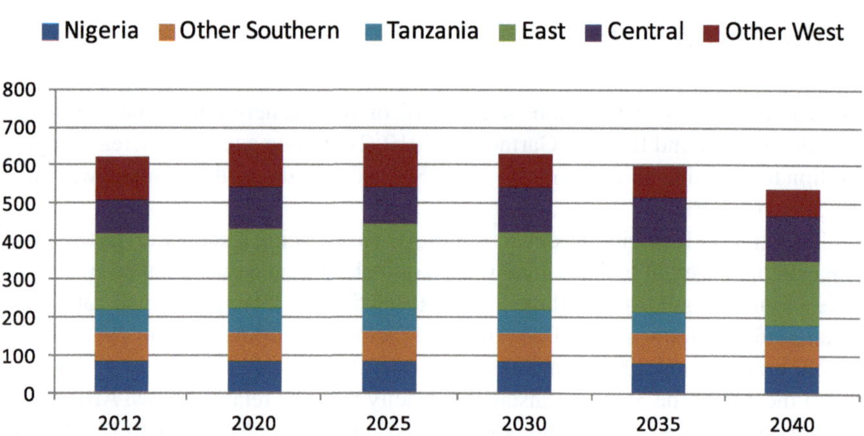

Fig. 10.1 Population without access to electricity

distribution of renewable energy, these sources are heavily underused, like solar energy, hydrogeneration in many countries, and wind on coasts. The smart cities can represent a resilient and dynamic organism.

Electricity consumption in Africa will double by 2030 to 1012 TWh (IEA 2010) and requires a concrete extension of electricity infrastructure (Parshall et al. 2009, 2395). The IEA estimates that for ensuring worldwide access to electrical energy by 2030, additional investments in energy sector of 35 billion dollars /year will be required, a large share being needed in Africa.

Smart Mobility in Africa and the Needs of Smart Cities

The global electric vehicle (EV) market is evolving and growing rapidly both thanks to new technologies (resulting in ease of use and reduced costs) and the urgent need to tackle climate change, which is leading countries to, globally, the elimination of vehicles with internal combustion engines (ICE). Consequently, automakers are increasingly designing EVs, with a multi-billion dollar sale, thanks also to tax incentives. By 2030, electric vehicles are projected to account for 35% of global vehicle sales.

The electric vehicle market across Africa is very promising, increasing investor interest and engagement in the sector and the momentum for growth. National and urban governments are including bold electric vehicle targets in their plans to decrease greenhouse gas emissions. Private sector investors and developers are exploring market opportunities in more countries, and global change will lead to more used electric vehicles entering the African market. At the moment, however, electric cars are still very little widespread in most of Africa. In South Africa, considered the largest electric vehicle market on the continent, only 1000 electric vehicles were purchased in 2019, out of more than 12 million vehicles on South Africa's roads, as can be seen from Fig. 10.2, and even fewer electric cars are in the market function in most other African countries.

The cities represent the main point for worldwide sustainability, and they need to get smart as fast as possible to reduce energy infrastructure costs. For the mobility sector, the energy need is linked to the diffusion of transportation in the cities' area and thus has high impact on the utilization of energy. The realization of smart cities requires optimization of electrical systems by integrating as much as possible renewable energy sources for increasing efficiency and security of supply. The lack of stability and capital investments and inadequate infrastructures currently limit innovations, while investments could be made, which, by attracting capital, would lead to the strengthening of the networks already built on smart grids, offering important possibilities for Africa. There are, however, several issues that African countries need to address to be able to reap the benefits of the electric vehicle movement.

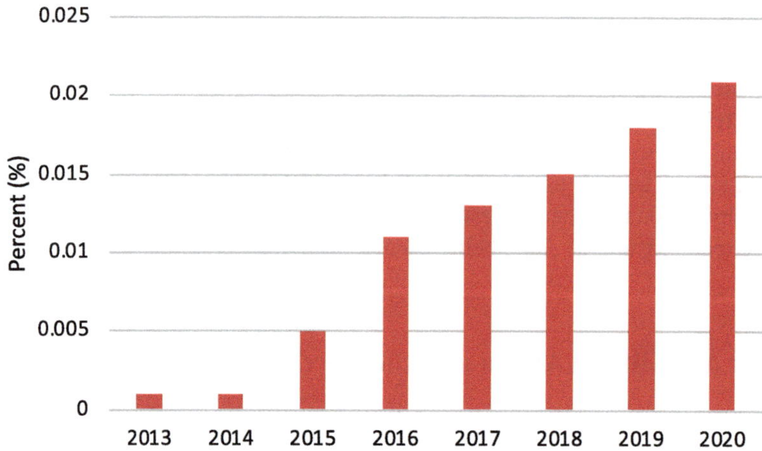

Fig. 10.2 Electric vehicle stock share evolution in South Africa

The main areas toward *African Transformation* are the following:

1. **"Smart governance":** this is the first element; without the political will to turn the page, Africa will not be able to seize the opportunity that technologies can offer: facilitating the rural electrification project, balancing the electricity costs, supporting the management and development of demand policies to realize the use of distributed resources and the interoperability of nowadays grids with the future ones, and bringing to the realization of smart grids.
2. **The redesign of distribution grids** to make use of smart metering and smart reclosers to improve security of supply and power quality reliability indices, fulfilling the requirements for village electrification with minimum costs toward smart villages (Fletcher and Strunz 2007, 791).
3. **Disruptive technologies**: it is necessary to start from the implementation of disruptive technologies for the optimization of interoperability and development of solutions for the needs of electrification (Welsch et al. 2013, 336). The distribution on the large territory will be different at regional and national level, in order to satisfy the requirements and goals of the various companies and the various local markets. ICT already contributes to about 7% of Africa's GDP, which is above the world average. ICTs will provide the ability to change the businesses and government in Africa by driving innovation and economic growth. The diffusion of the Internet is at 17% below the 28% world average. In addition, there are African countries where the population have access to a cell phone rather than water or electricity. The need for high-speed broadband and the introduction of smart meters for electricity exchange possibilities are parts of the goals to make smart cities in Africa.
4. **Renewable energy** leads above all to a reduction in pollution and cost (fuel costs), especially in remote areas on the continent (Teravaninthorn and Raballand 2009). Using locally available renewable energy sources improves the security of supply, which is critical to supporting the growth of enterprises and industrial

customers dependent on electricity. There are currently insufficient incentives in Africa to switch to a sustainable energy, and so photovoltaic production is not cost-effective for most consumers.

5. **Smartphone**: Africa has a mobile telephony market, after South Asia, which has developed better than in other globe areas, as for 300% per year in Cameroon and Kenya. On average, the distribution of smartphones has reached more than 72% across Africa. The physical infrastructure for telephony is lacking in almost all African countries. The World Bank and the African Development Bank indicate more than 650 million new mobile phone users in Africa, more than Europe.
6. **Smart urban infrastructures** are required to be sustainable, with the citizens involvement increasing. Sustainable urban infrastructure should not in-charge governments with debts almost impossible to return or end users with high costing tariff and must not negatively impact the ambient. Thus, the reduction of carbon emissions during construction and operation should be done. The transportation infrastructure should be resilient as much as possible. The different transportation systems are function of territory development, and the energy requirements must be tailored function of the cities and needs of the inhabitants.
7. **Smart grid (SG):** African electricity could be guaranteed by distributed resources, in particular renewable energy units. This capacity could be optimized by using smart meters, accounting for the variable production of renewable sources. Smart grids can reduce a deeply felt problem in Africa by identifying areas of illegal interception and network leaks, which end up burdening costs and therefore end users (Welsch et al. 2013, 336). With these intelligent grids, any connected electronic device can communicate the energy consumed or produced in real time. From the analysis of the information collected, decision support is created, allowing a more efficient energy system (Dlodlo et al. 2014, 95). A bidirectional system can be generated, strategically positioned, and connected to a cloud, capable of managing the data acquired from different sources, for the analysis and decision making of power flows in real time. The solar and wind production forecasts in Africa leads to a better utilization of "green" generators, and end users and industry can shift their gaily operation to periods when they can make maximum use of these resources. The complex development of smart grids in Africa is dependent on territory economical and political differences, the requirement to supply village areas.

Rethinking Smart Mobility in Africa

It is estimated that 80–90% of vehicles imported into Kenya are second-hand, and policy makers must work to ensure that Africa does not become a landfill for polluting vehicles but rather grows its electric vehicle market. However, this will not be enough since, despite the improvement in battery technology and charging, with an even greater distance traveled, and the reduction of associated costs, the main obstacle to the electric vehicle sector in Africa, as well as in Europe, is the lack of electricity and reliable infrastructure. Electric vehicle charging stations are rare and found only in large urban centers, so a well-planned network of charging stations with sufficient energy capacity is needed. Furthermore, in many African countries, the electricity grid is already under great pressure and will require significant investments in grid capacity and grid integration to manage the demand for electric mobility. Intelligent and electric transport services will not develop without an extensive and secure network of access to electricity in their area of business. Finally, it is highly unlikely that the demand for electric vehicles will increase if we consider the current electric grid model, which is inefficient and with frequent voltage drops.

Synergy must be understood between the incentives for the growth of smart grids, the use of renewable energy, and the use of electric vehicles; the growth of these sectors at the same time will allow broad-spectrum benefits in the African continent. There needs an enabling policy toward electric vehicles, targeted investments, and market incentives that will lead to the growth of smart grids and provide energy storage solutions for solar PV, wind, hydro, and other renewable and clean energies with the natural development of the electric vehicle market.

With the development of intelligent charging of electric vehicles via cloud-connected devices, excess energy from electric vehicle batteries can be fed back into the grid. Some companies in South Africa are already integrating vehicle-to-grid (V2G) technology which enables bi-directional charging, enabling electric vehicles to help stabilize the power grid during peak electricity demand; most notably, the national e-Mobility program plans to bring fast chargers and V2G functionality into the smart grid ecosystem.

Some interesting and successful projects are listed below:

(a) **Mobility for Africa** is a women-led start-up that provides small farmers with electric tricycles to transport their produce from farm to market and is partnering with Rift Valley Energy to harness energy from hydroelectric plants, and some charging points will use containerized photovoltaic solar systems.
(b) **Zembo** is an electric motorcycle start-up in Uganda. In just 2 years, more than 200 electric motorcycles have been sold to boda-boda taxi drivers and has created over 15 battery change and solar charging stations in Kampala. Zembo partnered with SafeBoda, a local racing call app with a reputation for trustworthy drivers, to facilitate their entry into the market.
(c) **NopeaRide (and EkoRent)** is the continent's first 100% electric taxi service. The company operates a small fleet of electric vehicles and charging stations located throughout the city, with drivers and customers accessing the platform via a mobile application. The new charging stations will be implemented efficiently and economically using data derived from Nopea applications (both drivers and passengers), Nopea vehicles, and charging stations and by monitoring traffic flow and popular travel routes. The EkoRent model is highly replicable, and it is expected that by enabling the company to expand its offering in Nairobi, the project will attract private investment in electrification of private and corporate transportation in the cities of the region.

Another central node for electric mobility in Africa is represented by the type of cables, which, with intelligent charging devices, will provide both power and communication. The IEC 62893 and EN 50620 standards specify and recommend the cable types and test methods that must be followed. This is a fundamental aspect, as the cables will have to withstand particularly stressful environmental conditions: most of the charging stations for electric vehicles are outdoors, exposed to considerable temperatures, and with cables that are subject to high volumes of energy; as a result, devices can get extremely hot. To ensure the imminent increase in charging demand for electrified vehicles, the use of approved cables at charging stations is required to protect charging infrastructure.

A further development key is the implementation of measurement. While smart metering is not always necessarily fit for purpose, in many cases it is an efficient and effective way to handle large load consumers. For other classes of consumers, prepaid meters have proven very effective in increasing revenue collection and reducing non-technical losses. Unless utility companies get revenue metering and management as part of their business law, they will continue to operate with cash flow insecurity, diverting attention from the business of providing safe, secure, and affordable energy for consumers.

Strategical Challenges for E-Mobility in Africa

In a city with high density of population, a good solution for sharing electric vehicles is to have a high number of vehicles for the inhabitants. In this way, the offer incidence as the number of vehicles over the number of kilometers to be served is sufficient to cover the time-shifting of the demand.

When this ration is more than 60%, up to 90%, the" free floating" can be introduced. This approach is not working for lower ratios as there are a limited number of vehicles available to customers, and the user cannot find a vehicle in the vicinity.

Considering the emissions of pollutants, a taxation based on contribution to pollution would lead to the adoption of electric vehicles. A medium size electric vehicle pollutes, from its production until the final use, less a fuel vehicle. In France, the sharing of electric vehicles is highly diffused as the charging infrastructure and reserved parking are available but at high costs. Many cities are currently not highly served by electric vehicles, but the transportation models based on electric vehicles will be different in the future.

Thus, for low-density environments, it would be useful to consider a seasonal strategy in particular in tourist-based areas as the demand is irregular. In case of increased demand for electric vehicle sharing, the charging infrastructure and transportation development must be done coherently. Thus, the operating environment and the management model must account for:

- The separation of pollutant energy sources
- The reduction of dependence on local energy grid

- Using communication systems to enable use of "electric vehicle sharing"

The strategical implementation of these actions can be done in three stages:

- Year 2021: to integrate communication model of the free flow managed by a main processor with intelligent algorithms for vehicle-to-vehicle (as illustrated in Fig. 10.3)
- Up to 2024: this strategical implementation is illustrated in Fig. 10.3 and can be implemented in rural and island locations. A model can be set up to realize the energy management such that to program the charging process to use clean energy for the electric vehicles sharing services and thus to improve the management algorithms and to benefit from vehicle-to-grid (V2G), grid-to-vehicle (G2V), Internet-to-grid (I2G), and Internet-to-vehicle (I2V). In this way, a further protection of the environment and a higher degree with respect to the main supply (influence of charging infrastructure on the network, optimization of the energy mix, power quality) can be achieved. The setting of information and communication technologies is a requirement for managing service models in medium and large density scenarios, involving vehicles, grids with renewable energy sources, and storage systems, charging stations' infrastructures. A real-time communication on different layers can be achieved for charging, traffic and parking information, and interoperability with other transportation means. A possible model of electric vehicle sharing is proposed and illustrated in Fig. 10.4.
- Up to 2035: the introduction of modular-open source vehicles, small capacity batteries and reduced sizes for future environment protection, wireless charging diffused on traffic routes and associated with fixed smart charging stations will

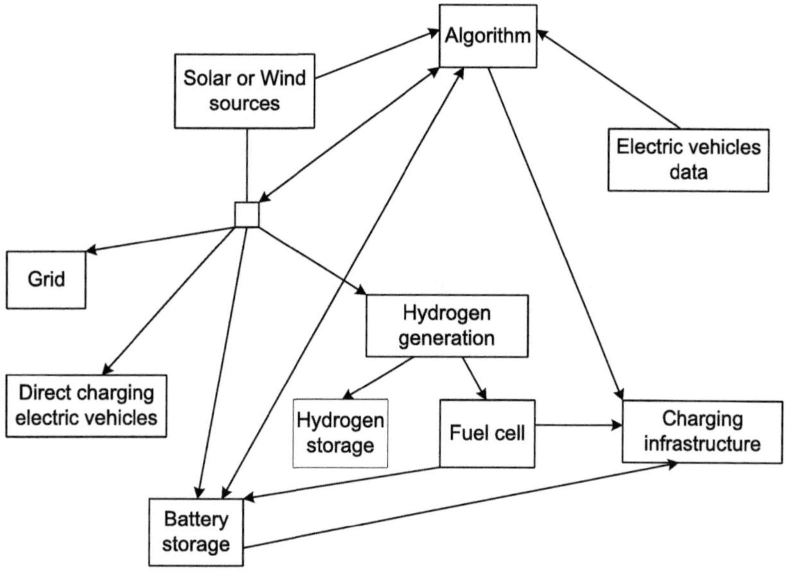

Fig. 10.3 Model of free flow managed by a central processing algorithm

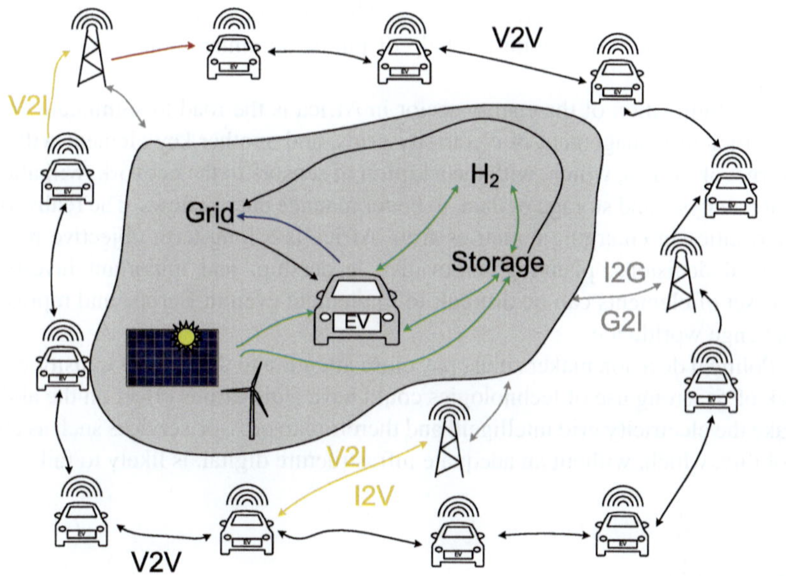

Fig. 10.4 Model for sharing electric vehicles

lead to a cost reduction, low impact on the main supply, and environmental protection (Roscia et al. 2016, 1120).

Conclusion

By 2030 the population of Africa will reach about 1 billion citizens, with six cities having more than ten million inhabitants. The jet points in Africa are the development of infrastructure, energy systems, technology, and mobility, with a vital role from the governments. The energy production and consumption, the displacement of materials and people, and traffic vary in time all over the territory and can be measured by adequate sensors and controlled through networks in real time. By interconnecting buildings, vehicles, power plants, and city lighting, the cities will become "smart."

Africa is growing exponentially, and the widespread use of smartphones will determine an improvement of services toward sustainable, augmented quality of life. The smart grids are a core element of smart cities and will facilitate voltage control, reactivation of power compensation, and reduction of distribution losses. A better planning integrated with disruptive technologies will change the cities not only in Europe but also in Africa. The final objective is to change the cities in a sustainable, well-being, secure and resilient. The future urbanization of cities and the urban influence on energy demands require a holistic approach for the management of the sustainable energy systems integrated with the various services offered in smart cities (e.g., electric mobility). This challenge will foster the economic improvements in Africa with the help of public administrations, universities, and companies.

The digitization of the energy sector in Africa is the road to significant changes and efficient management of electricity grids, and another key element will be the Internet of Things, which, with the addition of sensors to the network, will allow for better analysis and storage of data, to better manage energy flows. The realization of smart cities in emerging countries from Africa is a long-term objective requiring political decisions, planning, innovative leadership, and important investments. This set of elements can be difficult to implement even in Europe and represents a challenge worldwide.

Political decision makers must pay close attention to this energy transition, as the lack of or wrong use of technologies could have a disastrous effect on the ability to make the electricity grid intelligent and therefore to provide services such as electric mobility, which, without an adequate infrastructure digital, is likely to fail.

References

Anastasiadis P, Metaxas G (2010) Sustainable city and risk management. In 1st WIETE annual conference on engineering and technology education, 64–67

Deichmann U, Meisner C, Murray S, Wheeler D (2011) The economics of renewable energy expansion in rural Sub-Saharan Africa. Ener Pol 39:215–227

Dlodlo N, Mudumbe, JM, Jama Ndwe T (2014) The internet of things for a smart South African grid architecture. In International development informatics association conference, 95–107

Fletcher RH, Strunz K (2007) Optimal distribution system horizon planning–part I: formulation. IEEE Trans Power Syst 22:791–799

Foster V, Briceno-Garmendia C (2010) Africa's infrastructure: a time for transformation. Africa Development Forum. https://openknowledge.worldbank.org/handle/10986/2692. Accessed 18 May 2021

Herbert G (2004) Cities people planet: liveable cities for a sustainable world. Wiley

Parshall L, Pillai D, Mohan S, Sanoh A, Modi V (2009) National electricity planning in settings with low pre-existing grid coverage: development of a spatial model and case study of Kenya. Energy Policy 37:2395–2410

Roscia M, Lazaroiu GC, Mingrone L, Pignataro G (2016) Innovative approach of the sharing E-mobility. In 2016 international symposium on power electronics, electrical drives, automation and motion, 1120–1126

Satterthwaite D (1997) Sustainable cities and cities that contribute to sustainable development? Urban Stud 34:1667–1691

Sorensen A, Marcotullio PJ, Grant J (2014) Towards sustainable cities. In: Sorensen A, Marcotullio PJ, Grant J (eds) Towards sustainable cities. Routledge

Teravaninthorn S, Raballand G (2009) Transport prices and costs in Africa: a review of the international corridors. Directions in development; infrastructure. World Bank. https://openknowledge.worldbank.org/handle/10986/6610. Accessed 18 May 2021

Timothy E, McLaren D, Hillman M (1991) Reviving the city: towards sustainable urban development. Continuum International Publishing

UN (2019) World urbanization prospects. United Nations Department of Economic and Social Affairs. https://population.un.org/wpp/. Accessed 18 May 2021

Welsch M, Bazilian M, Howells M, Divan D, Elzinga D, Strbac G, Jones L, Keane A, Gielen D, Murthy Balijepalli VSK, Brew-Hammond A, Yumkella K (2013) Smart and just grids for sub-Saharan Africa: exploring options. Elsev Renew Sustain Ener Rev 20:336–352

WHO (2010) Hidden cities: unmasking and overcoming health inequities in urban settings. WHO:1–126

Part III
Social Involvement

Chapter 11
Sustainable Smart Cities: The European Case

María Teresa Nevado Gil, María Pache Durán, Luísa Cagica Carvalho, and Inna Sousa Paiva

Abstract Smart cities are emerging all over the world and are a promise to combat various problems currently faced by cities. This context makes smart cities a relevant topic for research. This study aims to identify which factors influence the development of sustainable smart cities. The empirical study uses a sample with 73 European cities and applies a multiple linear regression. The results suggest differences between smart cities in Europe, such as smarter cities are located geographically in the western region and are governed by women. This study provides an academic and empirical study on smart cities and contributes to a better explanation of a still under explored theme.

Keywords Smart cities · Europe · Sustainability · Factors

Introduction

In recent decades, the concept of 'smart city' (SC) has been gaining more and more importance in the literature regarding international policies (Albino et al. 2015). The unceasing growth of cities and improvements in quality of life have been key

M. T. N. Gil (✉) · M. P. Durán
Facultad de Empresa, Finanzas y Turismo, Universidad de Extremadura, Badajoz, Spain
e-mail: tnevado@unex.es, mpache@uma.es

L. C. Carvalho
Instituto Politécnico de Setúbal (CICE-IPS) & CEFAGE, Universidade de Évora, Évora, Portugal
e-mail: Luisa.c.carvalho@esce.ips.pt

I. S. Paiva
Instituto Universitário de Lisboa (ISCTE-IUL), Business Research Unit (BRU-IUL), Lisbon, Portugal
e-mail: Inna_Paiva@iscte-iul.pt

G. C. Lazaroiu et al. (eds.), *Holistic Approach for Decision Making Towards Designing Smart Cities*, Future City 18,
https://doi.org/10.1007/978-3-030-85566-6_11

elements for the future of society. Cities play a key role in social and economic aspects around the world and have a notable impact on environmental problems (Mori and Christodoulou 2012). For this reason, smart cities work in networks with the aim of promoting political and social efficiency, allowing economic, political, urban, cultural and social innovation (Carvalho 2017), nurtured by growing population flows that seek opportunities in cities for work and personal growth. In addition, they produce various business opportunities, allowing collaboration between different public and private sectors (Nevado et al. 2020). According to the United Nations (UN), in 2050, the world population will have increased by two billion people, with 70% living in urban areas. According to the UN, this increase in the population in cities may be a problem, since cities consume 75% of the world's energy and produce 80% of greenhouse gases. Despite this, cities favour innovation, knowledge and creativity, which is why a balance must be sought between social, economic and environmental aspects, as well as that of the citizens themselves. According to the European Parliament (2014), smart cities are classified according to six main dimensions: smart economy, smart governance, smart mobility, smart environment, smart people and smart living (Villarejo 2015).

Upon reviewing the literature, we found various studies on the definition, creation and use of indicators for the measurement and evaluation of smart cities (Nevado et al. 2018). Therefore, various approaches, objectives and methodologies can be found to diagnose and classify smart cities. Although there are no cities yet where urban systems and services are interconnected, many cities around the world have already become smart and sustainable. Information and communications technologies (ICTs) are responsible for accelerating the process for a city to become intelligent, thus fulfilling one of the 17 Sustainable Development Goals (SDGs) of the UN (De Guimarães et al. 2020), specifically SDG 11, whose objective is to achieve sustainable cities and communities. In this way, the aim is to create cities that reduce environmental impact, in addition to reflecting on models of access to resources, resource and energy management, transportation, etc. (Longo et al. 2014; Patsakis et al. 2015).

In 2012, the European Commission (EC) carried out a specific initiative for the development of smart cities called the 'Association for European Innovation on Smart Cities and Communities' (EIP-SCC) (https://eu-smartcities.eu/), which brings together cities, industries, small and medium-sized companies and financial and research entities with the aim of improving life in cities through more sustainable integrated solutions through concrete challenges of the city through different political areas and ICTs. For its part, the IESE Center for Globalization and Strategy is worthy of highlighting; it annually produces a ranking through the Cities in Motion index where it analyses the level of development of 174 cities in 80 countries in nine dimensions (Berrone and Ricart 2020). The smartest cities in the world are London, New York and Paris. Globally, Europe continues to dominate the ranking, with 27 cities among the 50 smartest in the world, followed by 14 cities in North America, five in Asia and four in Oceania.

In this context, this study focuses on SC, due to the importance that these cities have for the economy and sustainable development, and aims to identify the factors that may favour their development. To do this, and based on the objectives of the European Union (EU 2014), which include developing and redeveloping smart cities and communities, we use data from the Cities in Motion index (CIMI 2017) for 73 European cities.

To achieve the proposed objectives, a multiple linear regression of the levels of smart cities is proposed. The analysis of the factors influencing the development of sustainable SCs is at a relatively early stage of development. That is why our study contributes to the existing literature on regional development and smart cities and, in our case, in the European context. The work is structured as follows. After this introduction, a review of the literature on the regional development of SCs is carried out, from which the hypotheses of the study are drawn. Later, the methodology used, the descriptive analysis and the results obtained are given. The work concludes by summarizing the main aspects of the current problems and future challenges of smart cities.

Literature Review

Although there are many definitions of the 'smart city', to date there is no globally accepted definition (Al Nuaimi et al. 2015). According to Meijer and Bolívar (2016), there are three different types of definitions of 'smart city': smart cities understood as cities that use smart technologies themselves (technological approach), smart cities understood as cities with smart people (focus on human resources) and smart cities as cities with smart collaboration (governance approach). Therefore, there is no single model to frame a smart city or a one-size-fits-all concept (O'Grady and O'Hare 2012). In the literature review, multiple concepts can be found, among which the one defined by Caragliu et al. (2011, p. 70) stands out:

> 'We believe a city to be smart when investments in human and social capital and traditional … a city to be smart when investments in human and social capital and traditional (transport) and modern (ICT) communication infrastructure fuel sustainable economic growth and a high quality of life, with a wise management of natural resources, through participatory governance'.

The term appears for the first time in the 1990s. Here, the term pays special attention to the importance of ICTs with respect to modern infrastructures within cities (Meijer and Bolívar 2016). One of the first organizations to focus on the study of smart cities was the California Institute for Smart Communities, with the aim of seeing how a city could be designed to implement ICTs (Alawadhi et al. 2012). Years later, the Center of Governance at the University of Ottawa criticized this idea, as it was too technique-oriented (Albino et al. 2015). Despite this, the concept of 'smart city' spread in the early years of the twenty-first century and is a phenomenon that has been growing to this day. Thus, authors such as Nam and Pardo (2011) focused on investigating the meaning of the term 'smart' in this context, understanding this as a system adapted to the needs of users. Others like Harrison et al. (2010) stated that the term 'smart city' refers to 'instrumented, interconnected and intelligent city', understanding intelligence as the inclusion of complex analysis, modelling, optimization and visualization services to make better operational decisions. Some, for their part, postulate that the smart city concept often acts more like a label (Van den Bergh and Viaene 2015; Glasmeier and Nebiolo 2016).

Organizations such as the UN or the European Union (EU) coincide in pointing out the main feature of smart cities: the use of ICTs in managing the challenges of cities with the purpose of improving the living conditions of the citizens who live in them and promoting their personal development. Thus, the OECD Oslo Manual (EUROSTAT 2005) highlights the role of innovation in ICT sectors by providing a set of tools to identify consistent indicators, setting up an analysis framework for researchers on innovation in cities. Authors such as Hollands (2008) emphasize that smart cities require not only the use of ICTs but also the contribution of different groups of people, collecting what smart cities are (smart people and governance) and what they intend to achieve (economy, mobility, environment and smart life). For their part, the International Telecommunications Union (ITU) and the United Nations Economic Commission for Europe (ECE) also add the concept of 'sustainability': 'a smart and sustainable city is an innovative city that uses ICT to improve the quality of life, the efficiency of urban operations and services and competitiveness, while ensuring that the needs of present and future generations are met with respect to economic, social, environmental and cultural aspects'.

As a result of all the above, it can be affirmed that the concept is extremely complex, and, in addition to the physical infrastructure, includes the human and social factors which are essential for its existence (Aguilera et al. 2013). In this way, many researchers have separated the concept into many characteristics and dimensions (Albino et al. 2015). In the review of previous literature, we found different classifications of the dimensions associated with different aspects of human life that partly affect common characteristics of smart cities. Giffinger et al. (2007) identify four components of the smart city: industry, education, participation and technical infrastructure. For Komninos et al. (2011), smart cities are associated with four dimensions: the application of a wide range of electronic and digital technologies; the use of ICTs to transform life and work; the integration of ICTs into the infrastructure of cities; and the union of ICTs with people to generate innovation, learning and knowledge. Since then, the list has been expanding, identifying six main components (Giffinger and Gudrun 2010): smart economy, smart society, smart governance, smart mobility, smart environment, and smart well-being, which Lombardi et al. (2012) associate with different aspects of urban life (see Table 11.1).

Table 11.1 Dimensions of Smart Cities

Dimensions of a smart city	Related aspects of urban life	Objectives
Smart economy	Industry	Achieve greater competitiveness and productivity in cities, based on innovation and the creation of a strong, dynamic and sustainable economy, through the use of ICTs by companies, the promotion of entrepreneurial initiative and the promotion of creativity, in direct relation to the knowledge-based economy
Smart people	Education	Focus on society and human capital as a fundamental integral part of the city whose society has 'learned to learn', carrying out actions to prevent excessive energy consumption and pollution. The aim is to enhance the quality of life in the city, making it more competitive and sustainable
Smart governance	E-democracy	Centre the management of the city itself with participation and transparency towards the citizen. The aim is to improve the quality and efficiency of the services provided by public administrations through digitized processes, generating a transparent and rapid access to public information
Smart mobility	Logistics and infrastructure	Ensuring the accessibility and sustainable mobility of transport and communication through the use of ICTs. In addition, it must ensure public access to the internet for all its citizens
Smart environment	Efficiency and sustainability	Achieve environmental sustainability, optimizing energy consumption and managing resources efficiently
Smart living	Safety and quality	Offer a higher level and adequate use of services, improving the quality of life of the population, creating a virtuous circle of economic and social Well-being

Source: Own elaboration from Lombardi et al. (2012)

Design of the Investigation

Sample Selection

The selected population is made up of 73 European cities whose levels of smart cities for the year 2017 were studied and classified by the IESE Business School, one of the most important business schools in the world, which in recent years has published an index of cities which is objective, exhaustive, comprehensive and guided by the criteria of conceptual relevance and statistical rigour. The 2017 edition of CIMI includes 180 cities, 73 of them capitals, representing 80 countries. The breadth of the project establishes the CIMI as one of the city indexes with the broadest geographic coverage available today. In addition, to calculate the index, the authors included 79 indicators that provide a comprehensive view of each city.

CIMI has empirically validated the conceptual model developed from 79 indicators that provide a complete vision of each city, grouped into ten key dimensions to determine its efficiency: economy, human capital, technology, environment, international impact, social cohesion, mobility and transport, governance, urban planning and public management. All the indicators are linked to a strategic objective that leads to a new form of local economic development: the creation of a global city, the promotion of entrepreneurship and innovation, among others.

Methodology

To achieve the objective of this study and following Laswad et al. (2005), Moura et al. (2014) and Alcaraz-Quiles et al. (2015), among others, we propose a multiple linear regression of the CIMI and the ten dimensions that make up the index, based on three variables: gender, geographic location and political ideology. Thus, an explanatory analysis is carried out in order to identify the factors that affect smart city levels in European cities. All estimates will be made using SPSS software version 20.

Model Specification

To carry out the contrast of the hypotheses raised in previous sections, multivariate regression techniques were used. Through a multiple linear regression, using the ordinary least squares (OLS) method, the following models were estimated, one of the global index for the 73 European cities (CIMI) and the rest with the values of each of the ten dimensions of which the CIMI is composed.

$$\text{ICMI}_j = \beta_0 + \beta_1 \text{Gender}_j + \beta_2 \text{Localization}_j + \beta_3 \text{Ideology}_j + \varepsilon_j$$

$$\text{D1}_j = \beta_0 + \beta_1 \text{Gender}_j + \beta_2 \text{Localization}_j + \beta_3 \text{Ideology}_j + \varepsilon_j$$

$$\text{D2}_j = \beta_0 + \beta_1 \text{Gender}_j + \beta_2 \text{Localization}_j + \beta_3 \text{Ideology}_j + \varepsilon_j$$

$$\text{D3}_j = \beta_0 + \beta_1 \text{Gender}_j + \beta_2 \text{Localization}_j + \beta_3 \text{Ideology}_j + \varepsilon_j$$

$$\text{D4}_j = \beta_0 + \beta_1 \text{Gender}_j + \beta_2 \text{Localization}_j + \beta_3 \text{Ideology}_j + \varepsilon_j$$

$$\text{D5}_j = \beta_0 + \beta_1 \text{Gender}_j + \beta_2 \text{Localization}_j + \beta_3 \text{Ideology}_j + \varepsilon_j$$

$$\text{D6}_j = \beta_0 + \beta_1 \text{Gender}_j + \beta_2 \text{Localization}_j + \beta_3 \text{Ideology}_j + \varepsilon_j$$

$$D7_j = \beta_0 + \beta_1 \text{Gender}_j + \beta_2 \text{Localization}_j + \beta_3 \text{Ideology}_j + \varepsilon_j$$

$$D8_j = \beta_0 + \beta_1 \text{Gender}_j + \beta_2 \text{Localization}_j + \beta_3 \text{Ideology}_j + \varepsilon_j$$

$$D9_j = \beta_0 + \beta_1 \text{Gender}_j + \beta_2 \text{Localization}_j + \beta_3 \text{Ideology}_j + \varepsilon_j$$

$$D10_j = \beta_0 + \beta_1 \text{Gender}_j + \beta_2 Localization_j + \beta_3 Ideology_j + \varepsilon_j$$

Likewise, an error term (Ɛ) has been incorporated, which collects the incidence that the indices have on the effects not explained by the independent variables.

Results and Discussion

Table 11.2 shows the results related to the effect that the gender of the governing person, the geographical location of the cities and the political ideology of the representatives of the government of a city have on the smart city level of European cities, and on each of its dimensions. As can be seen in Table 11.2, the global model, where the influence of the proposed variables on the total smart city level is measured, has an explanatory power of 45.7% for a confidence level of 99% (value of $p < 0.01$). Of the three proposed variables, gender and geographic location are statistically significant for a confidence level of 95% and 99%, respectively, and are negatively associated with the dependent variable. With respect to the gender variable, this indicates that European cities governed by women reach better positions in the smart cities ranking than those governed by men. Although the literature suggests some differences between men and women in governance, studies on leadership and gender in CIs are still in their early stages. For this reason this study provides relevant and novel information on one of the factors that favours the development of smart cities, the gender of the governing party.

On the other hand, regarding the geographic location variable, the results confirm that European cities belonging to the western region occupy the best positions in the smart cities ranking. We can therefore affirm that there is a significant relationship between the levels of smart cities and the region to which European cities belong. These results are consistent with the literature since, as referenced above, it identifies Western Europe as more prosperous and more industrialized than Eastern Europe. For its part, the political ideology of the rulers of European cities lacks econometric relevance. However, these results are contradictory to the works of other authors such as those of Prado-Lorenzo et al. (2011), who obtained evidence that the political ideology of the ruling parties affected the sustainability of cities. Specifically, they concluded that a leftist ideology negatively affects the development of cities.

Table 11.2 Regression results

Explanatory variable	Global model		Model 1 Economy		Model 2 Human capital		Model 3 Technology		Model 4 Environment		Model 5 International projection	
	β	t (Sig.)	β	t (Sig.)	β	t (Sig.)	β	t (Sig.)	β	t (Sig.)	β	t (Sig.)
Constant	97.41	12.74 (0.000)	107.648	13.805 (.000)	95.573	10.045 (.000)	109.012	10.154 (.000)	57.784	6.236 (.000)	96.458	8.154 (.000)
Gender	−17.50	−2.129 (.037**)	−28.166	−3.358 (.001***)	−20.253	−1.979 (.052*)	.576	.050 (.960)	−17.020	−1.708 (.092*)	24.561	1.930 (.058*)
Region	−45.85	−5.821 (.000***)	−41.086	−5.111 (.000***)	−22.200	−2.263 (.027**)	−26.353	−2.381 (.020**)	−10.890	−1.140 (.258)	−43.616	−3.577 (.001***)
Ideology	6.85	1.082 (.283)	8.180	1.266 (.210)	−.269	−.034 (.973)	4.000	.450 (.654)	7.156	.932 (.355)	2.458	.251 (.803)
F-statistic	19.346		19.604		4.621		2.637		2.290		5.948	
F-sig	.000***		.000***		.005***		.056*		.086*		.001***	
R^2	.457		.460		.167		.103		.091		.205	

Explanatory variable	Model 6 Social cohesion		Model 7 Mobility and transportation		Model 8 Government		Model 9 Urban planning		Model 10 Public management	
	β	t (Sig.)	β	t (Sig.)	β	t (Sig.)	β	t (Sig.)	β	t (Sig.)
Constant	78.867	7.554 (.000)	101.222	8.285 (.000)	127.233	17.393 (.000)	77.215	8.865 (.000)	121.392	10.711 (.000)
Gender	−31.729	−2.826 (.006***)	−1.290	−.098 (.922)	−19.114	−2.430 (.018**)	−16.380	−1.748 (.085*)	−12.232	−1.003 (.319)

	Model 6		Model 7		Model 8		Model 9		Model 10	
	Social cohesion		Mobility and transportation		Government		Urban planning		Public management	
Explanatory variable	β	t (Sig.)	β	t (Sig.)	β	t (Sig.)	β	t (Sig.)	β	t (Sig.)
Region	−29.874	−2.776	−38.583	−3.064	−69.741	−9.248	−36.915	−4.111	−18.802	−1.609
		(.007***)		(.003***)		(.000***)		(.000***)		(.112)
Ideology	8.728	1.009	−4.056	−.401	−.377	−.062	17.523	2.427	4.004	.426
		(.317)		(.690)		(.951)		(.018**)		(.671)
F-statistic	8.265		3.698		41.571		13.143		1.898	
F-sig	.000***		.016**		.000***		.000***		.138	
R^2	.264		.139		.644		.364		.076	

Significance at the level 0.10*. Significance at the level 0.05**. Significance at the level 0.01***

In relation to the models of the levels of smart cities in each of the ten dimensions, it can be observed that, in general, they have a low explanatory capacity (between 10% and 20%), except for the economy, governance and urban planning, which reach high percentages (46%, 64.4% and 36.4%, respectively) for confidence levels of 99% (p values <0.01). On the other hand, in general terms, two of the three selected factors have an important effect on each of the dimensions of the IQ levels. The gender of the person who rules in European cities affects the dimensions represented by the economy, human capital, the environment, international projection, social cohesion, governance and urban planning. The same occurs with the variable 'Region', which affects eight of the ten dimensions. European cities concentrated in the western region obtain better results as smart cities in the dimensions that affect the economy, human capital, technology, international projection, social cohesion, mobility and transport, governance and planning urban.

Finally, the variable political ideology lacks explanatory capacity in all dimensions of the smart city level except urban planning, where it reaches a significant level at 95% confidence ($p = 0.018$). According to these results, a left-wing ideology in the governments of European cities would favour their positions in the ranking in terms of the sustainability of the environment and its inhabitants.

Taking the results obtained for these 11 models into account, we can conclude that the gender of the person who governs and the geographical situation of the city favour the SC levels of European cities.

Conclusions

This work has allowed us to obtain empirical evidence and contribute to the generation of knowledge about smart cities, given the lack of research with this orientation so far. Thus, an existing gap in the academic literature is covered, since the analysis of the factors that influence the development of smart cities is in a relatively early stage of development, making study of it necessary and interesting.

Regarding the analysis of the associated factors, evidence has been found that shows that the levels of smart cities are related to the geographical location of the cities and the gender of the governors. However, it has not been possible to verify any type of association with the political ideology of the governing team. We can conclude, therefore, that those cities located in the western region obtain better results as CI. The same occurs with cities governed by women, which achieve the best results in the smart cities ranking. Finally, the fact that city government representatives have conservative or progressive ideologies does not seem to be decisive for cities to reach high positions in the ranking.

Finally, we point out that this study highlights the need to develop future lines of research, among which we propose to expand the sample in order to carry out comparative analyses at the level of countries and continents, as well as to analyse the reasons and motivations that may lead to cities obtaining high of smart city levels.

References

Aguilera G, Galán JL, Campos JC, Rodríguez P (2013) An accelerated-time simulation for traffic flow in a smart city. FEMTEC 26

Al Nuaimi E, Al Neyadi H, Mohamed N, Al-Jaroodi J (2015) Applications of big data to smart cities. J Inter Serv Appl 6(1):1–15

Alawadhi S, Aldama-Nalda A, Chourabi H, Gil-Garcia JR, Leung S, Mellouli S et al (2012) Building understanding of smart city initiatives. In: international conference on electronic government. Springer, Berlin/Heidelberg, pp 40–53. https://doi.org/10.1007/978-3-642-33489-4_4

Albino V, Berardi U, Dangelico RM (2015) Smart cities: definitions, dimensions, performance, and initiatives. J Urban Technol 22(1):3–21. https://doi.org/10.1080/10630732.2014.942092

Alcaraz-Quiles FJ, Navarro-Galera A, Ortiz-Rodriguez D (2015) Factors determining online sustainability reporting by local governments. Int Rev Adm Sci 81(1):79–109

Berrone P, Ricart JE (2020) IESE cities in motion, Business School, University of Navarra. Available at https://media.iese.edu/research/pdfs/ST-0542.pdf

Caragliu A, Del Bo C, Nijkamp P (2011) Smart cities in Europe. J Urban Technol 18(2):65–82

Carvalho L (2017) Entrepreneurial ecosystems. Lisbon as a smart start-up city. In: Carvalho L (ed) Handbook of research on entrepreneurial development and innovation within smart cities. IGI Global, Pennsylvania, pp 1–24. ISBN-13: 978-1522519782

De Guimarães JCF, Severo EA, Júnior LAF, Da Costa WPLB, Salmoria FT (2020) Governance and quality of life in smart cities: towards sustainable development goals. J Clean Prod 253:119926. https://doi.org/10.1016/j.jclepro.2019.119926

European Parliament (2014) Mapping smart cities in the EU. Directorate General For Internal Policies. Policy Department

European Union (2014) Directorate General for Internal Policies. Policy Department A: Economic and Scientific Policy. Mapping smart cities in the EU

EUROSTAT (2005) Guidelines for collecting and interpreting innovation data (2005). A joint publication of OECD and Eurostat, Organization for Economic Co-Operation and Development. Statistical Office of the European Communities. Available at https://ec.europa.eu/eurostat/web/products-manuals-and-guidelines/-/OSLO

Giffinger, R. y Gudrun, H. (2010). Ranking de ciudades inteligentes: ¿un instrumento eficaz para el posicionamiento de las ciudades ?. ACE: arquitectura, ciudad y medio ambiente, 4 (12), 7–26

Giffinger R, Fertner C, Kramar H, Meijers E (2007) City-ranking of European medium-sized cities. Cent Reg Sci Vienna UT:1–12

Glasmeier AK, Nebiolo M (2016) Thinking about smart cities: the travels of a policy idea that promises a great deal, but so far has delivered modest results. Sustainability 8(11):1122. https://doi.org/10.3390/su8111122

Harrison C, Eckman B, Hamilton R, Hartswick P, Kalagnanam J, Paraszczak J, Williams P (2010) Foundations for smarter cities. IBM J Res Dev 54(4):1–16

Hollands RG (2008) Will the real smart city please stand up? Intelligent, progressive or entrepreneurial? City 12(3):303–320. https://doi.org/10.1080/13604810802479126

IESE Cities in Motion Index (CIMI) (2017). Available at https://media.iese.edu/research/pdfs/ST-0442-E.pdf

Komninos N, Schaffers H, Pallot M (2011) Developing a policy roadmap for smart cities and the future internet. In eChallenges e-2011 conference proceedings, IIMC International Information Management Corporation

Laswad F, Fisher R, Oyelere P (2005) Determinants of voluntary internet financial reporting by local government authorities. J Account Public Policy 24(2):101–121

Lombardi P, Giordano S, Farouh H, Yousef W (2012) Modelling the smart city performance. Innov Eur J Soc Sci Res 25(2):137–149. https://doi.org/10.1080/13511610.2012.660325

Longo M, Zaninelli D, Roscia M, Costoiu M (2014) Smart City to improve power quality. In: 2014 16th international conference on harmonics and quality of power (ICHQP). IEEE, pp 458–462

Meijer A, Bolívar MPR (2016) Governing the smart city: a review of the literature on smart urban governance. Int Rev Adm Sci 82(2):392–408. https://doi.org/10.1177/0020852314564308
Mori K, Christodoulou A (2012) Review of sustainability indices and indicators: towards a new City Sustainability Index (CSI). Environ Impact Assess Rev 32(1):94–106. https://doi.org/10.1016/j.eiar.2011.06.001
Moura A, Ribeiro V, Monteiro S (2014) Divulgação de informação na internet sobre Responsabilidade Social-Evidência empírica nos municípios portugueses. Trabajo presentado en las XXIV Jornadas Luso-Espanholas de Gestão Científica, Leiria, Portugal, 6
Nam T, Pardo TA (2011) Conceptualizing smart city with dimensions of technology, people, and institutions. In: Proceedings of the 12th annual international digital government research conference: digital government innovation in challenging times, pp 282–291
Nevado MT, Carvalho LC, Paiva IS (2018) Factores relevantes para convertirse en una Ciudad Inteligente Sostenible: un estudio empírico en Europa. In XXIII workshop on accounting and management control
Nevado MT, Carvalho LC, Paiva IS (2020) Determining factors in becoming a sustainable smart city: An empirical study in Europe. Econ Sociol 13(1):24–39. https://doi.org/10.14254/2071-789X.2020/13-1/2
O'Grady M, O'Hare G (2012) How smart is your city? Science 335(6076):1581–1582
Patsakis C, Laird P, Clear M, Bouroche M, Solanas A (2015) Interoperable privacy-aware e-participation within smart cities. Computer 48(1):52–58
Prado-Lorenzo J, García-Sanchez I, Cuadrado-Ballesteros B (2011) Sustainable cities: do political factors determine the quality of life? J Clean Prod 21:34–44
Van den Bergh J, Viaene S (2015) Key challenges for the smart city: turning ambition into reality. In: 2015 48th Hawaii international conference on system sciences. IEEE, pp 2385–2394
Villarejo H (2015) Smart cities: una apuesta de la Unión Europea para mejorar los servicios públicos urbanos. Revista de estudios europeos 66:25–51

Chapter 12
Public Governance of Digital Social Media

Sergei Kamolov and Alexandra Smagina

Abstract The process of digitalization is an inherent part of modern world society. Social relations face long-term deep changes fuelled by the spread of network paradigm and infrastructure of digital social media. The scope and penetration of the digital network society is so powerful and dramatic that any advanced public governance must have a clear understanding of this phenomenon and even elaborate comprehensive system of interinfluence for the sake of the citizens' interests – we call it a phenomenon of digital social twin.

Post-pandemic social architecture will rely greatly on social media which helps to maintain governmental control, to support business value and to provide for common people communication. Digital communication networks are becoming major mean to offset the risks of social distancing and other unseen challenges in social, economic, political and personal terms. Therefore, in the following chapter, we focus on the very dialectic nature of social media in smart cities concept through the analysis of key pros and contras of technologies interfering into social life, assessment of the public governance role with regard to social media and synthesis of applied logic that smart cities may use to retrieve maximum public value from digital communication platforms.

Keywords Digital social media · Digital social twin · Public governance · Smart communities

S. Kamolov (✉) · A. Smagina
Federal State Autonomous Educational Institution of Higher Education "Moscow State Institute of International Relations (University) of the Ministry of Foreign Affairs of the Russian Federation", Moscow, Russia
e-mail: s.kamolov@inno.mgimo.ru

G. C. Lazaroiu et al. (eds.), *Holistic Approach for Decision Making Towards Designing Smart Cities*, Future City 18,
https://doi.org/10.1007/978-3-030-85566-6_12

Introduction

Digital social media in its essence become social digital twin representing therefore not just a virtual phenomenon, but rather a complex substance, which needs to be governed as closely and effectively as any natural (analogue, offline) social communities. In this chapter, we state that digital social media in the era of smart cities cannot be studied as an infrastructural or communicative matter, i.e. functionally. It cannot be detached from its origins – the community itself. Moreover, we insist that the discussion of the possibility and approaches to the governance of the social media by public authorities has for a long time already surpassed the limits of prosaic issue of censorship. Traditional types of regulations (i.e. licensing) do not suffice the complexity of the matter which appeals for a thoughtful policy providing to the citizens both the freedom of self-fulfilment and the security in wide meaning of this term.

Unlike other types of digital twins (smart grids, BIM, intellectual control centres), digital social media (DSM) carry a property of less degree of predictability, but much greater degree of return influence, placing the matter of digital social media governance to a particular agenda of smart communities and demanding a particularly new set of measures, tools and professional competences from the public authorities.

We start our analysis with the study of the dialectical nature of the social media phenomenon. The systematization of theoretical fundamentals of network society and quantitative and qualitative analysis of digital social media points to the fact that DSM evolve into a primary matter of public governance in smart communities. In the second part, we present actual data indicating the level of the pervasiveness, power and complexity of DSM and derive the double-edge approach to this phenomenon. We also elaborate the constructive side of the DSM – the concept of digital social twin as the meaningful core of connected societies. The outline of the forms and methods of the DSM public governance, presented in the algorithmic way, suitable for applied use by local or regional authorities concludes this chapter.

Methods

The methodological basis of the study consists of paradigmatic, systemic, structural-functional and logical analysis methods, conceptualization and forecasting. When working with data, the method of content analysis and logical-linguistic analysis were used, followed by formalization of the obtained information.

Results

The very concept of smart society is built on the principle of digitalization – the introduction of an increasing number of technological tools into everyday life. When an information and technology come to the force, a change in relations and interconnections between people and surrounding objects inevitably occurs. Nowadays, DSM have become an effective means of social interaction since they preserve the essence of people's relations and at the same time represent a product of technical progress.

Quantitative analysis of current tendencies in DSM evolution reflects the trend towards the ubiquitous dissemination of online platforms in the near future. Qualitative examination shows that DSM wide functional coverage contributes to social transformation into online communities. This phenomenon explains the dialectical nature of online interactive platforms, which, on the one hand, may serve as convenient tool for facilitating public services provision. On the other hand, the advantages of DSM are used for ill-intentioned purposes such as cultivation of antisocial sentiments. In order to respond to existing challenges, both traditional forms and alternative methods of public governance should be applied. As dominating legal forms of regulation are ultimately reduced to licensing (prohibitions and censorship per se), new approaches are developed which presuppose the equal presence of the public governance in DSM with other users.

Despite the extensive list of methodological approaches of public administration in the sphere of social media, the current political agenda is narrowed to locally focused actions rather than comprehensive policies. The results of the research let us assert that the development of digital social media as a society's digital twin at the applied level means the extensive development of recommender systems of public governance – which is the matter of a separate study.

Theoretical Fundamentals

The world is now moving to a so-called network paradigm, which along with technical progress and digitalization of every sphere of life leads to emergence of smart community. The gradual shift towards a new paradigm is changing the very basic social processes including social interaction which is being performed through technologies serving as a mediator between people. The primary role of this mediator belongs to DSM that are an organic component of smart communities and are seen as a functional base in a knowledge and technology paradigm.

DSM are a multidimensional substance defined as online platforms, the content of which is formed in an interactive format by all its participants. Such platforms include social networks, online mass media, blogs, microblogs, thematic portals, instant messengers, video hosting and forums. Conceptually, our understanding of DSM is based on different assumptions:

- In terms of communication theory, DSM are determined as an ambiguous term that unites various online technologies on the Internet that allow users to communicate and interact with each other (Gillen 2009).
- Another approach highlights the principle of general participation, stating that DSM are any Internet projects in the web 2.0 format, the content of which is formed by the users themselves in social networks, blogs, podcasts, websites, Internet forums, Wiki, video hosting, print, online and mobile products (Chumikov 2014).
- From the point of social responsibility, the defining feature of DSM is so-called UGC – user generated content which is generated not by the creators of the platform but by its users which considerably changes the whole paradigm of social responsibility (Grygiel and Brown 2019).
- According to socio-cultural approach, DSM are communication technologies that allow maintaining connections between people, as well as personal and cultural networks to which they want to belong (Schejter and Tirosh 2012).

Basically, the discussion on DSM is built around their initial structure – network, which may be regarded from different scientific perspectives:

- In philosophical understanding, the concept of 'network' is examined in terms of 'knowledge resource', place, role and meaning of information flows in the social structures, considering them an important foundation of the 'coming post-industrial society' (Drucker 1993; Machlup 1962; Bell 2004). The considerable contribution to the issue were made by constructivists, who devoted their works to the problems of dissemination and perception of information, communication systems, self-organization, the principles of circularity and organizational isolation and autopoietic and cognitive systems (Tsokolov 2000).
- In the field of late cybernetics (or post-cybernetics), the concept of 'networks' is based on the postulate of information interaction and influence, where signs and information semantic systems play a crucial role (Solomatin 1989).
- The systemic approach argues that network is a technical phenomenon of new communication technologies, the morphology of which is associated with the idea of information as knowledge that generates constructive changes in the system, the characteristics of its capabilities and the probability of further events (Nazarchuk 2008).
- The idea of a network paradigm formed the basis of the network society theory (Van Djik 1991; Castells 1999; Craven and Wellman 1973; Simmel 1996). The definitions and concepts described in the writings of these authors are vividly embodied in 'digital life.'

Interactive social platforms are defined through their various features and given their vital role in terms of social interaction. The principal function of DSM as technical tool is to mediate social contacts in emerging smart society, which is theoretically founded on the network paradigm.

Discussion

DSM are characterized through a number of quantitative and qualitative features. Quantitative parameters reflect the scale of DSM proliferation, population penetration and geographic distribution, as well as users' characteristics (time users spend on platforms, age of users). These figures help evaluate the significance of digital social media at different social levels (global, national, individual).

In qualitative terms digital social media may be classified according to their origins and consequently their aims. Furthermore, DSM as a social phenomenon is characterized by features usually inherent to traditional society including social involvement. Users of the platforms are being engaged in certain collective processes similar to how people are involved in real-life activities. One of the most controversial matters is the representation of different actors on the online platforms with their corresponding goals.

Quantitative Analysis of Digital Social Media

Currently, there are hundreds of social media which account for over 4 billion users worldwide (Statista Global Consumer Survey 2020a, b). They range from local use applications and special solutions designed for particular company employees to global digital giants uniting nations across the world. The group of majors are represented by Facebook, YouTube, WhatsApp, WeChat, Instagram and TikTok with the first most popular having more than 2,7 billion active users every month (see Table 12.1). The second and the third leading places are split between YouTube and WhatsApp with 2 billion active users as of October 2020. The most of these companies were launched in the early 2000s and have been growing and gaining popularity since then. The exclusion has come from China, where TikTok was founded in 2016 getting almost 0,7 billion users for the past several years.

Every large DSM is operated in the web and has a mobile version. That is why the proliferation of social media is related to the availability of access to mobile Internet.

Table 12.1 Most popular social networks worldwide as of October 2020, ranked by number of active users (in millions)

№	Name of DSM	Number of users (mln)	Initial release
1.	Facebook	2701	2004
2.	YouTube	2000	2005
3.	WhatsApp	2000	2009
4.	Weixin/WeChat	1206	2011
5.	Instagram	1158	2010
6.	TikTok	689	2016

Open source: Statista Global Consumer Survey (2020a, b), Schejter and Tirosh (2015)

The most involved DSM users live in Asia (see Table 12.2). With about 926 million users from China in 2020, an estimated number in 2025 reaches 1,1 billion from this country alone and approximately 490 million from India. As for the other countries, the United States get the fourth position with 223 million in 2020 and 243 predicted in 2025.

Social networking is one of the most popular activities of citizens enabling them to keep in touch with friends as well as to catch up with the news. On average, Internet users spend over 140 min per day surfing social networks (Statista Global Consumer Survey 2020a, b).

In the recent years, Russia is one of the countries demonstrating high level of involvement in DSM. This country has the largest number of DSM users among all European countries that carry the following distinguished details: 78% of Russian Internet users have accounts on social networks, which is more than in the USA (75%), Germany (56%), Sweden (74%) and France (49%) (Lebedev 2018). In terms of demographic characteristics, the share of Internet users aged 12 to 24 in Russia amounted to 97,1% in February–November 2020. On average, 95,six million people, or 78,1% of the country's population over 12, used the Internet in Russia at least once a month. Daily, 87, 1 million people, or 71,1% of the Russian population, went online. Internet penetration in Russia among the younger population (up to 44 years old) in 2020 exceeded 90%, and among the youngest Russians (12–24 years old) approached 100%. Among the population 45–54 years old, 84.2% of Russians used the Internet at least once a month, and among the oldest residents of the country (55+ years old) only half – 49.7% use the Internet (Mediascope Web Index 2019).

Table 12.2 Comparison of DSM users number predictions

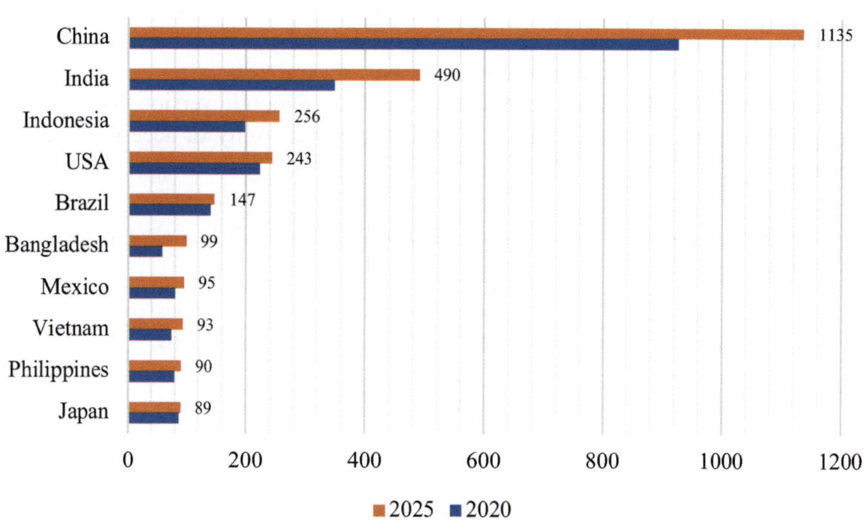

Open source: Statista Global Consumer Survey (2020a, b), Simmel (1996)

Qualitative Analysis of DSM

One of the core qualities of DSM is the establishing communicative and emotional connection among users. Online interactive platforms have considerable prerequisites for becoming the basis for lasting relations in the form of community. In fact, the scope and complexity of DSM structures, inter-influence that evolves between real and virtual communities, allow us to formulate a hypothesis of DSM as a digital twin of traditional societies.

Nowadays, social involvement as a phenomenon, which is a fundamental attribute of social media, is realized in a multidimensional system community. According to meso-level, a community consists of a certain number of individuals who are united by common interests. They also strive to achieve mutual goals and are to some extent willing to sacrifice their personal benefits (time, contacts) for the good of the group.

Interaction through DSM allows conversion of the physical contacts and meetings into a digital format. Nonetheless, everyone identifies the digital portrait (or digital twin) of a member with an original one. The definition of contacts is transforming and is currently understood as joint streaming, meetings and conferences in Zoom, Skype services, etc., whereas 'likes' and 'reposts' reflect and demonstrate group support and complicity in the life of an individual.

Similar to traditional reality, a community in virtual world has its boundaries and should be considered as a group: the so-called publics, channels and pages. Moreover, the access to these communities can be either closed or free, depending on the level of openness of the community to the 'others' and the 'society'.

DSM have significantly changed not only interaction but also decision-making procedures, as this process is now often carried out in digital format. Today social involvement in smart communities is basically exercised through digital social media. The provision of public services via social networks, for example, the debates on traffic in a district or on the environmental issues both at the municipal and federal levels, proves the effectiveness and efficiency of decision-making process online. The coronavirus pandemic forced some governments even to hold elections via Internet. The DSM are recognized as a convenient means in decision-making process, and their role in this sphere is rising fast.

Digital social media include a broad array of technologies of a different morphology. Basically, they may be classified to four types according to the functionality:

- Messengers (e.g. WhatsApp, Facebook Messenger, WeChat)
- Social networks (such as Facebook, Vkontakte)
- Analogue media (converting into digital format and usually associated to the phenomenon of prosuming – simultaneous production and consumption of content)
- Special-purpose platforms (for music content sharing, professional local networks, electronic governments)

DSM may differ according to the actors or entities using them. The variety of stakeholders is important, since it predefines the purposes social media serve for. At the moment, the most influential groups of social-economic actors are citizens, businesses, NGOs and governments. Social media may enable the formation of the networks among these actors, yet their role in the process has received little attention. However, this is the DSM which play the role of transnational knowledge network providing organizations, authorities and people with information useful in addressing shared problems (Wukich et al. 2017). The groups of the actors presented in DSM pursuing diverse goals are systematized in Table 12.3 below.

The analysis of quantitative parameters demonstrates that DSM are a global phenomenon involving all the population of the world having access to the Internet. The cheaper the technical devices, the more people start social networking. DSM imply a wide range of technologies with different functional coverage, which contribute to social involvement reflected in communities built online. Social involvement is the highest level of social maturity represented by citizens' networking, which is defined or calculated by the number of people involved in decision-making. All the features mentioned above precondition the controversial position of DSM in modern reality. On the one hand, they are considered to be a convenient tool for mitigating problems and improving living standards of the society. On the other hand, global character of proliferation of these platforms seems scary as they are not restricted by anyone.

Destructive Essence

Digital technology in identified as one of the major risks for the humanity (Global Risk Report 2021). This phenomenon embraces digital inequality, defense vulnerability, cyber security and socially dangerous behaviour promoted via DSM. Destructive behaviour of DSM users takes the roots in the digital format, crystalizes in digital communication, which eventually may lead certain people to

Table 12.3 DSM goals and actors matrix

	Core goal	Social goal	Individual goal	Applied goal
Citizens	Self-realization and communication	Find 'soulmates'	Organize leisure time	Use for work and business
Business	Generate profit	Collect and monetize big data	Develop and scale business	Search for talents
NGOs	Social impact resolution	Search for activists	Create issue-related networks	Propagate the ideas
Government	Achieve common societal goals	Collect and research big data to adjust the policy	'Measure' social sentiment	Facilitate the provision of common services

Compiled by the authors

actions in real-life dimension. With almost unrestricted content flow, digital platforms are becoming a tool for unacceptable materials distribution that may directly or indirectly contribute to social instability.

The so-called destructive content carries a negative message, implicitly pushing an individual towards generally dangerous actions or self-harm. It often starts in the form of rude humour – cruel jokes about death, murder and suicide. A targeted user can be exposed to the scenes of violence – physical, sexual, psychological – and images of illegal activities, such as drug use. Influenced by the destructive materials, the user becomes the object of direct psychological control by the distributors of these materials.

The involvement of users in socially destructive movements is a multi-phase process, and each of the stage may be examined separately. The primary impact on consciousness is carried out via certain techniques without a person noting that he or she is being involved in something extraordinary. At this moment, mental and psychological state of the individual is of vital importance. If a person favours an aggressive behaviour, he or she is more likely to pay attention to materials that are destructive in their essence and will be interested in membership in respective groups. Gradually, such person joins a number of violation-driven communities with the participants prone to deviant behaviour.

In marketing this method and mechanism is called 'the funnel of engagement' and reflects the sequence of stages of immersing a user in the environment of destructive content in virtual reality with the subsequent transition into objective, that is, demonstration of destructive behaviour in real life.

The graphic representation of the funnel of engagement is as follows (Fig. 12.1):

The level 1 of the funnel of engagement characterizes the stage at which a social media user joins groups of broad thematic coverage, publishing materials for the

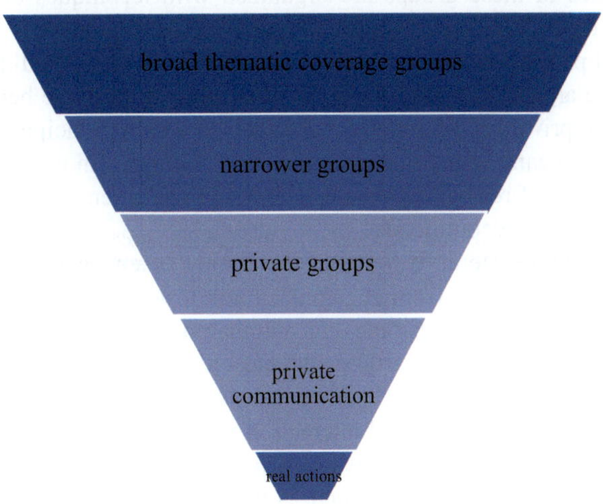

Fig. 12.1 The funnel of engaging in destructive behaviour in social media (Compiled by the authors)

widest possible audience. The content of such communities is usually accessible and understandable for everyone. However, among the jokes that seem harmless, materials with a destructive bias are concealed, for example, jokes about violence. Thus, the destructive content and the communities distributing it are attached with positive and attractive colouring. The share of such destructive messages may amount from 10% to 50% among in the total number of posts (Digital Hygiene 2019). There are no calls for any real action in these communities at this stage.

An interested user enters the second-level communities of engagement funnel. Such groups narrow the scope and plot of published content and focus on the preferences of their audience. Thus, they exert impact on the psyche of targeted audience. Each group is devoted to specific topics, and potential members join one of many according to the interests. The content is concentrated but has no direct calls to any actions yet.

After having been at the second level, gradually becoming the participants of the movement, users begin to perceive its culture and certain attitudes, follow the rules and understand a hierarchy established in such groups. They are considered ready to move to a new level, into private groups, where the regulation is more stringent and conducted by administrators. In addition, the procedure of joining implies the approval; it is not voluntary. Such groups heavily rely on recommender mechanism as they could potentially encourage users of open groups to explore relationships with unknown others by focusing on contact recommendation (Van Osch and Bulgurcu 2020). In order to remain a member of the community, it is necessary to demonstrate commitment on the regular basis participating in proposed activities, for example, challenges, voting, photo sharing, etc. Furthermore, there appear specific roles, hierarchy of power and signs of the subculture. The participants of such groups feel like being chosen; this impression is supported by different kinds of rewards.

The activities in these groups are organized with techniques and methods of manipulating users' minds to suggest specific ideas. The content is selected given the requests of participants and remains clear only to the 'dedicated' people. At this stage, it can be argued that users have propensity for destructive behaviour. There are also links to private chat rooms and channels in which participants are involved through various means of drawing attention and calling them to join private chats where instructions of real actions are distributed; most often dates and places of a crime-type action are recommended.

The fundamental difference between the funnel of engagement in destructive movements in social media from engaging in a destructive movement in objective reality is in the mechanism people get membered. In real life a person makes acquaintance with one of the participants and only then is accepted by the other members of the movement, whereas in virtual reality the user first adjoins a movement and then proceeds to personal interaction with followers.

This difference is important because new members are recruited faster and more effectively: the initial audience coverage is wider, which means that more people are likely to join as the large number of followers seems credible. Social media appear an effective platform for involving people into destructive behaviour online that gradually passes on to offline.

The society faces the problem of destructive content distribution via DSM that results in dangerous actions of people in the real life. 'The funnel of engagement' scheme implies gradual immersion which starts with drawing attention to detrimental jokes and results in action dangerous for the person and his or her surroundings. In fact, this is the other side of the coin of social involvement which is opposite to the smart one. Instead of uniting in communities aimed at well-being enhancing, DSM may contribute to society devastation.

Constructive Essence

Despite the risks accompanying DSM, they still must be regarded as opportunities especially in the frameworks of smart city concept. Digital social platforms are useful mechanisms of citizens' involvement in public problem detecting, agenda setting and participation in decision-making process. A profound approach towards DSM application suggests delegating a number of functions of the state to the platforms and even providing them with wider rights.

An interactive Internet space may serve as a thermometer of social wellbeing and tension for the government. The sentiments of citizens are broadcast on forums, blogs and social networks – on any platforms where involvement of the audience is available. Thus, the Internet reflects everything that happens in society (a concept of the so-called sociology one-to-one). Comprehensive monitoring systems of citizens' satisfaction (CMS) – smart systems of social networks monitoring and content analysing – serve for these goals (Kamolov and Smagina 2019). Such implementation is designed to show the correlation between local decisions and people's perception of them. CMS functionality includes:

- Measurement of the citizens' attitude towards the effectiveness of the implementation of public programs
- Classification of regions by level of satisfaction with the quality of public management
- Geo-tagging of problems within the implementation of public programs in the regions
- Automatic determination of responsible departments and the history of budget allocations related to specific geo-tag

As a result of the integration of social media and budget information within the CMS a universal tool for decision-making is shaped. Thus, clustering messages on a thematic and geographical basis allows focusing on relevant and important issues tied to their geographical location. The system allows decision-makers to determine the responsible authority and the causes for the citizens' dissatisfaction in real time. Analytical breakdown of the reviews into positive and negative lets identify a correlation between them and assess the quality of the implementation of decisions taken. The thematic distribution of messages allows geo-localization of the problems revealed by Internet users and immediate actions to protect the legitimate interests of citizens. The rating system of local authorities implemented in the system enhances the effect of users' liability towards it as they are more likely to believe the rated or at least assessed information (Kim et al. 2019).

Another advantage provided by DSM for the smart communities' governments is an opportunity to transfer several functions to the online mode. A number of specific applications are already developed in order to improve citizens-government integration: online receptions, tele-consulting, online education and digital health. Nowadays, the concept of digital government is on the agenda that includes a full-fledged integration of DSM into the network of self-interacting neuro-computer systems and Internet of things. The basic principle of the digital government implies elimination of the intermediary, which allows citizens and enterprises to directly connect to government systems, thereby eliminating the asymmetric distribution of information and significantly reducing corruption risks in the provision of public services. The so-called Facebook assessment index shows that communication strategy of local authorities on social networks among other things determines its popularity and citizens' liability to them (Miranda et al. 2018). Thus, DSM have become one of web-integrated services, implying the integration of electronic cross-services with a high level of transparency for citizens.

DSM as a social twin per se is becoming a part of a broader concept of digital twin of a nation often referred as a private virtual state (Fountain 2020). Digital platforms let governments operate on a global scale even with limited resources. Moreover, they contribute to electronic services provision to the citizens across the whole globe. Since digital platforms are represented almost in each country although without having their physical office, they may serve as embassies or consulates. The absence of territorial binding gives them an opportunity to act globally irrespective of jurisdiction.

Thus, DSM at the global level provide people with significant opportunities that can be advantageous for different stakeholders. Among others, the government gets the most considerable benefits which when combined upgrade the state to a new quality level. It is gradually transforming to the digital form transferring its activities and services to the online mode. Nowadays, the era of digital state is coming with social networks as the core element organizing interaction between the government, citizens and business. The appropriate social media strategy applied by authorities contributes to higher interaction rate, engagement and awareness (Prasetya et al. 2019). DSM are playing a significant role in the emergence of the private virtual state free of physical boundaries and jurisdictions. In fact, they enhance globalization eliminating common barriers and facilitating the government functions performance.

Traditional Forms of DSM Governance

From the moment DSM have emerged until developed into giants of the internet space, the state has been defining its attitude towards them. Against the backdrop of the growing platforms success, it seemed that they have stayed under control. Today the platform owners have at least technological capacity to dictate their own rules to the users (Isaak and Mina 2018), leaving the state without full-fledged immediate leverage to manage them.

The time lag between technological process and law-making has preconditioned the problem of lack of control. The speed of the Internet environment emergence has been outpacing the elaboration of corresponding regulatory frames. The platform owners have benefited from legal gaps and acted in the conditions of the absence of formal regulations. The faster the development of social media proceeded, the more the state 'lagged behind' in attempts to limit this development or at least influence it in some way.

The governance of the sphere of DSM is based on the set of forms and methods which are the subject of this part of the research. In the forms that are defined as external, typified expression of the practical activities of state bodies and officials in terms of formation and implementation of managerial goals and functions (Ohotskij 2016), there exist legal and organizational forms.

Currently, legal forms are dominating in the field of DSM governance. They may be divided into several categories depending on the sphere of regulation:

- Confidentiality
- Personal data protection
- Intellectual property rights
- Antitrust legislation
- Electronic public services
- Freedom of expression
- Public opinion shaping
- Political and electoral issues
- Protection for children on social media

A comprehensive legislation of supranational character aimed at the regulation of data collecting and operating has been actively developing during the past decade. The General Data Protection Regulation, often referred to as the GDPR, was adopted in 2018 based on the European Union Data Protection Directive 95/46/EU (1995) (General Data Protection Regulation 2018). The GDPR is regulating the activities of the companies established in the EU and processing personal data (i.e. subsidiaries located in the European Union). The GDPR is also applicable to organizations which activities is related to the provision of goods or services to EU personal data subjects. For example, websites available in the EU in the language of a member state are subject to this rule. The regulation is also applied to organizations related to monitoring the behaviour of personal data subjects.

However, the implementation of legal forms in the field of information security and personal data does not guarantee the comprehensive security. There is another sensitive problem which may be disclosed by all who once entered any social media. The GDPR and similar documents are aimed at protection of rather tangible matters as data, whereas the content of social media may impact directly on people. In the current conditions of unimpeded distribution of destructive content in the network, the security of users is challenged. Despite the number of different regulating acts, none of them may ensure that a user would not face uncomfortable posts logging in his or her account. The first steps in the regulation of the Internet space over the past two decades have already been taken by different countries. Awareness of the need to regulate human activity, as well as other actors in the open spaces of virtual reality, was developing gradually. This can be traced to the periods of the emergence of various legal acts in the jurisdictions of the countries of the world.

In the United States in 2000, the Children's Internet Protection Act 2000/ American Library Association Act (Children's Internet Protection Act (CIPA) 2020) was introduced to filter Internet content in schools; in the United Kingdom in 2003, the Law on Electronic Communications (Communications Act 2003) was implemented, which oversees the creation of an independent regulator of communication space; Japan passed the 'Act on Development of an Environment that Provides Safe and Secure Internet Use for Young People' (Act on Development of an Environment that Provides Safe and Secure Internet Use for Young People 2020), which controls providers of mobile Internet services to young people; the Russian Federation implemented the Federal Law 'On the Protection of Children from Information Harmful to their Health and Development' (Federal Law On the Protection of Children from Information Harmful to their Health and Development 2020), which provides for the categorization of information products and the ban on its dissemination among children; in South Korea in 2015 appeared the Enforcement Decree of the Amended Telecommunications Business Act (Enforcement Decree of the Amended Telecommunications Business Act 2020), obliging parents to install special Internet content filters on children's smart phones. In 2018, in a joint communiqué, the United Kingdom and France agreed on a joint action plan (United Kingdom – France Summit Communique) to increase the effectiveness of identifying and removing content that regarded as terrorist, radical or hateful.

The legal form of regulation is dominating; however, its shortcomings are getting clearer. The legal acts are reduced to censorship and mostly aimed at protecting content consumers rather than stopping destructive content production. An efficient management should be aimed at both reasons and consequences prevention. Thus, the legal form of regulation is to be complemented with institutional and organizational ones which will constitute a comprehensive approach of DSM governance.

Alternative Methods of DSM Governance

Based on the concept of digital social twin of traditional society, the DSM governance mechanism may be exercised on the basis of techniques of power-control influence aimed at streamlining public relations within established boundaries.

In contrast to traditional society, its digital twin requires specific governance methods due to several reasons. Firstly, DSM are too large, ephemeral and dispersed to be managed by common regulatory instruments. Traditional regulating processes are not able to provide an adequate response to modern complex global matters such as DSM. Globalization has led to the point when it is not always possible to adopt national laws regulating the activities of social media effectively in all countries. The inability of nation states to manage social media on a global scale is sometimes called 'regulatory vacuum effect' (Weber and Newell 2014.).

Secondly, DSM are a complex global network with a large number of multi-level nodes, the size of which makes impossible to exercise governance simultaneously and almost everywhere. The essence of social media is an aggregate that may be managed only with the help of aggregated approach that would reach each element of the network.

Thirdly, a close dependence of political agenda on the activity of social media users. One of the most significant features of DSM is considered user-generated content (UGC). The unification of people sharing content launches certain collective processes which stimulate a political potential which is capable to challenge the public governance system in terms of scope, coverage and access to citizens. The speed of information distribution in the network and the synchronization of many processes may cause unexpected effects with the government unable to response all the challenges.

These distinguishing features of DSM cause the discussions on the alternative regulating methods which may not be attached to a certain common methods classification and based on the principle of state involvement in DSM, equitable participation in digital life with other users of social networks. Such approach comprises the methods of monitoring the processes occurring online, collecting, aggregating and interpreting data from DSM, forecasting and modelling users' behaviour and stimulating digital platforms to impose their internal rules and codes of conduct.

First of all, such strategy involves monitoring of communication and content sharing on social networks and transferring interaction with the citizens to the online mode. In an ideal scenario, the government does not only monitor the processes but also participate in them as an entitled member, thus staying tuned and taking measures locally if necessary.

An extended strategy also includes working with the obtained data, modelling users' behaviour as well as forecasting the agenda based on the analysis of opinions. However, this set of methods has its own disadvantages primarily related to the ethics issues. On the one hand, monitoring can help to adapt the political course in accordance with the needs and requirements of citizens. On the other hand, social media monitoring involves analysing messages of individuals on virtual platforms that they can perceive as an intimate space (Bekkers et al. 2013).

In order to keep the processes occurring online under surveillance legally, the government has to come in to DSM. In the information society, the state transfers a part of its activities to the online mode, gradually converting into e-government. Some scientists insist on the new ontological understanding of this phenomenon, which they call web-ecology or ecosystem (Helmond 2013). The online government ecosystem also includes government social media account.

A 'self-regulation' mode of DSM is regarded as one of the effective methods to improve governance in this sphere. In this case, a digital platform acts as a legislator and compels everyone who wants to become its user to follow the rules. Such a prospect may be also beneficial for the platform. Firstly, it can build up its reputation and attract new users. Secondly, trust and loyalty to the platform will increase, and thirdly, it is likely to become a source of legalizing rules. It is especially important, since the policy changes implemented by a company may be potentially taken as a model at the national level. This seems to be significant advantage due to the differences in national legislations. At the same time, companies are able to determine the rules independently in accordance to the jurisdiction.

Another method is the principle of redistributive justice which facilitates the contribution of the public goods (Schejter and Tirosh 2015). Properties of social media should not be limited in any way; citizens must enjoy a wide access to the media. DSM become a source of the information never available before: for example, the population of remote regions or minority groups media can draw attention of the government to their problems through DSM. In this context, social networks sustain natural balance giving voice to those who have not been heard before the digital era.

In conclusion, traditional form of legal regulation adopted in DSM is applied: data protection, cybersecurity and prevention of sensitive content distribution. New approaches are developed which presuppose the equal presence of the state on DSM with other users: full-fledged participation of the government in the online life with the ultimate goal to forecast the agenda and to model people's behaviour based on the analysis of DSM content. Digital platforms themselves may be regarded as governing bodies since they are those who define the code of conduct and are able to force users to obey it. Finally, an approach based on redistributive justice theory presupposes the minimum control since it contradicts the very essence of information society nature. To conclude, it might be the system of check and balances that could be implemented for efficient governing of DSM, which is initially based on the legal form of regulation and would be further developed with the help of alternative governance methods.

References

Act on Development of an Environment that Provides Safe and Secure Internet Use for Young People. https://www8.cao.go.jp/youth/youth-harm/law/pdf/for_english.pdf. Accessed 12 Dec 2020

Bekkers V, Edwards A, de Kool D (2013) Social media monitoring: responsive governance in the shadow of surveillance? Gov Inf Q 30(4):335–342

Bell D (2004) The coming post-industrial society. Academia

Castells, M. 1999. Formation of a society of network structures

Children's Internet Protection Act (CIPA). https://www.fcc.gov/consumers/guides/childrens-internet-protection-act. Accessed 12 Dec 2020

Chumikov AN (2014) Who influences in new media and how contr-influence is performed: Russian applied experience. Kommunikologiya. T. 3. №. 1

Communications Act (2003) Chapter 21: The National Archives. http://www.legislation.gov.uk/ukpga/2003/21/pdfs/ukpga_20030021_en.pdf. Accessed 12 Dec 2020

Conference: Digital Hygiene organized by JSC Kribrum – Multifactorial Monitoring and Analysis System of Social Media (2019). http://digital-gigiena.ru

Craven P, Wellman B (1973) The Network City. Sociol Inq:57–88

Drucker P (1993) Post-capitalist society. New York

Enforcement Decree of the Amended Telecommunications Business Act. https://www.google.com/url?sa=t&rct=j&q=&esrc=s&source=web&cd=1&ved=2ahUKEwi_5Y6T0YvpAhWEyKYKHQfjApsQFjAAegQIAhAB&url=https%3A%2F%2Fkcc.go.kr%2Fdownload.do%3FfileSeq%3D24944&usg=AOvVaw3mnzDi-OgkmsJQUogKJspi. Accessed 12 Dec 2020

Federal Law On the Protection of Children from Information Harmful to their Health and Development. http://pravo.gov.ru/proxy/ips/?docbody&nd=102144583. Accessed 12 Dec 2020

Fountain JE (2020) Building the virtual state: information technology and institutional change. Brookings Institution Press, 2001. JSTOR. www.jstor.org/stable/10.7864/j.ctvcb59n3. Accessed 25 Feb 2020

General Data Protection Regulation 25.05.2018. https://gdpr-info.eu. Accessed 12 Dec 2020

Gillen P (2009) New agents of influence. Society for New Communications Research, USA

Global Risk Report 2021 – World Economic Forum. http://www3.weforum.org/docs/WEF_The_Global_Risks_Report_2021.pdf. Accessed 15 Feb 2021

Grygiel J, Brown N (2019) Are social media companies motivated to be good corporate citizens? Examination of the connection between corporate social responsibility and social media safety, Telecommunications policy. T. 43. №. 5, pp 445–460

Helmond, A. 2013. The algorithmization of the hyperlink. Computational Culture. №. 3

Isaak J, Mina J (2018) User data privacy: Facebook, Cambridge Analytica, and privacy protection. Computer 51(8):56–59

Kamolov S, Smagina A (2019) Digital innovation technologies in public governance: budget geolocation system. Institute of Scientific Communications Conference. Springer, Cham, pp 467–475

Kim A, Moravec PL, Dennis AR (2019) Combating fake news on social media with source ratings: the effects of user and expert reputation ratings. J Manag Inform Syst 36(3):931–968

Lebedev P (2018) The digital era. Lost on Social Media. https://www.ipsos.com/sites/default/files/ct/publication/documents/2019-01/lebedev_socseti.pdf. Accessed 15 Dec 2020

Machlup F (1962) The production and distribution of knowledge in the United States. Princeton University Press

Mediascope Web Index (2019). https://webindex.mediascope.net/research. Accessed 15 Dec 2020

Miranda FJ, Chamorro A, Rubio S (2018) The use of Facebook by local authorities: a comparative analysis of the USA, UK and Spain. Electron Gov Int J 14(2):200–216

Nazarchuk A (2008) Network society and its philosophical understanding. Problems of Philosophy 7:61

Ohotskij E (2016) Theory and mechanisms of contemporary public governance. Textbook, pp 209–215. [Teoriya i mekhanizmy sovremennogo gosudarstvennogo upravleniya: Uchebnik i praktikum dlya bakalavriata i magistratury. V dvuh chastyah. 3-e izdanie, pererabotannoe i dopolnennoe]. — M., Yurajt

Prasetya AD et al (2019) Social media strategy framework formulation and implementation: a case study of Indonesian government organization. Electron Gov Int J 15(4):417–438

Schejter A, Tirosh N (2012) Social media new and old in the Al-'Arakeeb conflict: a case study. Inf Soc 28(5):304–315

Schejter A, Tirosh N (2015) "Seek the meek, seek the just": social media and social justice. Telecommun Policy 39(9):796–803

Simmel G (1996) Selected. — M.: Yurist

Solomatin N (1989) Information semantic systems. K.: Vysshaya shkola

Statista Global Consumer Survey (2020a) Most popular social networks worldwide as of January 2021, ranked by number of active users. https://www.statista.com/statistics/272014/global-social-networks-ranked-by-number-of-users/. Accessed 20 Dec 2020

Statista Global Consumer Survey (2020b). Number of social network users in selected countries in 2020 and 2025. https://www.statista.com/statistics/278341/number-of-social-network-users-in-selected-countries/. Accessed 20 Dec 2020

Tsokolov S (2000) Discourse of radical constructivism: traditions of skepticism in modern philosophy and the theory of knowledge, Munich

Van Djik J (1991) The network society: social aspects of new media De netwerkmaastchappij Bohn Staflen Van Loghum, Houten, The Netherlands

Van Osch W, Bulgurcu B (2020) Idea generation in enterprise social media: open versus closed groups and their network structures. J Manag Inf Syst 37(4):904–932

Weber R, Newell A (2014) Corporate social responsibility as a gap-filling instrument?

Wukich C et al (2017) The formation of transnational knowledge networks on social media. Int Public Manag J 20(3):381–408

Chapter 13
Citizen-Centrism in Smart Cities: *Reality or Rhetoric?*

Nina P. David and Thomas S. Benson

Abstract Smart cities have often been portrayed as cities that effectively utilize information and communication technologies to tackle many contemporary urban issues, such as dwindling economic growth, lack of sustainability, and poor citizen participation. Following critiques of overly top-down and technocratic urban governance, smart cities have re-branded themselves as citizen-centric, promising enhanced citizen participation. However, few studies have examined whether this citizen-centrism rhetoric materializes into reality. In this chapter, we first provide a comprehensive review of extant literature on smart cities and citizen participation. Next, we develop a framework for evaluating whether smart cities are using citizen participation mechanisms in reality or if it is indeed just rhetoric. We apply this framework to a case study of a smart city, specifically Washington D.C., to demonstrate its usefulness.

Keywords Citizen-centrism · Citizen participation · E-participation · Smart cities · Urban governance · Washington DC

Introduction

In recent years, many cities have adopted strategies to transform themselves into "smart cities." To achieve such a transformation, cities often espouse the importance of citizen participation as central to urban-level decision-making and the formulation of smart city initiatives. In turn, this has led many cities to emphasize their "citizen-centric" approach to tackling urban issues, such as sustainability, housing, and infrastructure (Joss et al. 2017, 30; Pereira et al. 2017, 528). Given the dearth of literature on citizen-centrism, the rhetoric warrants investigation to examine whether

N. P. David (✉) · T. S. Benson
University of Delaware, Newark, DE, USA
e-mail: npdavid@udel.edu; tsbenson@udel.edu

G. C. Lazaroiu et al. (eds.), *Holistic Approach for Decision Making Towards Designing Smart Cities*, Future City 18,
https://doi.org/10.1007/978-3-030-85566-6_13

it follows through into reality. To this end, methodologically, this chapter reviews extant literature on citizen participation, e-participation, and smart cities. A framework is then devised to determine whether smart cities are truly citizen-centric, as they often proclaim.

Theoretically, this chapter contributes to urban studies by establishing a conceptual and evaluative framework to assess citizen-centric claims in smart cities. Empirically, this chapter employs the aforementioned framework in a case study of a smart city, specifically Washington DC from 2014 to 2021. Methodologically, to test the citizen-centric rhetoric, content and discourse analysis methods are used to examine Washington DC's Smarter DC website, as well as two official city documents: (i) Washington DC's *DC Smarter Business Challenge Strategic Plan* (2014) and (ii) Washington DC's *Smart DC Vision* (2016). Analysis of the website and two documents unveils the clear use of citizen-centrism as a narrative. The following analysis of e-participation mechanisms then illustrates that the rhetoric does not meet reality, and this is in line with current literature.

Theoretical Background

Smart Cities

Academics and professionals continue to debate what exactly constitutes a "smart city" and how to operationalize it. This debate largely revolves around the identifiable characteristics of smart cities (e.g., smart buildings, environmental protection, digitalization of government services, high Internet speed) (EasyPark 2019; Tomor et al. 2019, 13; Wong and Welch 2004, 275), the goals of smart cities (e.g., greater efficiency, interconnected infrastructure, sustainability, higher quality of life) (Washburn et al. 2010, 2; Ahvenniemi et al. 2017, 242; Meijer and Bolívar 2016, 401), the role of citizens in smart cities (e.g., big data, active participants) (Vanolo 2014, 887; Meijer and Bolívar 2016, 404), and how to conceptualize smart governance (e.g., co-production, collaborative, citizen-centrism) (Arnstein 1969, 289; Kneuer 2016, 673; Ruhlandt 2018, 2).

One guiding structure to identifying smart cities comes from Vanolo (2014, 887), who highlights: (1) a smart economy, (2) smart mobility, (3) smart governance, (4) a smart environment, (5) smart living, and (6) smart people as the key aspects of smart cities. Other scholars have identified similar dimensions, with focus on interconnected infrastructure for improved mobility (Washburn et al. 2010, 2), enhanced education as essential for smart citizens (Wiesmeth et al. 2018, 1), and environmental sustainability (Ahvenniemi et al. 2017, 242). Additional characteristics that help to identify smart cities include the establishment of a long-term, city-wide vision and the existence of several smart initiatives. In the case of the former, this discerns genuine smart cities from those that use the "smart city" brand to induce investment as part of a neoliberal strategy without aspiring to achieve the goals commonly associated with smart cities (e.g., enhance quality of life, sustainability). For the latter, this separates smart cities from short-term urban innovation projects.

Of the six characteristics identified by Vanolo (2014), two are of particular importance for this chapter: smart governance and smart people. Vanolo (2014, 887) suggests smart governance is "related to participation in decision-making processes, transparency of governance systems" and that smart people are linked to "participation in public life." In our conceptual and evaluative framework, we build upon these dimensions to incorporate additional academic literature to provide a fuller picture of smart governance and citizen participation in smart cities. However, prior to establishing this framework, it is appropriate to explore the "citizen-centrism" narrative that has dominated recent smart city literature since the "co-" turn ("collaboration, co-production, co-creation, and co-design") (Schliwa 2019, 154).

Finally, in light of the literature outlined above, we define a smart city as a city that uses a *citizen-centric urban development and governance strategy facilitated by technology to improve, overall, sustainability*. With this definition and Vanolo's identifying structure (including our additional characteristics), urban practitioners and scholars can perceive smart cities as being on a continuum of smartness, whereby cities vary in the degree to which they can be considered smart. By satisfying the above definition and structure, the smart city brand is understood as one that needs to be earned (Hollands 2008, 316).

Citizen Centrism, Participation, and Governance

The term "citizen-centrism" appeared in the early 2010s, and smart cities were often implicated as being "overly technocratic and top-down in orientation," "paternalistic," and "neoliberal" (Cardullo and Kitchin 2019, 2, 6). That is, in practice, smart cities continued to be governed by elites who made decisions on behalf of citizens, and citizens were more often than not treated as consumers and users. In response to these critiques, smart cities have engaged in a "re-branding exercises" creating initiatives and programs that claim to be "citizen focused" or "citizen engaged," though these have, in turn, been critiqued as an "empty signifier[s]" aimed at quieting critics without any fundamental changes to governance itself (Cardullo and Kitchin 2019, 2). There are opportunities, however. Through information communication technologies (ICTs), many citizens around the world have more opportunities than ever before to have their voices heard through "new channels of communication" (e.g., "smartphones," "social networks," and "web-based media sharing platforms"), and these channels can challenge the conventional "top-down paradigm of governance and agenda-setting in cities" (Razaghi and Finger 2018, 686–687). In conjunction with the fact that many people are moving into cities – and this trend is expected to continue – there is also an opportunity for urban governments to boost urban democracy through e-participation (Castelnovo et al. 2016,

725).[1] The turn to citizen centrism, in practice, places citizens at the heart of *smart* urban governance and recognizes that citizens are "bearers and users of local knowledge and expertise" (Tomor et al. 2019, 3).

Theoretically, citizen engagement has come to be recognized as one of the cornerstones of democracy and governance in general and smart governance in particular (David 2017, 73; Castelnovo et al. 2016, 731). Conceptual models of smart cities, today, include citizens as central players, a key component, or at the very least one of the dimensions of smart cities. For Albino, Berardi, and Dangelico (2015), for example, citizens are the protagonists of the smart city. For Chourabi et al. (2012, 2289), "people and communities" are one of eight factors that characterize smart cities. For Lombardi et al. (2012, 138), smart people and smart governance are two of the six key dimensions of smart cities. This leads Castelnovo et al. (2016, 729) to conclude that "[p]articipatory governance and citizen involvement…are key concepts in many smart city frameworks and even researchers that do not give governance such a central role in smart cities at least include it as one of the dimensions that should be targeted by smart city initiatives."

But what is the connection between citizen participation and governance, particularly smart governance? Vanolo (2014, 887) directly links citizen participation to smart governance, by defining smart governance as "participation in decision-making processes, transparency of governance of systems, availability of public services and quality of political strategies." Tomor et al. (2019), in an effort to articulate what smart governance for sustainability might look like, describe citizen participation as one of the three essential components – a building block – of smart governance. For Castelnovo et al. (2016, 3), collaborative governance is an essential component of smart governance. They describe collaborative governance as collective decision-making processes where stakeholders are directly involved in co-design[ing], co-decision [making], co-production, and co-evaluation. Similarly, Meijer and Bolívar (2016, 400) describe smart urban collaboration as the "widespread adoption of a more community-based model of governance" facilitated by new technologies and a network of actors (Meijer and Bolívar 2016, 400). Participation by stakeholders, including citizens, is considered active – rather than passive – and is argued to foster knowledge creation, exchange, and innovation, and open data (Meijer and Bolívar 2016, 402). Razaghi and Finger (2018, 681) frame smart urban governance as a problem-solving approach to fill-in the "incoherencies" of conventional urban governance structure, and this is done by acknowledging the need for collaborative governance as "no single actor can have the necessary knowledge, expertise, power, and resources to tackle [urban issues] … alone." Moreover, Ruhlandt (2018, 2) argues that smart urban governance entails "connecting all forces at work" to facilitate "decision-making" to achieve optimal urban outcomes in terms of "socioeconomic and environmental performance."

[1]The United Nations anticipates that, by 2050, about 66% of the world's population will live in urban areas (UN 2018).

If citizens are an essential component of smart governance, then how would government, in reality, better enable citizen participation in decision-making? We acknowledge that simply providing opportunities to participate might not result in participation itself. Citizens might need to be empowered to participate. There might be issues with equity, in terms of citizens' capacity to participate (e.g., access to technology, and information). Further, participation might not guarantee sustainable outcomes on the ground. Nevertheless, whether government is serious about citizen engagement is a good place to start. Here, we focus on citizens, not in their capacity as consumers of governmental services, but rather as participants in decision-making processes. To this end, this chapter focusses on two aspects: the extent to which governments invest in institutional development and concurrently provide opportunities for citizens to participate in decision-making.

Institutional Development

Institutional development can be defined as the design of citizen participation processes, where appropriate, to allow for genuine citizen participation (Bryson et al. 2013, 24; Ferilli et al. 2016, 97). This recognition of institutional development is both explicit and implicit in extant e-participation and e-government literature.[2] Current urban governance literature explores the importance of citizens placing trust in government institutions. This trust can be developed from increased transparency of decision-making processes, improved procedural justice, and citizen access to deliberation (David 2017, 75–76). To this end, it is crucial to "institutionalize policies and procedures" that will "promote" citizen participation (David 2017, 86; Bonsón et al. 2015, 59). Institutionalizing such policies also serves to guard against the "risk of technological determinism" as smart urban governments subsequently recognize that there is more to citizen participation (e-participation in particular) than technology (Castelnovo et al. 2016, 728). Therefore, institutionalizing citizen participation as part of smart city governance means that participation is emphasized, prioritized, articulated, and planned for in the formal documents that outline smart city-related visions and goals.

[2] E-government can be narrowly defined as "any way of using information and communication technology (ICT) to improve the relationship between governments and constituents, businesses, and other governmental agencies" (David 2017, 77; Bonsón et al. 2015, 53). E-participation can be defined as "the process of engaging citizens through ICTs in policy, decision-making, and service design and delivery in order to make it participatory, collaborative, inclusive, and deliberative" (David 2017, 78).

Opportunities for Participation

In terms of opportunities for citizen participation, we use Arnstein's *ladder of participation* and Cardullo and Kitchin's (2019, 5) *scaffold of smart citizen participation* as a foundation to develop a framework that enables us to effectively probe and assess citizen participation mechanisms. The framework is captured by Table 13.1.

At the bottom of our framework, both manipulation and therapy are non-participation (Arnstein 1969). Here citizens might be used as data points for problem definition, they might be users of governmental infrastructure and services, and the goal of participation might be to "educate." While it is problematic when non-participation is used as a substitute for participation, it can have significant benefits when thought of as one of the many rungs of the ladder with the aim of building civic capacity to participate. Use of ICTs might be in the form of Big Data, e-reporting mechanisms, webinars, and so on. At the middle of our framework are Arnstein's (1969) informing, consultation, and placation. Here, the government might be interested in keeping citizens informed, soliciting their feedback, and providing some opportunities of influence (e.g., virtual council and town meetings, online surveys, e-voting, dashboards, open data portals, etc.). ICTs could be used in this middle section, to expand access to information and expand opportunities to provide feedback and participate. Genuine citizen participation – which can legitimize urban governments – can be understood as the top three rungs of Arnstein's (1969, 289–290) ladder of citizen participation: partnership, delegated power, and citizen control. Here, ICTs can be used to promote collaborative problem-solving and decision-making (e.g., civic tech platforms, citizen-owned apps).

We acknowledge that technology alone cannot answer questions regarding privacy, anonymity, lack of digital skills, unequal access to social media, security of personal information, low levels of trust of governmental institutions among citizens, inadequate access to the Internet, or low citizen participation (David 2017, 90–92; Johnson and Halegoua 2014, 64–71). However, as shown in our framework, ICTs can play a role in all sections and rungs of the ladder. For example, even in the bottom and middle sections, ICTs can be used to improve the flow of information (Tomor et al. 2019, 11), and this improved flow of information can help build the public's capacity to participate (Tomor et al. 2019, 14). ICTs can be deployed to assist citizens in understanding "technical information," provide them with access to "information typically available only to experts," and make it "more convenient for citizens to access information from the comforts of their home or workplace" (Bryson et al. 2013; 30; Kim and Lee 2019, 1031; David 2017, 81).

ICTs also benefit urban governments enabling them to publish "huge amount[s] of information" for "relatively low cost[s]" (Bonsón et al. 2015, 52). Social media can provide a "secondary or supplemental source of information and communication," which Johnson and Halegoua (2014, 65) argue is appropriate for "today's hybrid, digital-physical public space" (Bonsón et al. 2015, 52). Vragov and Kumar (2013, 446–447) go as far as to claim that technology can "aid a peaceful transition from thin democracies" (limited citizen participation) to "strong democracies" (strong emphasis on citizen participation). Through the delivery of information,

Table 13.1 Framework – E-participation in smart cities

Form and level of participation		Citizen role and power	Discourse/ Framing	Process and communication	E-participation mechanisms
Citizen power	Citizen control	Leader, decision-maker	Rights, social/ political citizenship, commons	Process: Inclusive, bottom-up, collective, autonomy, consensus-building, negotiation, and bargaining Communication: Two-way	Citizen-owned and operated civic apps
	Delegated power	Decision-maker			Online forums, hackathons/civic hacking, living lab projects, smart city advisory committee, discussion forums, civic technology platforms (e.g., neighborly)
	Partnership	Co-creator Co-designer Co-producer	Participation, co-creation		
Tokenism	Placation	Proposer		Process: Top-down, civic paternalism, stewardship Communication: Can be two-way communication but is predominantly one-way communication	Smart city advisory committees, e-tools to propose alternative city initiatives

(continued)

Table 13.1 (continued)

Form and level of participation		Citizen role and power	Discourse/ Framing	Process and communication	E-participation mechanisms
	Consultation	Participant	Civic engagement		E-surveys, E-access to council meetings (with participation), E-feedback, E-voting, social media, emails
	Informing	Recipient, learner, consumer			Websites, E-access to council meetings (without participation), E-newsletters, E-access to council meeting agenda and minutes, social media, emails, blogs, dashboards, open data portals
Non-participation	Therapy	User, consumers, data-point, supporters	Stewardship, technocracy, paternalism	Process: Representation, communication: One-way	E-report urban issues, live in a smart building/ district, citizens as big data (consensual), use of ICTs (e.g., webinars) to educate citizens
	Manipulation				E-report urban issues, citizens as big data (non-consensual)

ICTs can also help facilitate "transparency" and "accountability" (Castelnovo et al. 2016, 726; David 2017, 79; Pereira et al. 2017, 529).[3] Therefore, ICTs can greatly facilitate the creation of the building blocks of legitimate participation: informing citizens of their "rights, responsibilities, and options" (Arnstein 1969, 285). ICTs

[3] Transparency can be defined as providing, at a minimum, "making available meeting minutes, ordinances, plans, updates on implementation of plans and ordinances, government investments…governmental activity at large…and data" (David 2017, 79). Accountability can be understood as including "aspects of fiscal transparency where government provides budgets and justifications for how money was spent, by whom, for whom, and the expected impacts of those investments" (David 2017, 79, 81).

can also allow governments to expand the range of participation mechanisms from those that require in-person attendance to those that are more flexible in terms of time and commitment. This, in turn, has the potential to bring into the public sphere those who might typically be underrepresented in policy processes (e.g., young people, working parents with children).

That said, despite the increased use of ICTs, many urban governments have yet to fully utilize it to enhance citizen participation. Instead, they have often focused solely on the provision of data and information – singular efforts to boost transparency and accountability – and this process treats citizens as passive observers, rather than citizens who are actively engaged in decision-making processes in advance of decisions being made (Pereira et al. 2017, 548; David 2017, 77). Passive observers become "mere recipients of information" without their views being taken into serious consideration by urban authorities (Tomor et al. 2019, 9). This results in citizens being "accustomed to being treated as clients and not partners," further underscoring the aforementioned notion that neoliberalism underpins the smart city concept (Bonsón et al. 2015, 59). This is likely the result of urban governments being unwilling to "share power with citizens" (giving up their autonomy) or their "lack of understanding, capacity, and expertise" to wholly incorporate citizens into urban decision-making processes (Tomor et al. 2019, 9). Nonetheless, without genuine citizen participation, citizen participation risks being relegated to "cosmetic forms of involvement" and "empty ritual[s]" (Ferilli et al. 2016, 99; Arnstein 1969, 282). This is increasingly problematic in smart urban governance as Tomor et al. (2019, 19) highlight: ambitions for genuine citizen participation in smart cities frequently remain "within the realm of rhetorical phrasing." This underscores the need for a critical examination of citizen-centrism rhetoric in smart cities to truly determine whether the rhetoric matches reality.

Methodology

Our methodology focusses on developing a holistic approach to assessing citizen-centrism claims in smart cities. Thus, the objective of the case study is to evaluate claims of citizen-centrism, and citizen participation mechanisms. Our approach employs content and discourse analysis – the latter of which forms a novel approach to assessing citizen-centrism rhetoric, especially in first asking whether such rhetoric exists, rather than presuming it is an inherent aspect of smart cities. The content analysis includes an examination of existing academic literature, city indexes and reports, and official city documents and webpages explicitly related to the case study.

The conceptual and evaluative framework (see Table 13.1) assists in highlighting key linguistic markers of interest (e.g., narratives, discourse) within the citizen-centrism phenomenon. For the purpose of this study, citizen-centrism forms the narrative, as it is a discursive process through which aspects of smart cities (e.g., citizen participation) are "foregrounded or backgrounded, filtered-in or filtered-out, and through which strategic targets are realized" (Zhang and Orbie 2019, 2). Such narratives are created and sustained by discourse and may be shaped by a range of actors, such as the media, the scientific and policy community, non-governmental organizations, citizens, and political officials and leaders (Zhang and Orbie 2019, 7). For this study, the citizen-centrism rhetoric as a narrative further conveys the significance of citizen participation as the purported foundation and glue that binds together the varying dimensions of smart cities.

Additionally, to explore the phenomenon of citizen-centrism in smart cities, a case study approach is used. This particular research method was selected because of its ability to allow sufficient analysis of the breadth and depth of citizen-centrism rhetoric and citizen participation mechanisms. Washington DC was selected given that it has a smart city strategy – *Smart DC* – as well as existing smart city initiatives that have been implemented. Washington DC is also considered to be an exemplar smart city by many recent city indexes and reports (see Table 13.2).

The content and discourse analyses are of official city documents and webpages that are publicly available on Washington DC's smart city website.[4] Analysis spans a period of 2014 (the year when Washington DC first floated the smart city concept) to 2021 (the time of writing). Thus, analysis of citizen-centrism rhetoric was conducted on two crucial documents: (1) *DC Smarter Business Challenge Strategic Plan* (2014) and (2) *Smart DC Vision* (2016). The *Comprehensive Plan* (2020) for Washington DC is currently being updated and, therefore, was not considered in this study.

In each document, a search was conducted for key terms that can be attributed to citizen-centrism. For example, "participation" (including variants, e.g., "participant," "participate," etc.) was used to denote the emphasis placed on citizen participation, and "citizen(s)" and "resident(s)" (aggregated) were used to depict the emphasis placed on the integral role played by citizens in urban governance (e.g., a bottom-up governance model) or as a target audience of urban government (e.g., a technocratic, top-down governance model). Additional keywords – often attributed to smart cities – included "inclusivity" (and its variants), "collaboration" (and its variants), and "stakeholder(s)." The context for each keyword was assessed to determine whether it was inclusive of citizen participation as opposed to, for example, collaboration strictly between the urban government and private enterprises. Thus,

Table 13.2 Washington DC as a Smart City – Rankings (2018–2020)

Publication date	City index/report	Global/domestic	Ranking
2018	Arcadis: Citizen centric cities	Global	39/100
2019	EasyPark: Smart cities index	Global	16/100
2020	IESE: Cites in motion	Global	15/174
2020	IMD: Smart City index	Global	12/109

Notes: "Domestic" is understood as within the USA and its overseas territories; Arcadis 2018; EasyPark 2019; IESE 2020; IMD 2020

[4] See https://smarter.dc.gov/.

it is argued that the greater the number of aforementioned keywords, the greater the rhetoric.

Then, to test whether the citizen-centrism narrative is rhetoric or reality in Washington DC, we turn to an assessment of e-participation mechanisms used in Washington DC's smart city strategy by using the 2014 and 2016 documents (e.g., contact city officials through social media, report urban issues electronically, attend city council and committee meetings virtually, etc., as demonstrated in the framework). To continue to evaluate citizen participation post-strategy (2016 onward), we analyze citizen participation in existing smart city initiatives by identifying the number of "public engagement" markers on smart city initiatives on the Smarter DC website.

Case Study: Washington DC

Washington DC has a population of 705,749 (2019) and a land area of 61.05 sq. mi (2010) (Census Bureau 2019). In terms of governance and city officials, Washington DC has a mayor (Muriel Bowser, 2015–present),[5] a city council, an Open Government Advisory Group, 40 Advisory Neighborhood Commissions, 12 committees, an open data portal, and numerous advisory committees (DC Council A 2021). Each of these governance components provide mechanisms for citizen participation (e.g., attending and participating in meetings, contacting city officials) and are not necessarily unique to smart cities. These aspects are also not unique to Washington DC, despite its status as the national capital of the USA. There are two exceptions, however. Unique to Washington DC, legislation approved by city council is subjected to congressional review whereby congress has the opportunity to disapprove of the legislation and prevent it from becoming law. That said, in the 47 years since Washington DC's first elected city council was established via the Home Rule Act of 1973, about 5000 legislative acts have been transmitted to congress and only 3 have been disapproved. Washington DC's budget also requires congressional approval. While these exceptions are noteworthy distinctions of the policy-making process, they do not affect our data or the scope of our analysis because we focus our chapter on the already adopted public-facing aspects of DC's smart city strategy to distinguish reality from rhetoric (a focus on "what" rather than "why") (117 Cong. Rec. HR411 2021; DC Council B 2021).

DC's characteristics including a large population size – the presence of a smart city vision, and, for the sake of comparable citizen participation features, the presence of a mayor, city council, and advisory committees, council meetings being open to the public, meeting minutes, and agendas (Wirth 1938, 116) – will allow future scholars to make for comparisons with other smart cities. Comparable smart cities have been identified by the smart city indexes and reports presented in Table 13.2.

[5] 'Present' being March 2021

Results and Discussion

Citizen-Centrism Rhetoric

The differences in frequency of the aforementioned keywords for each source appears in Table 13.3.

In the *DC Smarter Business* (2014) document, there is no mention of citizen participation – zero recorded instances of citizen/resident, other than a single reference each to "citizenship" and "residential homes" (irrelevant). Discussion of stakeholders, inclusivity, collaboration, and participation – language in line with smart urban governance literature – is, expectedly (given the title of the document), limited to businesses, landlords, non-profits, and urban government representatives (Greenspace and Eco-Coach 2014, 3, 5). The extent to which these stakeholders participated in the formulation of *Smart DC Vision* (2016) is unclear, and the "Action Plan" in the *DC Smarter Business* document was limited to "sector leaders," thus raising questions from the outset regarding the degree to which citizens would participate – if at all – in the development of Washington DC's smart city vision (Greenspace and Eco-Coach 2014, 17).

Collaboration with businesses within the *DC Smarter Business* document was envisaged as "forming critical partnerships with key stakeholders and media," "engaging major donors and sponsors," employing "gamification" strategies, clear "communications," and "creating an Advisory Council to provide input into the direction and content of the program" (Greenspace and Eco-Coach 2014, 3). Few details are provided about the advisory council, other than that businesses – not citizens – would be the participants and that the council would be established in Fall 2014 (Greenspace and Eco-Coach 2014, 13–14). No current details can be found for this, suggesting the council was established and subsequently disbanded or was never established. Nonetheless, these participation mechanisms and partnership efforts underscore exclusionary practices in the development of Washington DC as a smart city, with no opportunities for citizen to participate. To this end, the

Table 13.3 Differences in Word Frequency in 2014–2021 Washington DC Urban Government Sources

	DC Smarter Business Challenge Strategic Plan (2014)	Smart DC Vision (2016)	Smarter DC Website (aggregated) (2021)
Participation	102	6	2
Citizen(s) and resident(s)	0	34 (8 + 26)	5 (4 + 1)
Collaboration	6	13	6
Inclusivity	4	1	3
Stakeholder(s)	10	10	2

Note: The first numerical figure in parentheses represents the frequency of "citizen(s)," whereas the latter represents "resident(s)" citizen-centrism narrative is not supported as the rhetoric is not employed in the *DC Smarter Business* document. However, the exclusive focus on businesses does portray Washington DC as potentially utilizing the smart city brand as a neoliberal strategy to induce investment.

In the *Smart DC Vision* (2016), the citizen-centrism narrative is emphasized through the repeated use of "citizen(s)" (appearing eight times) and "resident(s)" (appearing 28 times). This finding is in line with extant literature that finds many self-proclaimed smart cities, use their smart city strategies, to stress citizen-centrism (Spil et al. 2017, 127; Levenda et al. 2020, 353; Przeybilovicz et al. 2020, 12). In Washington DC's vision, engagement is primarily paternalistic, top-down, and technocratic. This is demonstrated through a closer examination of the context-specific discourse analysis. Notably, "[s]mart citizens are empowered citizens," whereby citizens can engage in "new smart infrastructure" that will "enable resident and visitor feedback … [and] diffusion of information" (DDOT 2016, 3). Smart citizens are also referred to as "[s]mart users," thereby reflecting the transformation of citizens into "data products" and "consumer[s]," which is arguably underpinned by the "neoliberal conception of citizenship that favors consumption choice" (DDOT 2016, 3; Cardullo and Kitchin 2019, 2–6). The same notion is later repeated in the district department of transportation aspiring to "nimbly engage citizens" by "identify[ing] and proactively respond[ing] to the needs of its users" through a "regular flow of rich, meaningful data" (DDOT 2016, 4).

The strategy comprises multiple "Vision Element[s]," one of which is "Connected, Involved Citizens" (DDOT 2016, 17). The "Smart DC Approach" to this vision is "[c]rowdsourcing and *deep public engagement* in transportation management processes," which is reflective of the citizen-centrism narrative and indicative of potential efforts to enhance citizen participation mechanisms (DDOT 2016, 17; emphasis added). However, existing and proposed investments further stress how entrenched this narrative – within Washington DC's vision – is in rhetoric. For example, investments include open data, free public Wi-Fi, transit web portals, as well as "[u]ser-generated data" from these investments (DDOT 2016, 17). It is important to highlight here that user-generated data is not considered inherently "bad," but that the critique rests with treating citizens as data products, as opposed to citizens that *could* actively participate – given the opportunity – and share their knowledge, insights, and expertise with urban planners in a city that they live and/or work in.

In relation to the use of "participation" (and its variants), this discourse seldom appears in the context of citizens in the *Smart DC Vision*. There is an acknowledgement that citizens will not necessarily "need a smartphone to participate in the smarter city," but *how* is not outlined, which arguably illustrates additional rhetoric relating to inclusivity (DDOT 2016, 3). Another instance of ambiguity is in the mention of the government's open data portal, whereby "all the software and algorithms developed … will be open source and available for public participation" (DDOT 2016, 8). In this case, it appears that the government considers transparency, in and of itself, to amount to citizen participation. This is later clarified, in that the government "hopes" the open data will foster "greater public participation and collaboration" by being grounded in principles including "transparency, collaboration, openness, and discoverability" (DDOT 2016, 14). While transparency is undoubtedly a step toward genuine citizen participation, it is reflective of citizen participation – at least in the context of open data – being in its infancy in Washington DC.

Moreover, the vision highlighted an aspiration to "deploy a platform that will provide all *users* the ability to *provide feedback, test new technologies*, and improve movement of people, goods, and services within the city" (DDOT 2016, 26; emphasis added). Citizens are, again, referred to as "users," and the "broad [citizen] participation" here is *reactive* as citizens are providing feedback *after* the initial decision-making has taken place (DDOT 2016, 26). This reinforces the perspective that Washington DC, as a smart city, is underpinned by "a neoliberal ideology and corporate interests" as it "circumscribe[s] a particular role for citizens" – seemingly passive observers that act as data products (Cardullo and Kitchin 2019, 10).

Discussion of partnerships and stakeholders in the vision were limited to a triple stakeholder model, featuring only universities, regional institutions, and private and non-profit entities (DDOT 2016, 20–21). In terms of creating this vision, there are no mentions of workshops, meetings (open to the public), consultations, or any kind of citizen-facing communication. The single use of "inclusive" refers to collaboration across urban government agencies and sectors, thus further cementing the lack of citizen participation (DDOT 2016, 1). Somewhat ironically, this inclusivity is aimed at capitalizing on the "best practices for city governance and management," and the document later adds that citizens "will not be constrained by the capabilities of government agencies" (DDOT 2016, 1, 3). Therefore, it can be argued that the *Smart DC Vision* (2016) is underpinned by neoliberalism as it adopts a civic paternalism mode of governance. Washington DC's urban government is focused on providing services to residents who are treated as consumers and data products and play little to no role in the urban governance structure, including the formulation of the smart city vision itself.

Citizen Participation Mechanisms

While the analysis above focusses on citizen participation in the establishment and planned implementation of the 2014 and 2016 urban strategies, it is also important to assess the post-2014 (*DC Smarter Business*) e-participation mechanisms – given the centrality of technology, ICT-enabled governance, and e-government to the smart city concept – within the broader urban governance structure and setting. The availability of specific e-participation mechanisms is essential in determining the extent to which citizens can participate in Washington DC's smart urban governance structure. Table 13.4 reflects these participation mechanisms, which is in line with existing an literature that has assessed social media (Spil et al. 2017; Przeybilovicz et al. 2020), open data portals (Spil et al. 2017; Levenda et al. 2020), and e-surveys (Levenda et al. 2020). To add to this list, we examine whether citizens have the capacity to watch city council meetings electronically (live or not), whether they participate in city council meetings electronically (e.g., typed or verbal comment),

Table 13.4 E-participation mechanisms in Washington DC (2014–2021)

E-participation mechanisms	Form and level of participation	Communication (1 = one-way, 2 = two-way, 3 = both/either)	URLs and Citations
Social media (for the city council): Twitter, Facebook, YouTube, not Instagram	Tokenism (informing and consultation)	3 (dependent on responsiveness)	Twitter: https://twitter.com/councilofdc
			Facebook: https://www.facebook.com/councilofdc
			Instagram: N/A
			YouTube: https://www.youtube.com/channel/UCXTTGvPOwbn3QSov4FaQU-w/videos
Open data portal (with feedback)	Tokenism (informing and consultation)	3 (dependent on incorporation of feedback)	https://opendata.dc.gov/
Email (city council and mayor)	Tokenism (informing and consultation)	3 (dependent on responsiveness)	Council: https://dccouncil.us/councilmembers/
			Mayor: https://mayor.dc.gov/page/invite-mayor
Watch live meetings electronically	Tokenism (informing)	1	https://dccouncil.us/council-videos/
Watch archived meetings electronically	Tokenism (informing)	1	https://dccouncil.us/council-videos/
E-report urban issues	Non-participation (therapy)	1	https://311.dc.gov/citizen/home
Smart city website	Tokenism (informing)	1	https://smarter.dc.gov/
Citizens as big data	Non-participation (manipulation)	1	(DDOT 2016, 3).
E-access to current meeting agendas and minutes	Tokenism (informing)	1	https://ddc.dc.gov/page/dd-council-meetings
E-access to archived meeting agendas and minutes	Tokenism (informing)	1	https://ddc.dc.gov/page/dd-council-meetings

whether citizens can watch previous (archived) meetings, whether a smart city-specific website exists, whether feedback can be given on this website (by citizens) in relation to smart city initiatives, whether electronic access to meeting agendas and minutes exists, and whether previous meeting agendas and minutes exist (archived).

The capacity to contact city council members and the mayor via email provides a direct opportunity for citizens to engage with city officials. However, the availability of email, in addition to the capacity to contact city officials via social media (Twitter, Facebook, and YouTube), does not indicate the degree of responsiveness – a key aspect of transparency (Pereira et al. 2017, 532; David 2017, 80). Nevertheless, this is a form of e-participation available to Washington DC citizens. Further reflecting on transparency, Washington DC citizens enjoy access to city council meetings (live and archived), city council meeting agendas and minutes (current and archived), access to an open data portal (feedback can be provided), and access to a smart city-specific website that features previous, current, and planned smart initiatives. Although transparency – on par, in this instance, with Arnstein's "informing" – is a "good first step," the capacity for genuine citizen participation and meaningful dialogue is limited (Arnstein 1969, 280).

Applying the evaluative framework to e-participation mechanisms that are present in Washington DC – demonstrated by Table 13.4 – the majority of mechanisms adopt one-way modes of communication and are tokenistic in form. Our framework diverges from Cardullo and Kitchin's (2019, 8) framework as we not only acknowledge that social media can operate as a means for citizen "consultation" but also as a mechanism of inquiry, thereby also constituting "informing."

Email can serve the same twofold purposes and is – like social media – a one- or two-way mode of communication, depending on responsiveness. Other e-participation mechanisms demonstrate tokenism (informing) and one-way communication, such as watching live and archived meetings electronically, a smart city-specific website, and e-access to current and archived meeting agendas and minutes. These mechanisms allow citizens to keep updated with urban affairs but provide no opportunities for genuine citizen participation (citizen power). Furthermore, the open data portal is, in and of itself, another form of tokenism (informing) by granting citizens access to urban data. However, this portal can also be construed as another form of tokenism (consultation) by giving citizens the opportunity to provide feedback (two-way communication) on the portal. If feedback is refined, then the e-participation is noticeably minimal, but it could be a form on non-participation (manipulation) if citizens' feedback is not incorporated – an "illusory" form of participation that is uni-directional in communication (Arnstein 1969, 284).

In addition, there is evidence of non-participation. As the *Smart DC Vision* (2016) makes clear, citizens are treated as "users" of services, thereby relegating them to two forms of participation: first, citizens are consumers of urban services, such as transportation (consumerism: choice), and, second, citizens are a source of big data by which to non-consensually provide Washington DC with a "regular flow of rich, meaningful data" (non-participation: manipulation) (DDOT 2016, 4). The non-consensual aspect helps to distinguish citizens from those who consensually and willingly serve as data products (consumerism: choice), such as cyclists in Glasgow using an app to play the role of "sensors, supplying information to the service providers" to assist in reworking the city's "cycle lane network" (Przeybilovicz et al. 2020, 14). The capacity for citizens to provide feedback on the "new smart infrastructure" presents another tokenistic (consultation) mechanism for citizen participation (DDOT 2016, 3). Citizens can also use 311 – a service to report urban issues (by call or text), such as potholes, broken streetlights or meters, illegal dumping – which is a form of non-participation (therapy).

The lack of e-surveys (e.g., to consult citizens, receive feedback, develop ideas), the lack of any ability to participate in meetings electronically, and the lack of any mechanism to provide feedback on smart city initiatives on the smart city website (this is limited to businesses) demonstrates the limited ability for citizens to participate in the smart urban governance structure in Washington DC. This critique is not just limited to the lack of e-participation mechanisms, but also in relation to the lack of two-way communication and participation mechanisms regarding the development of the *DC Smarter Business* (2014) and *Smart DC Vision* (2016). Such a lack of genuine participation mechanisms suggests that the citizen-centrism narrative adopted by Washington DC in its *Smart DC Vision* (2016) is, indeed, rhetoric.

Additional e-participation mechanisms, as listed in the framework, that were noticeably absent in Washington DC include: (i) citizen-owned and operated civic apps (citizen control); (ii) a smart city advisory committee (depending on level of power, delegated power/partnership or placation); (iii) an e-tool to propose smart city initiatives (placation); (iv) e-newsletters (informing); (v) a smart building and/or district for residents (choice)[6]; (vi) hackathons; (vii) living lab project(s)[7]; (viii) an online forum[8]; and (ix) e-participation in city council meetings. Of all of these, a smart city advisory committee with "genuine bargaining influence" (partnership) could allow citizens to have a stake in the continued development of *Smart DC* (Arnstein 1969, 289). Through this advisory committee, citizens can work with city officials to create genuine participation mechanisms (electronic or otherwise), and although the committee is not, in and of itself, an e-participation mechanism, it can provide two-way communication and showcase e-participation mechanisms (e.g., meetings could be held electronically).

Further examination of the development of Washington DC's smart city brand and strategy reveals additional information on the lack of citizen participation. On the Smarter DC website, citizens cannot provide feedback on smart city initiatives, and of the 27 initiatives listed, only six featured markers for "Public/Market

[6] While a 'Smart EcoDistrict' has been planned, it remains in the conceptual stage of development. See https://smarter.dc.gov/Detail.aspx?Id=29

[7] An "Urban Living Labs" project exists but is only in the planning stage of development and has yet to involve any citizen participation, given the absence of a "Public/Market Engagement" marker. See https://smarter.dc.gov/Detail.aspx?Id=25

[8] A website for "Washington DC Message Boards" exists, but there is no indication that it is owned and operated by the Washington DC urban government. See https://www.dcmessageboards.com/

Engagement."[9] Despite the absence of any details regarding whether citizens participated (e.g., if it was only corporations) and the degree to which they were able to "participate" (e.g., a workshop, consultation, informational campaign), it is clear that input from citizens has – for the most part – not been requested nor expected. This is in spite of rhetoric used on the website that the smart city strategy is being "guided through community engagement" that is "inclusive" and "foster[s] a sense of shared ownership and collaboration."[10] In Washington DC's own admission, cultivating a "Smarter city requires transparent and engaging governance, culture of participation…and empowered and active citizenry"[11] – none of which, in relation to the smart city brand, seemingly rings true.

These findings – a self-proclaimed and accredited smart city that employs citizen-centrism rhetoric without achieving genuine citizen participation mechanisms (especially e-participation mechanisms) – are in line with the literature on citizen participation in smart cities and their respective smart urban governance structures. Spil et al. (2017, 127–129) find, overall, in three smart cities in Germany (Berlin, Hamburg, Enschede) that citizen participation is limited: open data portals are underutilized and more useful for businesses than citizens; collaboration between businesses, universities, and government often excludes citizens; and social media responsiveness is limited due to budgetary constraints. Levenda et al. (2020, 353) examine Calgary as a smart city and find that, despite the use of in-person consultations with citizens, workshops, and direct testimony to the city council, citizen participation remained largely tokenistic, and businesses were given more opportunities to participate. They also found that citizen-centrism rhetoric often remains rhetoric in smart cities as the neoliberal stronghold over local government continues to be firmly in place, and participation is frequently tokenistic (Levenda et al. 2020, 355).

Moreover, Goodman et al. (2020) assessed citizen participation in the City of Guelph, the Region of Waterloo, and the Mohawk Council of the Akwesasne (MCA) and found evidence of citizen-centrism rhetoric in the first two. They noted that opportunities for citizen participation are usually limited to informational exercises and pre-determined courses of action as the top-down governance structure persists (Goodman et al. 2020, 421). Przeybilovicz et al. (2020, 12) analyzed three cities: Curitiba, Glasgow, and Utrecht. They found that the urban governments largely treated citizens as data products, provided few opportunities for participation, and sought feedback (reactive) on smart initiatives. Although some feedback was incorporated, it is worth acknowledging that citizens are infrequently given the opportunity to participate from the outset in terms of developing smart initiatives, thereby further depicting an ongoing theme – top-down, technocratic governance and civic paternalism (Przeybilovicz et al. 2020, 14). Other scholars have recognized the technocratic nature of smart urban governance and neoliberal underpinning of the

[9] See https://smarter.dc.gov/FeaturedProjects.aspx

[10] See https://smarter.dc.gov/

[11] See https://smarter.dc.gov/

smart city brand, despite recent efforts to shift away from these critiques and toward a more citizen-centric model of governance (Joss et al. 2017, 32, 40; Schliwa 2019, 155; Cardullo and Kitchin 2019, 6; Ghose and Johnson 2020, 340).

However, there is some cause for optimism. Levenda et al. (2020, 349) identified partially successful e-participation in Calgary with the establishment of its "Engage platform" – an "online portal for citizen participation" – and this "increased the number of comments on any project," and the enhanced participation contrasted greatly with few attendees of similar in-person events. They also find that investing in particular participation mechanisms, as opposed to a broad range of mechanisms, yielded better rates of participation as it became less of a "tickbox" strategy and more about ensuring meaningful dialogue (Levenda et al. 2020, 352). E-surveys can also be useful for urban governments for examining citizen opinions and priorities, and these can, subsequently, be turned into policy and law (Levenda et al. 2020, 354). For Goodman et al. (2020, 424–425), citizen participation in MCA was interactive (e.g., citizens placing stickers on a map to indicate their desires), informational, and accessible (e.g., monthly newsletter with "forthcoming opportunities for engagement" and contact information), provided inclusive access (e.g., a rideshare program, and the provision of hot meals), and refined ideas for an app based on citizen input.

Conclusion

This chapter has analyzed citizen participation in Washington DC and has dissected the use of rhetoric in two official city documents and the city's official smart city website, as well as several e-participation mechanisms. As demonstrated – through application of the conceptual and evaluative framework – Washington DC did not offer genuine citizen participation mechanisms in line with the smart city rhetoric and narrative of citizen-centrism. The e-participation mechanisms are limited to the provision of information electronically, including city-level data, and watching meetings live – none of which constitute genuine citizen participation. Further, civic paternalism and top-down, technocratic governance remain prevalent due to the neoliberal underpinning of the city. This "paternalism," for Arnstein, is one of the "most significant roadblocks to achieving genuine levels of participation" (Arnstein 1969, 284).

Second, e-participation mechanisms like watching meetings live and accessing the open data portal treats citizens as passive consumers of information (transparency) that may create better-informed citizens, but the crucial question is, to what end? Without genuine bargaining influence to challenge the status quo or propose alternative ideas, citizen participation will remain limited, tokenistic, and unidirectional. Citizens ought not to be treated as "mere recipients of information," and the re-branding exercise of smart cities with the citizen-centrism narrative ought not to be employed to "silence detractors or bring them into the fold while keeping the central mission of capital accumulation and technocratic governance intact" (Cardullo and Kitchin 2019, 2, 10).

Third, the findings presented here are, as aforementioned, in line with the literature on citizen participation in smart cities. However, whereas most current research examines in-person citizen participation mechanisms (e.g., consultations, workshops, council meetings), this chapter focuses largely on e-participation mechanisms. This chapter also makes headway in that there is very little literature on citizen participation in Washington DC, despite its highly ranked status as a smart city, in addition to being capital of USA. Nevertheless, in-person citizen participation mechanisms in Washington DC also warrant additional analysis to further probe the extent to which Washington DC allows for genuine citizen participation, as well as the notion of "consent" in e-participation mechanisms that employ the use of citizens as data products.

Finally, in addition to the common government-sanctioned mechanisms for citizen participation highlighted in this chapter, there has been a notable increase in citizen-led demonstrations and protests in recent years across USA. Notable examples include Black Lives Matter, the Women's Marches, March for Our Lives, the Keystone XL and Dakota Access Pipeline protests, White Nationalism, and Climate Marches (including school strikes). How these demonstrations and protests were facilitated or co-produced by technology – such as social media – is becoming an increasingly important question in the digital age, especially in the context of smart cities where city officials need to navigate cyberspace regulations and privacy rights (Wilkins et al. 2019; Boulianne et al. 2020). To shine a light on this, scholars ought to explore how technology is playing a role in the co-production of demonstrations and protests.

Bibliography

Ahvenniemi H, Huovila A, Pinto-Seppa I, Airaksinen M (2017) What are the differences between sustainable and smart cities? Cities 60:234–245

Albino V, Umberto B, Rosa MD (2015) Smart Cities: Definitions, Dimensions, Performance, and Initiatives. Journal of Urban Technology 22(1): 3–21. https://doi.org/10.1080/10630732.2014.942092

Arcadis (2018) Citizen centric cities: the sustainable cities index 2018. Arcadis https://www.arcadis.com/media/1/D/5/%7B1D5AE7E2-A348-4B6E-B1D7-6D94FA7D7567%7DSustainable_Cities_Index_2018_Arcadis.pdf. Accessed 10 Jan 2021

Arnstein S (1969) A ladder of citizen participation. In: LeGates RT, Stout F (eds) The city reader, 6th edn. Routledge

Bonsón E, Royo S, Ratkai M (2015) Citizens' engagement on local governments' Facebook sites. An empirical analysis: the impact of different media and content types in Western Europe. Gov Inf Q 32:52–62

Boulianne S, Lalancette M, Ilkiw D (2020) "School strike 4 climate": social media and the international youth protest on climate change. Media Commun 8(2):208–218

Bryson JM, Quick KS, Slotterback CS, Crosby BC (2013) Designing public participation processes: theory to practice. Public Adm Rev 73(1):23–34

Cardullo P, Kitchin R (2019) Being a 'citizen' in the Smart City: up and down the scaffold of smart citizen participation in Dublin, Ireland. GeoJournal 84:1–13

Castelnovo W, Misuraca G, Savoldelli A (2016) Smart cities governance: the need for a holistic approach to assessing urban participatory policy making. Soc Sci Comput Rev 34(6):724–739

Census Bureau (2019) QuickFacts: District of Columbia. US Census Bureau https://www.census.gov/quickfacts/fact/table/pittsburghcitypennsylvania,bostoncitymassachusetts,miamicityflorida,denvercitycolorado,philadelphiacitypennsylvania,DC/PST045219. Accessed 18 Feb 2021

Chourabi H, Nam T, Walker S, Gil-Garcia JR, Mellouli S, Nahon K, Pardo TA, Scholl HJ (2012) Understanding smart cities: an integrative framework. In: 45th Hawaii international conference on system sciences, pp 2289–2297

David N (2017) Chapter 4: Democratising government: what we know about e-government and civic engagement. In: Bolívar MPR, Alcaide-Muñoz L (eds) International e-government development: policy, implementation and best practices. Springer, Berlin

DC Council A (2021) The council. DC Council https://dccouncil.us/the-council/. Accessed 1 Jan 2021

DC Council B (2021) How a bill becomes a law. DC Council https://dccouncil.us/how-a-bill-becomes-a-law/. Accessed 9 Mar 2021

District Department of Transportation (DDOT) (2016) Smart DC: making the district a smart city. Washington DC. https://ddot.dc.gov/sites/default/files/dc/sites/ddot/page_content/attachments/District%20of%20Columbia%20Smart%20City%20Application%20Part%201.pdf. Accessed 30 Dec 2020

EasyPark (2019) Smart cities index 2019. EasyParkGroup. https://www.easyparkgroup.com/smart-cities-index/. Accessed 10 Jan 2021

Ferilli G, Sacco PL, Blessi GT (2016) Beyond the rhetoric of participation: new challenges and prospects for inclusive urban regeneration. City Cult Soc 7:95–100

Ghose R, Johnson PA (2020) Introduction to the special section: smart citizens creating smart cities: locating citizen participation in the Smart City. Can Geogr 64(3):340–343

Goodman N, Zwick A, Spicer Z, Carlsen N (2020) Public Engagement in Smart City Development: Lessons from Communities in Canada's Smart City Challenge. The Canadian Geographer 64(3): 416–432

Greenspace and Eco-Coach (2014) DC smarter business challenge strategic plan: FY 2015 – FY 2019. DC Gov Dept of Energy & Environment https://doee.dc.gov/sites/default/files/dc/sites/ddoe/service_content/attachments/SBC%20Strategic%20Plan%20Final%209%2030%202014.pdf. Accessed 19 July 2020

Hollands RG (2008) Will the real smart city please stand up? Intelligent, progressive or entrepreneurial? City 12(3):303–320

IESE (2020) IESE: cities in motion index 2020. IESE https://media.iese.edu/research/pdfs/ST-0542-E.pdf. Accessed 10 Jan 2021

IMD (2020) Smart city Index 2020. IMD https://www.imd.org/smart-city-observatory/smart-city-index/. Accessed 10 Jan 2021

Johnson BJ, Halegoua GR (2014) Potential challenges for social media in the neighbourhood context. J Urban Technol 21(4):51–75

Joss S, Cook M, Dayot Y (2017) Smart cities: towards a new citizenship regime? A discourse analysis of the British smart city standard. J Urban Technol 24(4):29–49

Kim S, Lee J (2019) Citizen participation, process, and transparency in local government: an exploratory study. Policy Stud J 47(4):1026–1047

Kneuer M (2016) E-democracy: a new challenge for measuring democracy. Int Polit Sci Rev 37(5):666–678

Levenda AM, Keough N, Rock M, Miller B (2020) Rethinking public participation in the smart city. Can Geogr 64(3):344–358

Lombardi P, Giordano S, Farouh H, Yousef W (2012) Modelling the smart city performance. Innov Eur J Soc Sci Res 25(2):137–149

Meijer A, Bolívar MPR (2016) Governing the smart city: a review of the literature on smart urban governance. Int Rev Adm Sci 82(2):392–408

Pereira GV, Cunha MA, Lampoltshammer TJ, Parycek P, Testa MG (2017) Increasing collaboration and participation in smart city governance: a cross-case analysis of smart city initiatives. Inf Technol Dev 23(3):526–553

Przeybilovicz E, Cunha MA, Geertman S, Leluex C, Michels A, Tomor ZC, Webster WR, Meijer A (2020) Citizen participation in the smart city: findings from an international comparative study. Local Gov Stud:1–25

Razaghi M, Finger M (2018) Smart governance for smart cities. Proc IEEE 106(4):680–689

Ruhlandt RWS (2018) The governance of smart cities: a systematic literature review. Cities 81:1–23

Schliwa G (2019) Smart cities by design? Interrogating design thinking for citizen participation. In: The right to the smart city. Emerald Publishing Ltd, pp 151–164

Spil TAM, Effing R, Kwast J (2017) Smart city participation: dream or reality? A comparison of participatory strategies from Hamburg, Berlin & Enschede. Int Feder Inf Process 10595:122–134

Tomor Z, Meijer A, Michels A, Geertman S (2019) Smart governance for sustainable cities: findings from a systematic literature review. J Urban Technol 26(4):3–27

United Nations (UN) (2018) World urbanisation prospects: the 2018 revision. United Nations Economic and Social Affairs https://population.un.org/wup/Publications/Files/WUP2018-KeyFacts.pdf. Accessed 8 Jan 2021

Vanolo A (2014) Smartmentality: the smart city as disciplinary strategy. Urban Stud 51(5):883–898

Vragov R, Kumar N (2013) The impact of information and communication technologies on the costs of democracy. Elect Comm Res Appl 12(6):440–448

Washburn D, Sindhu U, Balaouras S, Dines R, Hayes N, Nelson L (2010) Helping CIOs understand "Smart City" initiatives. Forrester Research, Cambridge, MA

Wiesmeth H, Fiala O, Stegareva E, Häckl D, Weinhold I (2018) Smart institutions for smart cities. Int Conf Sustain Cities Earth EnvironSci 177:1–7

Wilkins DJ, Livingstone AG, Levine M (2019) Whose tweets? The rhetorical functions of social media use in developing the black lives matter movement. Br J Soc Psychol 58:786–805

Wirth L (1938) Urbanism as a way of life. In: LeGates RT, Stout F (eds) The city reader, 6th edn. Routledge

Wong W, Welch E (2004) Does E-government promote accountability? A comparative analysis of website openness and government accountability? Gov Int J Pol Admin Inst 17(2):275–297

Zhang Y, Orbie (2019) Strategic narratives in China's climate policy: analysing three phases in China's discourse coalition. Pac Rev:1–28

117. Cong. Rec. HR411 (2021, Jan 21) Statement of Rep. Norton

Chapter 14
Smart City, the Citizen Response and the Social and Human Need: *Some Sociological Worries*

Vasile Sebastian Dancu

Abstract Most often we refer to the smart city technologies as ways to increase quality urban life. But, is this the main interest? We can assume the idea that smart city is really the need of the future urbanistic, but there are still few questions about the place it occupies and the citizen in this technologically integrated system, in addition to the fact that he will be the final beneficiary: 1. Can technology now meet all human needs? 2. How will this digital cohabitation work? The author followed in several sociological surveys on national urban population samples, conducted in 2018 and 2021, the measurement of indicators related to the use of smart city technologies and governance in Romania aiming to identify how smart government ideologies are correlated with the satisfaction of individuals and their prioritization in terms of urban quality of life. One of the conclusions of the studies shows that when we talk about smart cities, we focus too much on technology. However, rebuilding cities means from a sociological perspective bringing people together and restoring trust and participation in common life. This can be the final goal, and we should see how we put technology at the service of this goal.

Keywords Governance and planning · Citizen participation · Smart city · Quality of urban life · Environmental awareness and behavior · Attitudes and beliefs · Attitudes toward technologies · Conviviality · Human smart cities · Smart communities

V. S. Dancu (✉)
University of Bucharest, Bucharest, Romania
e-mail: vasile.dancu@sas.unibuc.ro

G. C. Lazaroiu et al. (eds.), *Holistic Approach for Decision Making Towards Designing Smart Cities*, Future City 18,
https://doi.org/10.1007/978-3-030-85566-6_14

Introduction

In the literature, most references to the concept of smart city are focused on how the elements of technology increase the quality of life. But, can technology now meet all human needs? We don't believe that, at least for the argument that needs and expectations change quite quickly. Another essential question is how will this digital cohabitation work? For now, we know that social networks are everywhere, but they do not create sociability but rather determine an isolation of individuals.

The smart city topic needs to be contextualized in relation to important topics to assess the broader framework of quality of life, such as the values, beliefs, norms, and behaviors related to the environment and technology; satisfaction with dwelling and the neighborhood and mobility habits; knowledge about the smart city concept and the local projects, attitudes toward local projects, and the expected projects outcomes; and attitudes toward technologies and actual behavior.

Speculations about the effects that the COVID-19 pandemic will have on the future development of cities seems, at first sight, to give a greater chance only to policies based on digital applications, without any precautions. Social distancing will require digital applications to avoid crowds in urban spaces or at work. On the other hand, urban sociology studies on the effects of the crisis on the population show that the biggest lack felt in isolation or lockdown was precisely the face-to-face social contact, the relationship with the other or the hugs.

A study conducted in Romania,[1] one year after the onset of the pandemic, revealed that the effects on the social relationship and sociality were most severely felt, although the penetration of mobile communications technology is very high. Against the background of restrictions and social distancing, the need for socialization was strongly affected in the first pandemic year: over half of Romanians (54%) felt more the need to contact a loved one, and 43% felt the need to talk with someone dear. Almost three quarters (71%) of the study participants said that during this period they met much less with friends. Analyses show that the need for contact is present in greater proportions among women, young people and people up to 50 years, and low-educated respondents but also those with higher education or among those living in cities. The impact of COVID-19 on the psycho-emotional state of Romanians is significant. According to the sociological survey, one of ten Romanians went through anxiety in the last 7 days, almost daily, 13% felt sad or hopeless, 14% could not control their worries, and 14% felt that they no longer have the interest or pleasure to do something. We still do not know what influence the data collected during the pandemic for post-COVID-19 development will have, but we can suspect that, beyond digital surveillance applications and those of

[1] The Romanian Institute for Evaluation and Strategy – IRES, "One year of COVID-19 in Romania. The impact of the pandemic on the Romanians quality", 2021, February 12–15. Representative sample for the adult population in urban Romania: 1.115 individuals. Maximum tolerated error ± 2,9%. Available at: https://ires.ro/uploads/articole/ires_1-an-de-covid-19_impact-calitatea-vietii_sondaj-de-opinie_c.pdf?fbclid=IwAR3EQ-JfJHAOUh0lO3DYxvv3Y4hBjDCmIs7sqvNTpyxgE7Ov6VsnV72nC88

organizing spaces and urban mobility flows, there will be many implications to humans of the smart city.

Certainly, to a large extent, the pandemic will be a lubricant for smart cities (Kunzmann, 2020), but the sustainability of development we believe will bring the requirement of strong connection to the specific needs of communities, not just for technological urbanization projects. Urban sociology and various pandemic studies show that public power gained a growing role during this period, to the detriment of urban planners or bottom-up initiatives. Research has shown that resource management during COVID-19 by policymakers led to a resurgence of symbolic thinking, with a pillar of "institutional magic" and rethinking the emotion-risk couple in government decisions (Brown, 2020).

Transforming Cities

When analyzing modernity, most analysts believe it is the result of urbanization, or that urban civilization brings with it a new civilization, a new culture, and new values. Even the emancipation of the individual was a utopia that not only early communism but many others used to link it to urban space and to the birth of the industry and the proletariat.

In a worldwide successful work, Urban Express (Nordström and Schlingmann 2015), Kjell Nordström, and Per Schlingmann launched the idea that the future belongs to the cities and to women. The city has become the natural environment of the people, the authors say, and it will be the most important form of organization in the future because it already has greater power than the nation states. Obviously, these types of postmodern utopias are based on many data and statistics. Two cities in Russia account for 30% of Russia's GDP, London for a quarter of UK GDP,[2] and Bucharest and the adjacent area for almost 30% of Romania's GDP.[3] In 2050 it is estimated that nearly 75% of the world's population will live in cities, and the concentration will be astonishing: 50% of the population will occupy only 1% of Earth's area.

The social consequences of smart city operations were not still studied by sociologists or psychologists. Massive migration to jobs (Silaghi and Ghatak 2011) has emptied rural areas and killed traditional rural communities (Sandu and de Jong 1998). However, large urban concentrations are increasingly difficult to manage and organize rationally. The creation of the great peripheries of the cities of millions of inhabitants in many parts of the world seems to be a good model for the decomposition of the Western city model (Dear and Flusty 1998) and the emergence of a dystopian end for the civilized world: huge neighborhoods with mountains of rubbish

---

[2] See https://ceoworld.biz/2018/12/28/gdp-rankings-of-the-worlds-largest-economies-2019/

[3] See https://cnp.ro/user/repository/prognoze/prognoza_profil_teritorial_decembrie_2016.pdf

and thousands of people on the streets, delinquency and violence all the way, a landscape that does not show civilization or modernity.

In the centers of cities where the poor were pushed to the outskirts and to the out-of-town industry, today are only the bright showcases of the big brands and alongside the tourists, an invasion of foreigners who seem to make the heart of the cities even colder for most of their inhabitants (Bärnthaler et al. 2020).

The city is no longer a topos of dwelling, but a ball of traffic flows (Aalbets and Gibb 2014*).*

Connectivity becomes the main feature and premise of urban life and governance (Olazabal et al. 2018). However, smart city connections and technologies will not make cities more homogeneous and neither more egalitarian nor cohesive, with more consolidated identities.

The sea of connections and possible virtual spaces will increase inequalities, as there will be a problem of access to them (Mausom and Choudhary 2017). A young man in the poor neighborhood can have a laptop and a smartphone, but these cannot replace the education and diploma or other resources he or she needs to build a career or to make a step forward in their family; sociologists consider that it should be the case (Pantazis et al. 2006).

The city is a segmented world (Harriso 2010*), with many speeds and so it will remain.*

If we look, for example, in a model city like Singapore, we see that intelligent mechanisms will not increase the number of jobs for the poor or democracy, even if some authors seek to demonstrate that citizens will be able to propose solutions and will be able to transmit them continuously, supervising in some way all of the spaces (Seta et al. 2017).

The possibilities of limiting individual liberty (Colema 2003*) and overseeing individuals (Coleman* 2005*) will increase exponentially, and thus individual autonomy and freedom of choice will be reduced.*

Already today, in many cities around the world, when going through a store area, window displays warn you on your phone over the offer, or when entering a mall, the phone can connect to intelligent devices that manage to monitor your route and access different personal data (Deakin et al. 2011).

The social consequences of smart city operations have not yet been studied by sociologists or psychologists, even in recent articles (Sandel 2017*).*

Generally speaking, smart cities are talking about ways to increase the quality of urban life (Lombardi et al. 2012). But, is this the main interest? We can move on to the idea that smart city is really the necessity of the urban future, but there are still few questions about the place the citizen occupies in this technologically integrated system (Cardullo and Kitchin 2019), besides being a beneficiary.

However, we can assume, as some sociologists do today, that we have a rather commercial strategy (Pellicer et al. 2013) but also a power strategy (Castells, 2016). The interest of service providers is unquestionable, and those who decide what is needed will have more and more power over the urban system. Proximity policy or participatory practices are replaced by an entrepreneurial (Calzada and Cowie 2017), technocratic design, a "top-down" kind.

Smart Cities = Happier People?

The fascination for technological innovation will turn citizens into technology consumers, but are they the only needs of the citizen to achieve the feeling of well-being or happiness?

The need for intelligent services is not the first option of those who appreciate the quality of living or the temptation to choose another city. In Romania, for example, if half of urban residents do not want to move at any price, the first option of the other half is to move to the country and not to a big city where there are much better services. Similarly, among the inconveniences of city dwelling mentioned by Romanian citizens, there are many things impossible to integrate in smart city optimization: increased cost of living (as a rule, technology increases costs), stress, lack of cleanliness, lack of green spaces, the difficulty of finding a place to live.

Centralized numerical control of infrastructure, equipment and services, urban planning optimization, intelligent controlled infrastructure, heating, and urban lighting are things that become more and more real, but it does not exhaust the human requirements for choosing a city.

Can technology now respond to all human needs?

We do not believe that, at least for the argument that needs and expectations change quite quickly (Van Cott 1985).

An essential question is how will this digital cohabitation work? For now, we know that social networks are everywhere, but they do not create sociability and rather lead to isolation of individuals. It's true, the Internet and everything it brought comes with a numerical utopia that has changed in a very short period of time a part of our world (Williams and Edge 1996). However, it is still a utopia. At the heart of this utopia is the idea that information can fundamentally change cities in safer and more rationally organized spaces, but the essential element is services. Smart city is almost synonymous with system optimization (Mohamed et al. 2017). But, will the logic of system optimization take on the whole logic of development without human creativity, just maximizing functional performances?

Slowly, cities will be less and less differentiated and become more and more similar; they will lose their "specialization" and their charm that makes them unique places with unique experiences. Becoming "cities without character" (De Frantz 2007), a sort of "system of systems," will no longer need the input of the inhabitants to give them the charm and the uniqueness of community and cultural experiences, suffering a cultural and individual deactivation (McCornack 2012). The city also lives through social memory imagery; it is imagined and imaginable, and so it cannot only be reduced to the quality of services. At no time, the stiffness of super-optimized service systems will have the power to strengthen or "optimize" the sense of belonging to the community.

Numerical technologies seek to unify, to bring everyone and everything to the same denominator, to the same formulas. It leads to impoverishment and to the blocking of the transformation force that the diversity and tensions of its manifestations in a community have.

There will be technical changes that will increasingly lead to the need for standardization through computer programs to reduce the elements of the random and weakness of human individuals, so that the temptation to push social management toward technocratic utopia will be growing (Zukin 1998). More and more changes in collective behavior will be required, so that the culture will become senseless, and the system of values will suffer modifications. The control over the individual and the violation of privacy will increase (Allam and Newman 2018). Concerns about Big Brother have so far supported the fantasy of a great writer or the totalitarian practices of fascism or communism. Now there is also technical support, but also an ideology that postulates that digitalization and technology are generally neutral. In addition, terrorism has brought back the need for security and surveillance not only of public spaces but also of individuals (Tzezana 2016). Personal data gathered through the surveillance cameras or the continued geolocalization of the individual may be a threat to freedom. Data can be used for profiling but also for surveillance. It is necessary to build procedures for mandatory anonymization but also for a strict regime of use of big data and open data systems.

Numerical technologies are already being seen to limit direct relationship and social interactions (Choudrie et al. 2013). Starting from informatization of the administration ("contactless" technology) and through surrogate participation through social networks or online consultation of citizens, the city loses its socializing force and direct interaction practices that create and sustain culture, identity, and the desire to communicate to transform social practices or even services through social innovation and co-participation.

We are hearing more and more often, with good reason, discussions about the harmfulness of electromagnetic waves on health through lots of cases of electro-hypersensitivity (Vizcaya et al. 2017). The city is becoming more and more a place bombed from all directions by such cannons with magnetic waves.

The more digital the city, the more vulnerable it is from the perspective of cyber-terrorism, knowing that the most integrated and strongest systems are the most vulnerable (Kitchin & Dodge 2019*).*

In a world where hackers from Râmnicu Vâlcea or Arad, small cities in Romania, enter into NATO or Pentagon systems (Alcantara 2013), there is a worry that road traffic guidance systems can be pirated and a simple altered display of traffic signs or signaling can cause disasters in the chain, so that we can only give a simplistic example; it is always possible, notwithstanding the disruption of public lighting systems or electricity supply systems.

Beyond that, the new lifestyle in cities will accentuate what today is called nomophobic syndrome, that is, addiction to smartphone (Gupta 2019*).*

If today's studies show that over half of smart technology owners experience this syndrome, that the fear of being out of the network, staying without battery, or losing the phone, so feeling the fear of leaving the connection, a greater dependence on the technology will be able to "remodel" a number of other components of social and individual behavior.

There are many other questions we have to answer, without, however, rejecting the technology.

Who Are the City Owners?

Do we want inclusive cities or cities that practice exclusion (Beall 2002)?

Analyses show that less and less the mass of the inhabitants are owners. The habitat is pushed further out of the cities – for the benefit of corporations and office buildings (Brenner et al. 2009). What place does the habitat occupy in the general economy of a city (Bastons and Armengou 2016), in the minds of the citizens and in their expectations regarding the future of the city?

When we talk about smart city, we put more emphasis on technology. However, rebuilding cities means from a sociological perspective bringing people together and restoring trust and participation in the common life (Lee et al. 2014). This may be the main target, and we should see how we put technology in the service of this goal.

Another fundamental question is, "who serves this type of intelligence"? Does it serve the people or the decision-makers or the administrators (Khan & Dambruch & Peters-Anders & Sackl & Strasser & Fröhlich & Templer & Soomro, 2017)? We believe that at this point in time, the way we think the intelligent city focuses on administration rather than on empowerment of the inhabitants, the citizens. Through smart city, we might think about how to return the cities to the people, how to get them out of the hands of their mayors and politicized projects, corporate oligarchy (Sadowski and Pasquale 2015), or even invading tourists (Concilio et al. 2016).

In this way, cities will not only cease to resemble each other, regain the charm and plurality of solutions, but also the city's governance will be easier. Solving citizens' requests online can be fascinating, but the total disappearance of the clerk will not facilitate rapid solutions to problems that cannot be coded in software procedures and solutions. The counter at the desk had a role other than automatic. Let's see what people's aspirations (Aljoufie and Tiwari 2017) are and start from them to build the smart cities of the future.

There is no question that the cities of the future need intelligence and technology, but they will become truly intelligent when they can accumulate the intelligence of the people in the city, as Saskia Sassen, professor of sociology at Columbia University, asserted.[4]

Cities will, in particular, need the living intelligence of their inhabitants and their emotions, their sense of living at home, and being responsible in any behavior more than artificial intelligence (Capdevila and Zarlenga 2015).

The happy ones in the small towns. The need for conviviality.

People in small towns say they are happier than those in big cities. It seems a paradox because in smaller cities there are lower wages, urban infrastructure is weaker, cultural infrastructure is less developed, and smart city elements are almost non-existent in Romania, at least.

It is true that in large cities happiness is a variable dependent on income level but also on human capital (Florida et al. 2013); studies of metropolitan economic performance have found that human capital is a strong predictor of regional economic outcomes including incomes.

Half of Romanians in urban areas do not want to move from the city where they currently live, the degree of conservatism being higher among residents of small towns, women, and the elderly,[5] and the other half would accept a migration to rural areas. The big, technological city, with many smart city elements, does not attract. The disadvantages of living in the city concern concrete and non-functional aspects such as pollution, lack of parking spaces, increased cost of living, and lack of cleanliness. Also, traffic and road infrastructure and waste management are aspects of the quality of urban life assessed negatively by Romanian citizens.

When asked if the administrations in their cities are open to accept proposals from citizens or if they have projects for the development of the city, opinions are rather diffuse: most respondents have positive opinions about the attitude of the administration or are placed in the area of neutrality. More than half of Romanians admit that they have not had any initiative or proposal aimed at improving the quality of life in the city where they live.

About a third of urban Romanians have heard about the notion of smart city, and four out of 10 are optimistic about the chances that the city where they live will become a smart city in the coming 10 years. Although 6 out of 10 Romanians in urban areas have a smartphone, few use it to access applications or features that can make life easier in cities.

In the IRES study conducted in 2018, even if 84% of Romanians believe that cities will be more digitalized in ten years, still 59% hope that they will be better thought out so that people can live together. However, digitization will not achieve this requirement; what we see at least so far is that cities are increasingly exclusive, not inclusive.

---

[4] See https://www.livemint.com/Specials/m21w1rzMM8KpbE9KO1iFVK/Redefining-notions-of-urban-intelligence.html

[5] See https://www.researchgate.net/publication/325319651_Perceptii_si_atitudini_privind_calitatea_vietii_urbane_in_Romania

Even though technology has been considered a panacea for the global and local challenges of urban development today, sociological research is increasingly showing that it is not enough (Santoso and Kuehn 2013). Conviviality strengthens social cohesion and harmonizes coordination between individuals, groups, and institutions in web communities, for example, in digital cities (Caire 2009). Conviviality is also a condition for social interactions but also a tool for the internal regulation of social systems.

Conclusions

Today, there are already practices of using social intelligence design to model conviviality for digital cities (Amsterdam, Helsinki), but it is necessary to integrate in the design of digital city applications, the necessary protection mechanisms against potential negative parts of conviviality, such as deception, domination, group fragmentation, and reductionism. Best practices and guidelines for the design of social intelligence systems should include issues such as ensuring the views of all parties, in order to avoid crushing one party to another. These experiments try to build a virtual conviviality (Ameripour et al. 2010), but the subjects of our research refer to real conviviality, as meeting and sharing experiences face to face. All the more important is the need for direct social contact, as the COVID-19 pandemic has strengthened this need to reunite people in common spaces and reduce spatial distances.

If the definitions for smart cities, but also the practices, are diverse, we believe that it is normal for the types of queries, when we make projects, to be as integrated as possible and to bring specialists from many fields. Our interrogative approach was placed somewhere in the area of the need for the participation of social sciences specialists in order to be able to achieve at least two qualities of the future smart cities: human smart cities and sustainability.

We believe that the "intelligence" of a city is not a property in itself and none that is immanent to the technology used, but it is built in the triad governance – citizen participation – technology. Technology dependence can deprive citizens of the skills that make a community resilient to unexpected risks.

Now technology and its followers are developing their own dynamics in which the technological imagination works only by virtue of pragmatic and utilitarian desideratum, without asking for feedback from the beneficiaries who are citizens. Thus, the social dimension of technologicalization is little studied, although it includes important elements: culture and cultural capital, social capital, relationship infrastructure, identity capital, creative practices, cultural traditions and practices, and other elements.

The city is a complex socio-technical structure in which it is impossible to harmonize everything. Here can come the role of governance and citizens' participation. Especially in this crisis caused by the COVID-19 pandemic, there is an important need to rethink socioeconomic development models to make them more related to the new social needs, especially related to territorial viability and social inclusion (Bencardino and Greco 2014), but also the need for social contact to combat isolation stress (Kim and Jung 2021). The paradigm of the human intelligent city (Concilio et al. 2014) projects the idea that future cities must bring more sustainable services, based on the collaborative nature of anthropocentric networks leading to the emergence of new partnerships of social actors interested in solving a problem unsatisfied societal.

Social scientists can provide data to show how we can, through smart technologies, build more inclusive, efficient, and sustainable projects, but especially models that stimulate social innovation. That is not just smart cities, but smart communities.

References

Aalbers MB, Gibb K (2014) Housing and the right to the city: introduction to the special issue. Int J Hous Policy 14(3):207–213. https://doi.org/10.1080/14616718.2014.93617

Alcantara LL (2013) Cybercrime and cybersecurity in the global south. J Glob Inf Technol Manag 16(4):72–74. https://doi.org/10.1080/1097198X.2013.10845650

Aljoufie M, Tiwari A (2017) People's aspirations from smart city technologies: what solutions they have to offer for the crucial challenges city of Jeddah is facing. Curr Urban Stud 5:466–482. https://doi.org/10.4236/cus.2017.54026

Allam Z, Newman P (2018) Redefining the smart city: culture, metabolism and governance. Smart Cities 1(1):4–25. https://doi.org/10.3390/smartcities1010002

Ameripour A, Nicholson B, Newman M (2010) Conviviality of Internet social networks: an exploratory study of Internet campaigns in Iran. J Infor Technol 25(2):244–257. https://doi.org/10.1057/jit.2010.14

Bärnthaler R, Novy A, Stadelmann B (2020) A Polanyi-inspired perspective on social-ecological transformations of cities. J Urban Aff:1–25

Bastons M, Armengou J (2016) Human habitat, space and place. J Agric Environ Ethics 29:559–570. https://doi.org/10.1007/s10806-016-9617-8

Beall J (2002) Globalization and social exclusion in cities: framing the debate with lessons from Africa and Asia. Environ Urban 14(1):41–51. https://doi.org/10.1177/095624780201400104

Bencardino M, Greco I (2014) Smart Communities. Social Innovation at the Service of the Smart Cities. TeMA - Journal of Land Use, Mobility and Environment. https://doi.org/10.6092/1970-9870/2533

Brenner N, Marcuse P, Mayer M (2009) Cities for people, not for profit, 13:2–3, 176–184. https://doi.org/10.1080/13604810903020548

Brown P (2020) Studying COVID-19 in light of critical approaches to risk and uncertainty: research pathways, conceptual tools, and some magic from Mary Douglas. Health Risk Soc 22(1):1–14. https://doi.org/10.1080/13698575.2020.1745508

Caire P (2009) Designing convivial digital cities: a social intelligence design approach. AI Soc 24:97. https://doi.org/10.1007/s00146-009-0201-x

Calzada I, Cowie P (2017) Beyond smart and data-driven city-regions? Rethinking stakeholder – helixes strategies. Region Magazine 308(4):25–28. https://doi.org/10.1080/13673882.2017.11958675

Capdevila I, Zarlenga MI (2015) Smart city or smart citizens? The Barcelona case. J Strateg Manag 8(3):266–282. https://doi.org/10.1108/JSMA-03-2015-0030

Cardullo P, Kitchin R (2019) Being a 'citizen' in the smart city: up and down the scaffold of smart citizen participation in Dublin, Ireland. GeoJournal 84:1–13. https://doi.org/10.1007/s10708-018-9845-8

Castells M (2016) A sociology of power: my intellectual journey. Annu Rev Sociol 42:1–19. https://doi.org/10.1146/annurev-soc-081715-074158

Choudrie J, Ghinea G, Songonuga V, Surfers S (2013) E-government and the Digital Divide: An Exploratory Study of UK Local Authority Websites and Older Citizens, Interacting with Computers, Volume 25, Issue 6, November 2013, Pages 417–442, https://doi.org/10.1093/iwc/iws020

Coleman R (2003) Surveillance, power and social order: a case study of CCTV in Liverpool. In: Doctorate (PhD) submitted to Liverpool John Moores University. The British Library, London

Coleman R (2005) Surveillance in the city: primary definition and urban spatial order. Crime Media Cult 2:131–148. https://doi.org/10.1177/1741659005054018

Concilio G, Deserti A, Rizzo F (2014) Exploring the interplay between urban governance and smart services codesign. ID&A Interaction Design Architecture Journal, v. 20, p. 33–47

Concilio G, Marsh J, Francesco M, Rizzo F (2016) Human smart cities: a new vision for redesigning urban community and citizen's life. In: Skulimowski A, Kacprzyk J (eds) Knowledge, information and creativity support systems: recent trends, advances and solutions. Advances in intelligent systems and computing, vol 364. Springer, Cham. https://doi.org/10.1007/978-3-319-19090-7_21

De Frantz M. The City without Qualities. Political Theories of Globalization in European Cities, EUI SPS, 2007/04. Retrieved from Cadmus, European University Institute Research Repository, available at: http://hdl.handle.net/1814/6732

Deakin M, Waer A, Husam (2011) From intelligent to smart cities. Intell Build Int 3(3):140–152. https://doi.org/10.1080/17508975.2011.586671

Dear M, Flusty S (1998) Postmodern Urbanism. Ann Assoc Am Geogr 88(1):50–72. https://doi.org/10.1111/1467-8306.00084

Florida R, Mellander C, Rentfrow PJ (2013) The happiness of cities. Reg Stud 47(4):613–627. https://doi.org/10.1080/00343404.2011.589830

Gupta R (2019) Nomophobia: a smartphone addiction. Int J Ind Psychol 7(1):969–979. DIP:18.01.110/20190701. https://doi.org/10.25215/0701.110

Harrison J (2010) Networks of connectivity, territorial fragmentation, uneven development: the new politics of city-regionalism. Polit Geogr 29(1):17–27., ISSN 0962-6298. https://doi.org/10.1016/j.polgeo.2009.12.002

Kim H H-s, Jung JH (2021) Social isolation and psychological distress during the COVID-19 pandemic: A cross-national analysis. Gerontologist 61(1) February 2021:103–113. https://doi.org/10.1093/geront/gnaa168

Kitchin R, Dodge M (2019) The (in)security of smart cities: vulnerabilities, risks, mitigation, and prevention. J Urban Technol 26(2):47–65. https://doi.org/10.1080/10630732.2017.1408002

Kunzmann KR (2020) Smart cities after Covid-19: ten narratives, disP. Plan Rev 56(2):20–31. https://doi.org/10.1080/02513625.2020.1794120

Lee JH, Gong Hancock M, Hu M-C (2014) Towards an effective framework for building smart cities: lessons from Seoul and San Francisco. Technol Forecast & Soc Change 89:80–99., ISSN 0040-1625. https://doi.org/10.1016/j.techfore.2013.08.033

Lombardi P, Giordano S, Farouh H, Yousef W (2012) Modelling the smart city performance. Innovation: Eu J Soc Sci Res 25(2):137–149. https://doi.org/10.1080/13511610.2012.660325

Mausom M, Choudhary A (2017) Urban poverty: trends, assessment, and inclusive planning as a solution. In: Seta F, Sen J, Biswas A, Khare A (eds) From poverty, inequality to smart city. Springer Transactions in Civil and Environmental Engineering. Springer, Singapore. https://doi.org/10.1007/978-981-10-2141-1_2

McCornack A (2012) The spirit of cities: why the identity of a city matters in a global age. J Int Affair; Spring/Summ 65:2

Mohamed N, Lazarova-Molnar S, Al-Jaroodi J (2017) "Cloud of things: optimizing smart city services", 7th International Conference on Modeling, Simulation, and Applied Optimization (ICMSAO): 1-5. https://doi.org/10.1109/ICMSAO.2017.7934875
Nordström K, Schlingmann P (2015) Urban express: 15 rules of the new world, in which the main roles of cities and women
Olazabal M, Chelleri L, Sharifi A (2018) Is connectivity a desirable property in urban resilience assessments? In: Yamagata Y, Sharifi A (eds) Resilience-oriented urban planning. Lecture notes in energy, Volume 65, Springer, Cham. https://doi.org/10.1007/978-3-319-75798-8_11
Pantazis C, Townsend P, Gordon D (2006) The necessities of life. In: Pantazis C, Gordon D, Levitas R (eds) Poverty and social exclusion in Britain, the millennium survey, vol 4. The Policy Press, Bristol, pp 89–122
Pellicer et al (2013) A global perspective of smart cities: a survey. 7th International Conference on Innovative Mobile and Internet Services in Ubiquitous Computing, Taichung (439–444)
Sadowski J, Pasquale F (2015) The spectrum of control: a social theory of the smart city. First Monday 20(7). https://doi.org/10.5210/fm.v20i7.5903
Sandel D (2017) Smart cities are 90% sociology and 10% infra- structure. In: Kumar V (ed) Smart economy in smart cities. Springer, Singapore, pp 1033–1036. https://doi.org/10.1007/978-981-10-1610-3
Sandu D, de Jong GF (1998) Political change, ideology and migration intention. Romanian Journal of Sociology 9(1)
Santoso S, Kuehn A (2013) Intelligent urbanism: convivial living in smart cities. iConference 2013 Proceedings (pp. 566–570). https://doi.org/10.9776/13278
Seta F, Sen J, Biswas A, Khare A (eds) (2017) From poverty, inequality to smart city. Springer transactions in civil and environmental engineering. Springer, Singapore. https://doi.org/10.1007/978-981-10-2141-1
Silaghi MI, Ghatak S (2011, March) Why do not they move from rural to urban areas? Inter-regional migration in Romania. Journal for Economic Forecasting, Institute for Economic Forecasting 0(1):143–158
The Romanian Institute for Evaluation and Strategy – IRES, "Perceptions and attitudes regarding the quality of urban life in Romania". 2018, March 16–19. Representative sample for the adult population in urban Romania: 800 individuals. Maximum tolerated error ± 3.5%. Available at: https://www.researchgate.net/publication/325319651_Perceptii_si_atitudini_privind_calitatea_vietii_urbane_in_Romania
The Romanian Institute for Evaluation and Strategy – IRES, "One year of COVID-19 in Romania. The impact of the pandemic on the Romanians quality", 2021, February 12–15. Representative sample for the adult population in urban Romania: 1.115 individuals. Maximum tolerated error ± 2,9%. Available at: https://ires.ro/uploads/articole/ires_1-an-de-covid-19_impact-calitatea-vietii_sondaj-de-opinie_c.pdf?fbclid=IwAR3EQ-JfJHAOUh0lO3DYxvv3Y4hBjDCmIs7sqvNTpyxgE7Ov6VsnV72nC88
Tzezana R (2016) Scenarios for crime and terrorist attacks using the internet of things. European Journal of Futures Research 4(1):18. https://doi.org/10.1007/s40309-016-0107-z
Van Cott HP (1985) High technology and human needs. Ergonomics 28:1135–1142
Vizcaya M, Felipe-Duran F, Acevedo-Mosqueda E, Sanchez-Meraz M, Acevedo Mosqueda M (2017) A web page design for monitoring electromagnetic pollution in a smart city. Paper presented at 2016 IEEE 36th Central American and Panama Convention, CONCAPAN 2016. https://doi.org/10.1109/CONCAPAN.2016.7942366
Williams R, Edge D (1996) The social shaping of technology. In: Dutton WH (ed) Information and communication technologies - visions and realities. Oxford University Press, Oxford, pp 53–67
Zukin S (1998) Urban lifestyles: diversity and standardisation in space of consumption. Urban Stud 35:825–839

Index

G. C. Lazaroiu et al. (eds.), *Holistic Approach for Decision Making Towards Designing Smart Cities*, Future City 18,
https://doi.org/10.1007/978-3-030-85566-6

GPSR Compliance

The European Union's (EU) General Product Safety Regulation (GPSR) is a set of rules that requires consumer products to be safe and our obligations to ensure this.

If you have any concerns about our products, you can contact us on ProductSafety@springernature.com

In case Publisher is established outside the EU, the EU authorized representative is:

Springer Nature Customer Service Center GmbH
Europaplatz 3
69115 Heidelberg, Germany

Batch number: 10732090

Printed by Printforce, the Netherlands